SIGNAL MEASUREMENT, ANALYSIS, and TESTING

SIGNAL MEASUREMENT, ANALYSIS, and TESTING

edited by

Jerry C. Whitaker

CRC Press

Boca Raton   London   New York   Washington, D.C.

Library of Congress Cataloging-in-Publication Data

Signal measurement, analysis, and testing / Jerry C. Whitaker, editor.
    p. cm.
    Includes bibliographical references.
    ISBN 0-8493-0048-7 (alk. paper)
    1. Signal processing—Digital techniques. 2. System analysis.
3. Fourier analysis. 4. Phase distortion (Electronics)
I. Whitaker, Jerry C.
TK5102.9.S537 1999
621.382′2—dc21   99-44876
                                                                                                                                            CIP

    This book contains information obtained from authentic and highly regarded sources. Reprinted material is quoted with permission, and sources are indicated. A wide variety of references are listed. Reasonable efforts have been made to publish reliable data and information, but the author and the publisher cannot assume responsibility for the validity of all materials or for the consequences of their use.

    Neither this book nor any part may be reproduced or transmitted in any form or by any means, electronic or mechanical, including photocopying, microfilming, and recording, or by any information storage or retrieval system, without prior permission in writing from the publisher.

    All rights reserved. Authorization to photocopy items for internal or personal use, or the personal or internal use of specific clients, may be granted by CRC Press LLC, provided that $.50 per page photocopied is paid directly to Copyright Clearance Center, 222 Rosewood Drive, Danvers, MA 01923 USA. The fee code for users of the Transactional Reporting Service is ISBN 0-8493-0048-7/00/$0.00+$.50. The fee is subject to change without notice. For organizations that have been granted a photocopy license by the CCC, a separate system of payment has been arranged.

    The consent of CRC Press LLC does not extend to copying for general distribution, for promotion, for creating new works, or for resale. Specific permission must be obtained in writing from CRC Press LLC for such copying.

    Direct all inquiries to CRC Press LLC, 2000 N.W. Corporate Blvd., Boca Raton, Florida 33431.

    The material in this book was taken from The Electronics Handbook, edited by Jerry D. Whitaker.

    Trademark Notice: Product or corporate names may be trademarks or registered trademarks, and are used only for identification and explanation, without intent to infringe.

© 2000 by CRC Press LLC

No claim to original U.S. Government works
International Standard Book Number 0-8493-0048-7
Library of Congress Card Number 99-44876
Printed in the United States of America 1 2 3 4 5 6 7 8 9 0
Printed on acid-free paper

Preface

Signal Measurement, Analysis, and Testing is intended for engineers and technicians involved in the design, production, installation, operation, and maintenance of electronic devices and systems. This publication covers a broad range of technologies with emphasis on practical applications. In general, the level of detail provided is limited to that necessary to design electronic systems based on the interconnection of operational elements and devices. References are provided throughout the book to direct readers to more detailed information on important subjects.

The purpose of this book is to provide in a single volume a comprehensive reference for the practicing engineer in industry, government, and academia. It deals with the prominent aspects of this process for any electronic design, including a thorough treatment of computer-aided analysis emphasizing MATLAB®. Most modern signal analyses make use of computer-aided analysis in one form or other, including peak detection, time base measurement, Fourier analysis, and display. While most sections concentrate on distortion mechanisms and analysis, other sections present extensive tables and data of properties of materials, frequency bands and assignments, international standards and constants, conversion factors, general mathematics, and abbreviations of communications terms.

Contributors

Samuel O. Agbo
California Polytechnic University
San Luis Obispo, California

W.F. Ames
Georgia Institute of Technology
Atlanta, Georgia

George Cain
Georgia Institute of Technology
Atlanta, Georgia

Robert D. Greenberg
Federal Communications
 Commision
Washington, DC

Jerry C. Hamann
University of Wyoming
Laramie, Wyoming

John W. Pierre
University of Wyoming
Laramie, Wyoming

James F. Shackelford
University of California
Davis, California

Jerry C. Whitaker
Technical Press
Beaverton, Oregon

Rodger E. Ziemer
University of Colorado
Colorado Springs, Colorado

Contents

Signal Measurement, Analysis, and Testing *Jerry C. Whitaker*

1 Introduction *Rodger E. Ziemer*

2 Audio Frequency Distortion Mechanisms and Analysis *Jerry C. Whitaker*
    2.2 Introduction 3
    2.2 Level Measurements 4
    2.3 Noise Measurement 6
    2.4 Phase Measurement 7
    2.5 Nonlinear Distortion Mechanisms 8
    2.6 Multitone Audio Testing 11
    2.7 Considerations for Digital Audio Systems 15

3 Video Display Distortion Mechanisms and Analysis *Jerry C. Whitaker*
    3.1 Introduction 19
    3.2 Video Signal Spectra 19
    3.3 Measurement of Color Displays 23
    3.4 Assessment of Color Reproduction 24
    3.5 Display Resolution and Pixel Format 26
    3.6 Applications of the Zone Plate Signal 27
    3.7 CRT Measurement Techniques 32

4 Radio Frequency Distortion Mechanisms and Analysis *Samuel O. Agbo*
    4.1 Introduction 38
    4.2 Types of Distortion 39
    4.3 The Wireless Radio Channel 42
    4.4 Effects of Phase and Frequency Errors in Coherent Demodulation
          of AM Signals 45
    4.5 Effects of Linear and Nonlinear Distortion in Demodulation of Angle Modulated Waves 49
    4.6 Interference as a Radio Frequency Distortion Mechanism 53

5 Digital Test Equipment and Measurement Systems *Jerry C. Whitaker*
    5.1 Introduction 57
    5.2 Logic Analyzer 57
    5.3 Signature Analyzer 60
    5.4 Manual Probe Diagnosis 60
    5.5 Checking Integrated Circuits 60
    5.6 Emulative Tester 61
    5.7 Protocol Analyzer 62
    5.8 Automated Test Instruments 62
    5.9 Digital Oscilloscope 63

6 Fourier Waveform Analysis *Jerry C. Hamann and John W. Pierre*
- 6.1 Introduction 70
- 6.2 The Mathematical Preliminaries for Fourier Analysis 71
- 6.3 The Fourier Series for Continuous-Time Periodic Functions 74
- 6.4 The Fourier Transform for Continuous-Time Aperiodic Functions 75
- 6.5 The Fourier Series for Discrete-Time Periodic Functions 76
- 6.6 The Fourier Transform for Discrete-Time Aperiodic Functions 76
- 6.7 Example Applications of Fourier Waveform Techniques 77

7 Computer-Based Signal Analysis *Rodger E. Ziemer*
- 7.1 Introduction 82
- 7.2 Signal Generation and Analysis 82
- 7.3 Symbolic Mathematics 86

Conversion Factors, Standards, and Constants *Jerry C. Whitaker*

8 Properties of Materials *James F. Shackelford*
- 8.1 Introduction 92
- 8.2 Structure 92
- 8.3 Composition 92
- 8.4 Physical Properties 93
- 8.5 Mechanical Properties 93
- 8.6 Thermal Properties 95
- 8.7 Chemical Properties 95
- 8.8 Electrical and Optical Properties 95
- 8.9 Additional Data 96

9 Frequency Bands and Assignments *Robert D. Greenberg*
- 9.1 U.S. Table of Frequency Allocations 118

10 International Standards and Constants
- 10.1 International System of Units (SI) 214
- 10.2 Physical Constants 216

11 Conversion Factors *Jerry C. Whitaker*
- 11.1 Introduction 222
- 11.2 Conversion Constants and Multipliers 238

12 General Mathematical Tables *W.F. Ames and George Cain*
- 12.1 Introduction to Mathematics Chapter 242
- 12.2 Elementary Algebra and Geometry 242
- 12.3 Trigonometry 247
- 12.4 Series 251
- 12.5 Differential Calculus 256
- 12.6 Integral Calculus 262
- 12.7 Special Functions 266
- 12.8 Basic Definitions: Linear Algebra Matrices 273
- 12.9 Basic Definitions: Vector Algebra and Calculus 278
- 12.10 The Fourier Transforms: Overview 282

13 Communications Terms: Abbreviations 288

Index 303

I

Signal Measurement, Analysis, and Testing

1. **Introduction** *Roger E. Ziemer* 2
2. **Audio Frequency Distortion Mechanisms and Analysis** *Jerry C. Whitaker* 3
   Introduction • Level Measurements • Noise Measurement • Phase Measurement • Nonlinear Distortion Mechanisms • Multitone Audio Testing • Considerations for Digital Audio Systems
3. **Video Display Distortion Mechanisms and Analysis** *Jerry C. Whitaker* 19
   Introduction • Video Signal Spectra • Measurement of Color Displays • Assessment of Color Reproduction • Display Resolution and Pixel Format • Applications of the Zone Plate Signal • CRT Measurement Techniques
4. **Radio Frequency Distortion Mechanisms and Analysis** *Samuel O. Agbo* 38
   Introduction • Types of Distortion • The Wireless Radio Channel • Effects of Phase and Frequency Errors in Coherent Demodulation of AM Signals • Effects of Linear and Nonlinear Distortion in Demodulation of Angle Modulated Waves • Interference as a Radio Frequency Distortion Mechanism
5. **Digital Test Equipment and Measurement Systems** *Jerry C. Whitaker* 57
   Introduction • Logic Analyzer • Signature Analyzer • Manual Probe Diagnosis • Checking Integrated Circuits • Emulative Tester • Protocol Analyzer • Automated Test Instruments • Digital Oscilloscope
6. **Fourier Waveform Analysis** *Jerry C. Hamann and John W. Pierre* 70
   Introduction • The Mathematical Preliminaries for Fourier Analysis • The Fourier Series for Continuous-Time Periodic Functions • The Fourier Transform for Continuous-Time Aperiodic Functions • The Fourier Series for Discrete-Time Periodic Functions • The Fourier Transform for Discrete-Time Aperiodic Functions • Example Applications of Fourier Waveform Techniques
7. **Computer-Based Signal Analysis** *Rodger E. Ziemer* 82
   Introduction • Signal Generation and Analysis • Symbolic Mathematics

1
Introduction

Rodger E. Ziemer
University of Colorado,
Colorado Springs

THE CHARACTERIZATION, analysis, trouble shooting, and repair of modern complex electronic systems requires sophisticated techniques. This section addresses these aspects of the electronics engineer's profession. Different techniques are required for different regimes of operation. A convenient categorization for this purpose is audio, video, and radio frequency. In addition, it is convenient to subdivide further into analog and digital.

Computers have become pervasive throughout society, and the electronics engineer perhaps finds computer-based analysis more important than any other branch of engineering. Indeed, most modern signal analysis instruments now make use of computer-aided signal analysis in one form or other, including peak detection, time base measurement, Fourier analysis, and display.

This section consists of six major chapters. Four deal with instrumentation aspects of signal measurement, analysis, and testing. Two deal with the mathematics of signal analysis.

Chapter 130 is concerned with audio distortion mechanisms and analysis, mainly signal level, phase, and frequency, or combinations of these, such as signal-to-noise ratio.

Chapter 131 deals with video signal distortion mechanisms and analysis. Because of the unique characteristics of video display systems, special considerations are required for characterization of distortion in them. For example, number of lines scanned per second is an important consideration in this area.

Chapter 132 addresses radio frequency distortion mechanisms and analysis. The considerations are similar to those for audio distortion mechanisms and analysis, except the frequency ranges are higher and are bandpass.

Chapter 133 considers digital test equipment and measurement systems. A universal instrument in this category is the logic analyzer, and this chapter gives an easily understood overview of this important instrument and its capabilities.

The important area of Fourier waveform analysis is dealt with in Chapter 134. In the not too distant past, Fourier analysis was strictly an analytical tool. The discovery of the fast Fourier transform (FFT) in the late 1960s and the development of fast hardware to perform the FFT in the 1980s have made Fourier analysis a mainstay of signal analysis instrumentation, and it is difficult to tell an oscilloscope (with digital signal processing capability) from a spectrum analyzer with FFT-based spectral analysis capability.

Many software tools are now available for computer-based signal analysis, which is the subject of Chapter 135. This chapter concentrates on only one general purpose analysis tool called MATLAB®, which is a vector-based high-level programming language. It has many capabilities including general number crunching, signal analysis, graphical interface, and symbolic manipulation.

This section, written by several experts in the disciplines represented, provides important information for any electronic design. Whether a component in an overall system or a stand-alone box, distortion mechanisms and their characterization are important in the design of high-quality instrumentation.

2
Audio Frequency Distortion Mechanisms and Analysis

Jerry C. Whitaker
Editor-in-Chief

2.1 Introduction 3
    Purpose of Audio Measurements
2.2 Level Measurements 4
    Root-Mean-Square Measurements • Average-Response Measurements • Peak-Response Measurements • Decibel Measurements
2.3 Noise Measurement 6
2.4 Phase Measurement 7
2.5 Nonlinear Distortion Mechanisms 8
    Harmonic Distortion • Intermodulation Distortion • Addition and Cancellation of Distortion Components
2.6 Multitone Audio Testing 11
    Multitone Versus Discrete Tones • Operational Considerations • FFT Analysis
2.7 Considerations for Digital Audio Systems 15

2.1 Introduction

Most measurements in the audio field involve characterizing fundamental parameters.[1] These include signal level, phase, and frequency. Most other tests consist of measuring these fundamental parameters and displaying the results in combination by using some convenient format. For example, signal-to-noise ratio (SNR) consists of a pair of level measurements made under different conditions expressed as a logarithmic, or decibel, ratio. When characterizing a device, it is common to view it as a box with input terminals and output terminals. In normal use a signal is applied to the input and the signal, modified in some way, appears at the output. Instruments are necessary to quantify these unintentional changes to the signal. Some measurements are *one-port* tests, such as impedance or noise level, and are not concerned with input/output signals, only with one or the other.

Purpose of Audio Measurements

Measurements are made on audio circuits and equipment to check performance under specified conditions and to assess suitability for use in a particular application. The measurements may be used to verify specified system performance or as a way of comparing several pieces of equipment for use in a system. Measurements may also be used to identify components in need of adjustment or repair. Whatever the application, audio measurements are a key part of audio engineering.

Many parameters are important in audio devices and merit attention in the measurement process. Some common audio measurements are frequency response, gain or loss, harmonic distortion, intermodulation distortion, noise level, phase response, and transient response.

---

[1]Portions of this chapter were adapted from: Benson, K.B. and Whitaker, J.C. 1990. *Television and Audio Handbook for Technicians and Engineers*. McGraw–Hill, New York. Used with permission.

Measurement of level is fundamental to most audio specifications. Level can be measured either in absolute terms or in relative terms. Power output is an example of an absolute level measurement; it does not require any reference. SNR and gain or loss are examples of relative measurements; the result is expressed as a ratio of two measurements. Although it may not appear so at first, frequency response is also a relative measurement. It expresses the gain of the device under test as a function of frequency, with the midband gain, typically, as a reference.

Distortion measurements are a way of quantifying the amount of unwanted components added to a signal by a piece of equipment. The most common technique is total harmonic distortion (THD), but others are often used. Distortion measurements express the amount of unwanted signal components relative to the desired signal, usually as a percentage or decibel value. This is another example of multiple level measurements that are combined to give a new measurement figure.

2.2 Level Measurements

The simplest definition of a level measurement is the alternating current (AC) amplitude at a particular place in the audio system. In contrast to direct current measurements, however, there are many ways of specifying AC voltage. The most common methods are average, root-mean-square (RMS), and peak response. Strictly speaking, the term *level* refers to a logarithmic, or decibel, measurement. However, common parlance employs the term for an AC amplitude measurement, and that convention will be followed in this chapter.

Root-Mean-Square Measurements

The RMS technique measures the effective power of the AC signal. It specifies the value of the DC equivalent that would dissipate the same power if either were applied to a load resistor. This process is illustrated in Fig. 2.1 for voltage measurements. The input signal is squared, and the average value is found. This is equivalent to finding the average power. The square root of this value is taken to translate the signal from a power value back to a voltage. For the case of a sine wave the RMS value is 0.707 of its maximum value.

Consider the case when the signal is not a sine wave, but rather a sine wave and several of its harmonics. If the RMS amplitude of each harmonic is measured individually and added, the resulting value will be the same as an RMS measurement on the signals together. Because RMS voltages cannot be added directly, it is necessary to perform an RMS addition, as follows:

$$V_{rms} = \sqrt{V_{rms1}^2 + V_{rms2}^2 + V_{rms3}^2 + V_{rmsn}^2}$$

Note that the result is not dependent on the phase relationship of the signal and its harmonics. The RMS value is determined completely by the amplitude of the components. This mathematical predictability is powerful in practical applications of level measurement, enabling measurements made at different places in a system to be correlated. It is also important in correlating measurements with theoretical calculations.

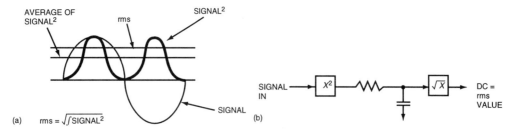

FIGURE 2.1 Root-mean-square (RMS) voltage measurements: (a) the relationship of RMS and average values, (b) RMS measurement circuit.

Audio Frequency Distortion Mechanisms and Analysis

WAVEFORM		rms	Avg.	rms avg.	CREST FACTOR
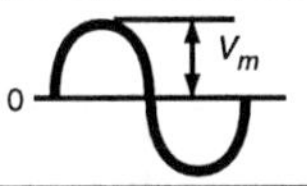	SINE WAVE	$\frac{V_m}{\sqrt{2}}$ 0.707 V_m	$\frac{2}{\pi} V_m$ 0.637 V_m	$\left|\frac{\pi}{2\sqrt{2}}\right| = 1.111$	$\sqrt{2} = 1.414$
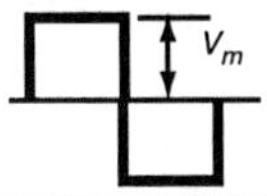	SYMMETRICAL SQUARE WAVE OR DC	V_m	V_m	1	1
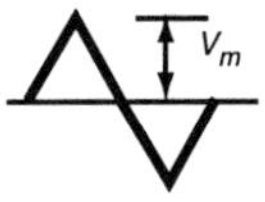	TRIANGULAR WAVE OR SAWTOOTH	$\frac{V_m}{\sqrt{3}}$	$\frac{V_m}{2}$	$\frac{2}{\sqrt{3}} = 1.155$	$\sqrt{3} = 1.732$

FIGURE 2.2 Comparison of RMS and average voltage characteristics.

Average-Response Measurements

Average-responding voltmeters were common in audio work some years ago principally because of their low cost. Such devices measure AC voltage by rectifying and filtering the resulting waveform to its average value, which can then be read on a standard DC voltmeter. The average value of a sine wave is 0.637 of its maximum amplitude. Average-responding meters are usually calibrated to read the same as an RMS meter for the case of a single-sine-wave signal. This results in the measurement being scaled by a constant K of 0.707/0.637, or 1.11. Meters of this type are called average-responding, RMS calibrated. For signals other than sine waves, the response will be different and difficult to predict. If multiple sine waves are applied, the reading will depend on the phase shift between the components and will no longer match the RMS measurement. A comparison of RMS and average-response measurements is made in Fig. 2.2 for various waveforms. If the average readings are adjusted as described previously to make the average and RMS values equal for a sine wave, all of the numbers in the *average* column should be increased by 11.1%, whereas the *RMS-average* numbers should be reduced by 11.1%.

Peak-Response Measurements

Peak-responding meters measure the maximum value that the AC signal reaches as a function of time. (See Fig. 2.3.) The signal is full-wave rectified to find its absolute value and then passed through a diode to a storage capacitor. When the absolute value of the voltage rises above the value stored on the capacitor, the diode will conduct and increase the stored voltage. When the voltage decreases, the capacitor will maintain the old value. Some means for discharging the capacitor is required to allow measuring a new peak value. In a true peak detector, this is accomplished by a switch. Practical peak detectors usually include a large resistor to discharge the capacitor gradually after the user has had a chance to read the meter.

The ratio of the true peak to the RMS value is called the **crest factor**. For any signal but an ideal square wave the crest factor will be greater than 1, as illustrated in Fig. 2.4. As the measured signal become more peaked, the crest factor will increase.

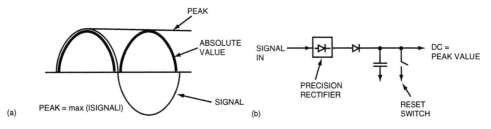

FIGURE 2.3 Peak voltage measurements: (a) illustration of peak detection, (b) peak measurement circuit.

By introducing a controlled charge and discharge time, a *quasipeak* detector is achieved. The charge and discharge times may be selected, for example, to simulate the ear's sensitivity to impulsive peaks. International standards define these response times and set requirements for reading accuracy on pulses and sine wave bursts of various durations. The gain of a quasipeak detector is normally calibrated so that it reads the same as an RMS detector for sine waves.

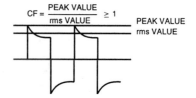

FIGURE 2.4 Illustration of the crest factor in voltage measurements.

Another method of specifying signal amplitude is the **peak-equivalent sine**, which is the RMS level of a sine wave having the same peak-to-peak amplitude as the signal under consideration. This is the peak value of the waveform scaled by the correction factor 1.414, corresponding to the peak-to-RMS ratio of a sine wave. This is useful when specifying test levels of waveforms in distortion measurements.

Decibel Measurements

Measurements in audio work are usually expressed in decibels. Audio signals span a wide range of level. The sound pressure of a live concert band performance may be one million times that of rustling leaves. This range is too wide to be accommodated on a linear scale. The logarithmic scale of the decibel compresses this wide range down to a more easily handled range. Order-of-magnitude changes result in equal increments on a decibel scale. Furthermore, the human ear perceives changes in amplitude on a logarithmic basis, making measurements with the decibel scale reflect audibility more accurately.

A decibel may be defined as the logarithmic ratio of two power measurements or as the logarithmic ratio of two voltages:

$$dB = 20 \log \frac{E_1}{E_2} \text{ for voltage measurements}$$

$$dB = 10 \log \frac{P_1}{P_2} \text{ for power measurements}$$

There is no difference between decibel values from power measurements and decibel values from voltage measurements if the impedances are equal. In both equations the denominator variable is usually a stated reference. A doubling of voltage will yield a value of 6.02 dB, whereas a doubling of power will yield 3.01 dB. This is true because doubling voltage results in a factor-of-four increase in power.

Audio engineers often express the decibel value of a signal relative to some standard reference instead of another signal. The reference for decibel measurements may be predefined as a power level, as in decibels referenced to 1 mW (dBm), or it may be a voltage reference, as in decibels referenced to 1 V (dBV). When measuring dBm or any power-based decibel value, the reference impedance must be specified or understood. Often it is desirable to specify levels in terms of a reference transmission level somewhere in the system under test. These measurements are designated dBr, where the reference point or level must be separately conveyed.

2.3 Noise Measurement

Noise measurements are simply specialized level measurements. It has long been recognized that the ear's sensitivity varies with frequency, especially at low levels. This effect was studied in detail by Fletcher and Munson and later by Robinson and Dadson. The Fletcher–Munson hearing-sensitivity curve for the threshold of hearing and above is given in Fig. 2.5. The ear is most sensitive in the region of 2–4 kHz, with rolloffs above and below these frequencies. To predict how loud something will sound it is necessary to use a filter that duplicates this nonflat behavior electrically. The filter *weights* the signal level on the basis of frequency, thus earning the name *weighting filter*. Various efforts have been made to do this, resulting in several standards for noise measurement.

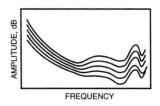

FIGURE 2.5 The Fletcher–Munson curves of hearing sensitivity vs frequency.

Some of the common weighting filters are shown overlaid on the hearing threshold curve in Fig. 2.6.

The most common filter used in the United States for weighted noise measurements is the A-weighting curve. An average-responding meter is often used for A-weighted noise measurements, although RMS meters are also used for this application.

European audio equipment is usually specified with a CCIR filter and a quasipeak detector. The CCIR curve is significantly more peaked than the A curve and has a sharper rolloff at high frequencies. The CCIR quasipeak standard was developed to quantify the noise in telephone systems.

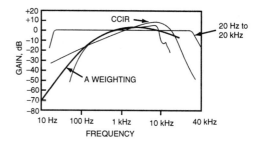

FIGURE 2.6 Response characteristics of several common weighting filters for audio measurements.

The quasipeak detector more accurately represents the ear's sensitivity to impulsive sounds. When used with the CCIR filter curve, it is supposed to correlate better with the subjective level of the noise than A-weighted average-response measurements do.

Some audio equipment manufacturers specify noise with a 20 Hz–20 kHz bandwidth filter and an RMS-responding meter. This is done to specify noise over the audio band without regard to the ear's differing sensitivity with frequency. The International Electrotechnical Commission (IEC) defies the audio band as all frequencies between 22.4 Hz and 22.4 kHz. Measurements over such a bandwidth are referred to under IEC standards as unweighted.

2.4 Phase Measurement

Phase in an audio system is typically measured and recorded as a function of frequency over the audio range. For most audio devices phase and amplitude responses are closely coupled. Any change in amplitude that varies with frequency will produce a corresponding phase shift. A fixed time delay will introduce a phase shift that is a linear function of frequency. This time delay can introduce large values of phase shift at high frequencies that are of no significance in practical applications. The time delay will not distort the waveshape of complex signals and will not be audible. There can be problems, however, with time delay when the delayed signal is used in conjunction with an undelayed signal.

When dealing with complex signals, the meaning of phase can become unclear. Viewing the signal as the sum of its components according to Fourier theory, we find a different value of phase shift at each frequency. With a different phase value on each component, the question is raised as to which one should be used as the reference. If the signal is periodic and the waveshape is unchanged passing through the device under test, then a phase value may still be defined. This may be done by using the shift of the zero crossings as a fraction of the waveform period. If there is differential phase shift with frequency, however, the waveshape will be changed. It is then not possible to define any phase-shift value, and phase must be expressed as a function of frequency.

Group delay is another useful expression of the phase characteristics of an audio device. Group delay is the slope of the phase response. It expresses the relative delay of the spectral components of a complex waveform. If the group delay is flat, all components will arrive together. A peak or rise in the group delay indicates that those components will arrive later by the amount of the peak or rise. Group delay ϕ is computed by taking the derivative of the phase response vs frequency:

$$\phi = \frac{-(\alpha_{f2} - \alpha_{f1})}{f_2 - f_1}$$

where α_{f1} is the phase at f_1 and α_{f2} phase at f_2. This requires that phase be measured over a range of frequencies to yield a curve that can be differentiated. It also requires that the phase measurements be performed at frequencies close enough together to provide a smooth and accurate derivative.

2.5 Nonlinear Distortion Mechanisms

Distortion is a measure of signal impurity. It is usually expressed as a percentage or decibel ratio of the undesired components to the desired components of a signal. There are several methods of measuring distortion, the most common being harmonic distortion and several types of intermodulation distortion.

Harmonic Distortion

The transfer characteristic of a typical amplifier is shown in Fig. 2.7. The transfer characteristic represents the output voltage at any point in the signal waveform for a given input voltage; ideally this is a straight line. The output waveform is the projection

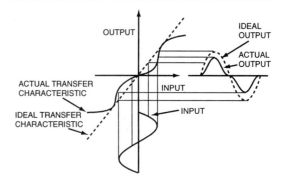

FIGURE 2.7 Illustration of total harmonic distortion (THD) measurement of an amplifier transfer characteristic.

of the input sine wave on the device transfer characteristic. A change in the input produces a proportional change in the output. Because the actual transfer characteristic is nonlinear, a distorted version of the input waveshape appears at the output.

Harmonic distortion measurements excite the device under test with a sine wave and measure the spectrum of the output. Because of the nonlinearity of the transfer characteristic, the output is not sinusoidal. By using Fourier series, it can be shown that the output waveform consists of the original input sine wave plus sine waves at integer multiples (harmonics) of the input frequency. The spectrum of the distorted signal is shown in Fig. 2.8 for a 1-kHz input, and output signals consisting of 1, 2, 3 kHz, etc. The harmonic amplitudes are proportional to the amount of distortion in the device under test. The percentage harmonic distortion is the RMS sum of the harmonic amplitudes divided by the RMS amplitude of the fundamental.

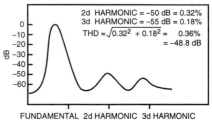

FIGURE 2.8 Example of reading THD from a spectrum analyzer.

Harmonic distortion may also be measured with a spectrum analyzer. As shown in Fig. 2.8, the fundamental amplitude is adjusted to the 0-dB mark on the display. The amplitudes of the harmonics are then read and converted to linear scale. The RMS sum of these values is taken, which represents the THD. This procedure is time consuming and can be difficult for an unskilled operator.

Notch Filter Analyzer

A simpler approach to the measurement of harmonic distortion can be found in the notch-filter distortion analyzer. This device, commonly referred to as simply a distortion analyzer, removes the fundamental of the signal to be investigated and measures the remainder. A block diagram of such a unit is shown in Fig. 2.9. The fundamental is removed with a notch filter, and the output is measured with an AC voltmeter. Because distortion is normally presented as a percentage of the fundamental signal, this level must be measured or set equal to a predetermined reference value. Additional circuitry (not shown) is required to set the level to the reference

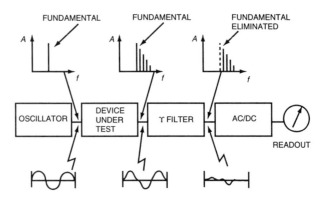

FIGURE 2.9 Simplified block diagram of a harmonic distortion analyzer.

value for calibrated measurements. Some analyzers use a series of step attenuators and a variable control for setting the input level to the reference value. More sophisticated units eliminate the variable control by using an electronic gain control. Others employ a second AC-to-DC converter to measure the input level and compute the percentage by using a microprocessor. Completely automatic units also provide autoranging logic to set the attenuators and ranges, and tabular or graphic display of the results.

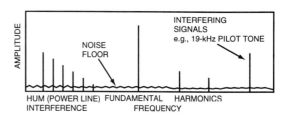

FIGURE 2.10 Example of interference sources in distortion and noise measurements.

The correct method of representing percentage distortion is to express the level of the harmonics as a fraction of the fundamental level. However, most commercial distortion analyzers use the total signal level as the reference voltage. For small amounts of distortion these two quantities are equivalent. At large values of distortion the total signal level will be greater than the fundamental level. This makes distortion measurements on these units lower than the actual value. The errors are not significant until about 20% measured distortion.

Because of the notch-filter response, any signal other than the fundamental will influence the results, not just harmonics. Some of these interfering signals are illustrated in Fig. 2.10. Any practical signal contains some noise, and the distortion analyzer will include these in the reading. Because of these added components, the correct term for this measurement is *total harmonic distortion and noise* (THD+N). Although this fact does limit the reading of very low-THD levels, it is not necessarily bad. Indeed, it can be argued that the ear hears all components present in the signal, not just the harmonics.

Additional filters are included on most distortion analyzers to reduce unwanted hum and noise. These usually consist of one or more high-pass filters (400 Hz is almost universal) and several low-pass filters. Common low-pass filter frequencies are 22.4, 30, and 80 kHz. A selection of filters eases the tradeoff between limiting bandwidth to reduce noise and the reduction in reading accuracy that results from removing desired components of the signal. When used in conjunction with a good differential input stage on the analyzer, these filters can solve most noise problems.

Intermodulation Distortion

Many methods have been devised to measure the intermodulation (IM) of two or more signals passing through a device simultaneously. The most common of these is **SMPTE IM**, named after the Society of Motion Picture and Television Engineers, which first standardized its use. IM measurements according to the SMPTE method have been in use since the 1930s. The test signal is a low-frequency tone (usually 60 Hz) and a high-frequency tone (usually 7 kHz) mixed in a 4:1 amplitude ratio. Other amplitude ratios and frequencies are used occasionally. The signal is applied to the device under test, and the output signal is examined for modulation of the upper frequency by the low-frequency tone. The amount by which the low-frequency tone modulates the high-frequency tone indicates the degree of nonlinearity. As with harmonic distortion measurement, this test may be done with a spectrum analyzer or with a dedicated distortion analyzer.

The modulation components of the upper signal appear as sidebands spaced at multiples of the lower-frequency tone. The RMS amplitudes of the sidebands are summed and expressed as a percentage of the upper-frequency level.

The most direct way to measure SMPTE IM distortion is to measure each component with a spectrum analyzer and add their RMS values. The spectrum analyzer approach has a drawback in that it is sensitive to frequency modulation of the carrier as well as amplitude modulation. A distortion analyzer for SMPTE testing is quite straightforward. The signal to be analyzed is passed through a high-pass filter to remove the low-frequency tone, as shown in Fig. 2.11. The high-frequency tone is then demodulated as if it were an amplitude modulated signal to obtain the sidebands. The sidebands pass through a low-pass filter to remove any remaining high-frequency energy. The resulting demodulated low-frequency signal will follow the envelope of the high-frequency tone. This low-frequency fluctuation is the distortion component and is displayed as a percentage of the amplitude of the high-frequency tone. Because low-pass filtering sets the measurement bandwidth, noise has little effect on SMPTE IM measurements.

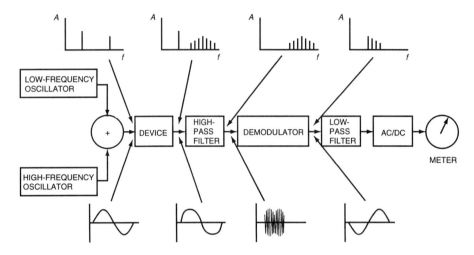

FIGURE 2.11 Simplified block diagram of a SMPTE intermodulation analyzer.

CCITT IM

Twin-tone intermodulation or CCITT difference frequency distortion is another method of measuring distortion by using two sine waves. The test signal consists of two closely spaced high-frequency tones as shown in Fig. 2.12. When the tones are passed through a nonlinear device, IM products are generated at frequencies related to the difference in frequency between the original tones. For the typical case of signals at 14 and 15 kHz the IM components will be at 1, 2, 3 kHz, etc., and 13, 16, 12, 17 kHz, etc. Even-order, or asymmetrical distortions produce low-frequency difference-frequency components. Odd-order, or symmetrical nonlinearities produce components near the

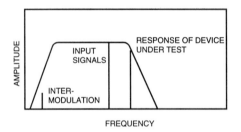

FIGURE 2.12 CCITTE intermodulation test of transfer characteristic.

input signal frequencies. The most common application of this test measures only the even-order components because they may be measured with a multipole low-pass filter. Measurement of the odd-order components requires a spectrum analyzer or DSP-based signal analyzer.

The CCITT test has several advantages over either harmonic or SMPTE IM testing. The signals and distortion components may almost always be arranged to be in the passband of the device under test. This method ceases to be useful below a few hundred hertz when the required selectivity in the spectrum analyzer or selective voltmeter becomes excessive. However, fast Fourier transform- (FFT-) based devices can extend the practical lower limit substantially below this point.

The distortion products in the CCITT IM test are usually far removed from the input signal. This positions them outside the range of the auditory system's masking effects. If a test that measures what the ear might hear is desired, the CCITT test is the most likely candidate.

Addition and Cancellation of Distortion Components

A common consideration for system-wide distortion measurements is that of distortion addition and cancellation in the devices under test. Consider the example given in Fig. 2.13. Assume that one device under test has a transfer characteristic similar to that diagrammed at the top of Fig. 2.13(a) and another has the characteristic diagrammed at the bottom. If the devices are cascaded, the resulting transfer-characteristic nonlinearity will be magnified as shown. The effect on sine waves in the time domain is illustrated in Fig. 2.13(b). The distortion component generated by each nonlinearity is in phase and will sum to a component of twice the original magnitude. If the second device under test has a complementary transfer characteristic, however, quite a different result is

Audio Frequency Distortion Mechanisms and Analysis

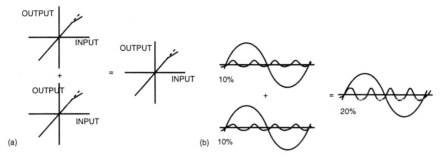

FIGURE 2.13 Illustration of the addition of distortion components: (a) addition of transfer-function nonlinearities, (b) addition of distortion components.

obtained. When the devices are cascaded, the effects of the two curves cancel, yielding a straight line for the transfer characteristic. The corresponding distortion products are out of phase with each other, resulting in no measured distortion components in the final output.

2.6 Multitone Audio Testing

Multitone testing operates on the premise that audio equipment can be stimulated to the same extent by a simultaneous combination of sine waves as it can be by a series of discrete tones occurring one at a time.[2] The advantage of multitone analysis is clear: it provides complete system evaluation in one step, eliminating the time-consuming process of applying a series of individual tones, allowing a settling time after each, making adjustments after each, and then repeating the process to check for interactions.

Multitone test techniques use carefully designed mixtures of tones applied simultaneously to the device under test (DUT). The individual tone elements have well-defined frequency, phase, and amplitude relationships. The frequencies of multitone components are selected to avoid mathematically predictable harmonic and intermodulation products that would fall on or near any of the fundamentals. The phase of each tone is fixed, but randomly selected relative to each of the other tones. Amplitude relationships may vary, depending on the device under test.

Multitone signals may be created with an inverse FFT for highly accurate and stable signal parameters. Figure 2.14 shows a spectral display of a multitone signal designed for testing wide-band audio recording equipment. Because the frequency, phase, and amplitude relationships of the elements of a multitone signal are well defined, digital signal processor (DSP) techniques allow easy detection of changes in these relationships. Thus, frequencies not present in the original multitone are the result of distortion and noise. Similarly, changes in phase relationships of the multitone elements can be used to derive phase response information. It follows that deviation from the known amplitude relationships will provide frequency response information.

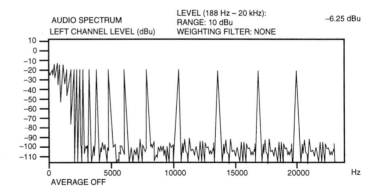

FIGURE 2.14 An audio spectrum display showing the spectral distribution of the numerous fundamental frequencies in a multitone signal. (*Source:* Tektronix, Beaverton, OR.)

[2]This section was adapted from Thompson, B. 1992. Multitone Audio Testing. Tektronix Application Note 21 W-7182, Tektronix, Beaverton, OR.

Although implementations of multitone analysis vary from one vendor to the next, plots of level vs frequency, level difference between channels, phase difference between channels vs frequency, and noise and/or distortion vs frequency are among those typically provided. Figures 2.15 and 2.16 show two of these multitone displays, levels vs frequency (frequency response) and level/phase differences between channels plotted against frequency. An audio spectrum display, such as the one shown in Fig. 2.14, is also a valuable tool when combined with the multitone signal.

Multitone Versus Discrete Tones

Multitone testing differs from traditional singletone testing in several ways. As previously mentioned, multitone analysis reduces testing and calibration times. This speed advantage stems from several properties:

- All tones are applied at once.
- The entire spectrum is measured/adjusted at once.
- Only one settling time is needed for the generator, DUT, and measurement device because a single signal acquisition is made.
- Level, phase, and noise and distortion calculations are made from one FFT record.

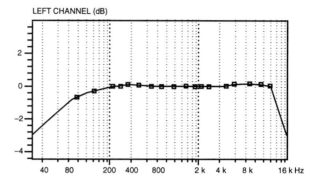
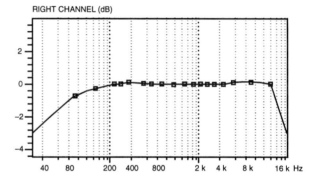

FIGURE 2.15 Level vs frequency plots resulting from audio system measurement with a multitone signal analyzer. (*Source:* Tektronix, Beaverton, OR.)

Not only does multitone eliminate the step-and-repeat process, but it can also provide immediate feedback for equipment adjustments. Watching the entire audio spectrum respond to interactive adjustments with little or no time lag further simplifies audio equipment calibration.

When compared to a single sine wave tone, looking at a multitone signal in the time domain (Fig. 2.17) illustrates how closely a multitone waveform resembles program audio. Two factors contribute to this resemblance. First,

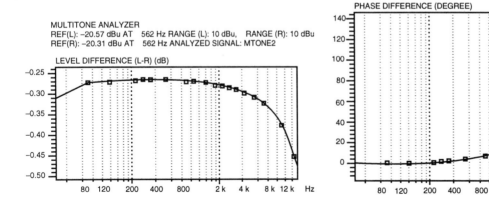

FIGURE 2.16 Level and phase difference measurements taken with a multitone signal analyzer. (*Source:* Tektronix, Beaverton, OR.)

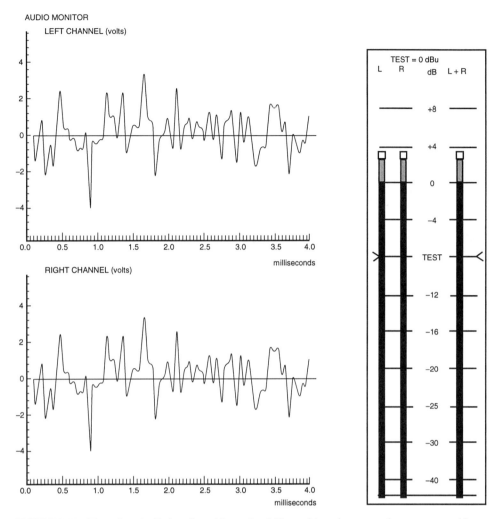

FIGURE 2.17 Time-domain display of a multitone signal. The multitone bears a much stronger resemblance to program material than a single sine wave tone. (*Source:* Tektronix, Beaverton, OR.)

multitones fill more of the spectrum than single sine wave tones. Second, the crest factor of multitone is similar to that of music or voice signals.

With multitone's similarity to program material, it can be argued that the adjustments performed and measurement results obtained through multitone give more realistic noise and distortion values than discrete tone testing. (In this case, more realistic means a value better representing the noise and distortion present with program material.) In the past, noise measurements were made on audio equipment while no input signal was present. Therefore, the quantization noise and other level-related anomalies were not a part of the noise measurement result. Furthermore, signal processors often operate at different gain ranges depending on the input signal level. So, a no-input noise measurement would be even less representative of actual operating conditions.

The different type of stimulation provided by multitone can, however, lead to measurement results that do not exactly match results obtained with conventional tones or tone sequences. With this in mind, multitone testing should not be used interchangeably with tests made using single-tone input signals. The strengths of multitone testing are speed and convenience, and so it excels in the roles of calibration, production line testing, and equipment performance tracking.

Standardized tone sequence tests require settling, acquisition, and measurement times for each tone in the sequence. All of the advantages of multitone previously mentioned apply here as well. And since multitone is a

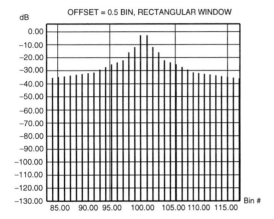

FIGURE 2.18 Computer simulation of FFT without windowing. The 588.87-Hz fundamental frequency in this model does not correspond exactly to an FFT bin. The result is leakage that provides a misleading noise floor display. (*Source:* Tektronix, Beaverton, OR.)

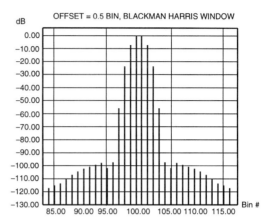

FIGURE 2.19 Computer simulation of FFT with Blackman–Harris windowing. The application of windowing prior to the FFT drastically reduces the effects of leakage. Within 23.44 Hz on either side of the fundamental, leakage is more than 90 dB down. (*Source:* Tektronix, Beaverton, OR.)

DSP-based technology, it is always backed up with the computing power necessary for automation. To fully mimic standard tone sequences, a multitone sequence of two or three different levels of multitone can be performed in one-tenth the time of conventional sequences.

Operational Considerations

Multitone is a strong candidate for rapid performance tests of common carrier and broadcast audio transmission linksÑwhile on the air or otherwise in use. A short burst of multitone, typically about 1 s, is all that is needed for analyzers to do a complete audio path characterization. Although most multitones do not sound particularly symphonic, they are not abrasive either, especially in the short bursts necessary for measurement.

Given the usual distance between ends of a transmission system during split-site testing (i.e., transmitter and studio), it is not practical to lock the sampling clocks of the generator and analyzer, resulting in slight frequency shifts in the multitone elements. Split-site testing is one application where frequency shifts in the multitone can occur, and windowing prior to the FFT analysis should be employed to ensure accurate measurement results.

FFT Analysis

A closer look at multitone shows the need for caution when processing the signal. Figures 2.18 and 2.19 illustrate how different processing techniques produce drastically different results, only some of which are meaningful. Both figures are based on a simulated 8192-point (8 K) FFT with a 48-kHz sampling frequency and a 588.87-Hz tone. FFT bin numbers (e.g., bin 100 out of 4096 bins) label the horizontal axes of each graph. The fundamental frequency of 588.87 Hz falls halfway between bin numbers 100 (585.94 Hz) and 101 (591.80 Hz).

Transforming time-domain information into the frequency domain with an FFT requires that the period of the time-domain signal be an integer multiple of the period of the FFT. When such a periodic relationship occurs, a fundamental frequency will fall into a single FFT bin. But this requires that the same sampling clock be used for both signal generation and subsequent resampling of the signal, and further, that no shift in the tone's pitch occurs while passing through the DUT. Because shifts in pitch can occur and because some testing applications require that the generator and analyzer be at separate locations, it is not safe to assume perfect tone-to-bin correspondence.

The result of this imperfect tone-to-bin correspondence is *leakage*, which simply means energy that should fall into a single bin is spread out over a wide frequency range. To reduce the effect of leakage, the time-domain data is multiplied with a window function.

The graph in Fig. 2.18 clearly shows the effects of leakage. If the fundamental was at 585.94 Hz, a single line at bin 100 with no energy in adjacent bins would result. A rectangular window, which is equivalent to no window, was used in Fig. 2.18. The graph in Fig. 2.19 shows the effect of applying a Blackman–Harris window to the same data. After only four bins either side of the fundamental the leakage is down more than 90 dB. In actual application, an algorithm can determine the exact frequency and amplitude of the fundamental. To compensate for significant shifts in pitch, the spectra surrounding each expected multitone fundamental may be searched ±5% of the frequency of each fundamental.

Extraneous signals introduced into the signal path will also have a serious impact on a nonwindowed analysis. Interfering carriers, hum, and sync buzz are a few types of noncoherent signals that fit into this category. These noncoherent signals do not fall within the periodicity requirements for a nonwindowed FFT and will affect the noise floor of a graphical display in the same manner as will a shift in pitch of the multitone fundamentals. With windowing, 60-Hz hum will cause a noise spike at 60 Hz. Without windowing, the same hum energy will be spread across the audio spectrum, raising the entire noise floor.

For noise and distortion measurements, the eight bins surrounding each fundamental in a multitone are set to zero. The remaining bins are plotted to represent the noise (and distortion) floor of the device under test.

2.7 Considerations for Digital Audio Systems

Perhaps the single most common operating problem in a digital audio system is **clipping**, which occurs when the input range of the A/D converter is exceeded.[3] It is important to detect this situation because it can degrade audio quality and introduce distortion artifacts. Figure 2.20 shows an audio analyzer display of input signal clipping. Because of its importance, some form of clipping analysis is necessary for audio work. The behavior of the clipping detector in an analyzer can often be modified according to the audio program material and preferences of the audio engineer. For example, if the mix and material demand that no full-scale samples be present in the media, then the trigger point for the clip indicator (and counter) can be set to its most sensitive position. At this point, just one full-scale sample will trigger the clip indicator. On the other hand, when some full-scale samples are permitted, the sensitivity of the clip detector can be reduced by specifying the number of consecutive full-scale samples that must occur before the clipping detector is triggered.

When compact discs (CD) are mastered, engineers attempt to adjust audio levels so that the clip point is just reached. This practice obtains the maximum dynamic range of the 16-bit resolution of CD media. In this situation, the engineer might choose a clip detector trigger point of two or more consecutive samples before a clip registers. A similar capability may be provided for mute detection, with the sensitivity of the trigger adjusted to indicate a single muted sample or some number of consecutive samples.

An important capability of some analyzers is the compilation of statistics associated with a digital audio program or path. A display of such statistics is shown in Fig. 2.21. The following parameters are logged in the unit shown:

- *Highest true peak* retains the largest value indicated by the peak hold indicator.
- *Highest bar reading* retains the largest value of bar graph level. In the event that the user selects either volume unit (VU) or PPM behavior for the bar graphs, the highest bar reading will always be less than the highest true peak. Conversely, when true peak behavior is selected for the bar graphs, then highest true peak and highest bar reading will have the same value.
- *Number of clips* is a counter that accumulates the number of times the clip detector is triggered. Of course, the clip is only triggered when the sensitivity conditions (number of consecutive full scale samples) are satisfied.
- *Number of mutes* is a counter that accumulates the number of times the mute detector is triggered.
- *Invalid samples* is a counter that accumulates the number of samples that include a high validity bit.
- *Parity errors and code violations* are counters that accumulate the number of times that each of these interface errors are detected.

---

[3]This section is based on Tektronix. 1995. 764 Digital Audio Monitor. Tektronix Application Note 21W7269, Tektronix, Beaverton, OR.

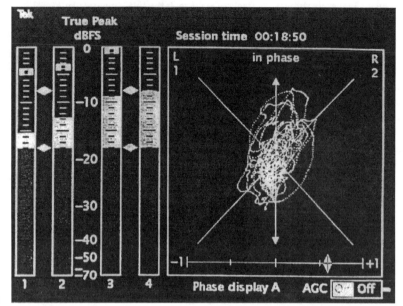
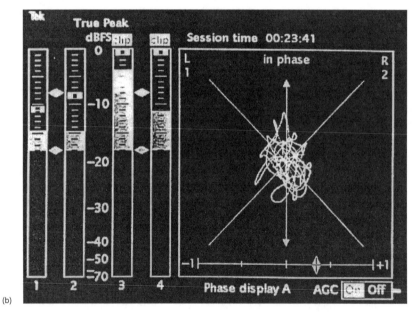

FIGURE 2.20 Detection of clipping in an A/D converter: (a) audio monitor display of normal program material below the clipping threshold, (b) display showing the effects of clipping. (*Source:* Tektronix, Beaverton, OR.)

- *Active bits* display the number of bits in a sample that are active or changing states.
- *DC offset* displays the DC offset contained in the encoded audio. This capability enables fine tuning of the DC offset characteristics of A/D converters.
- *Sample rate* displays the sample rate measured by the analyzer.

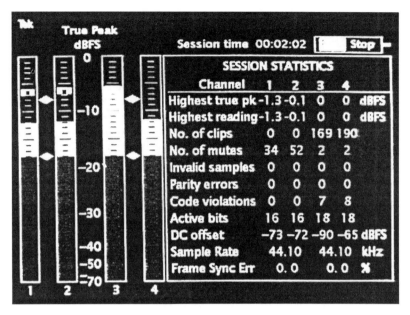

FIGURE 2.21 Audio monitor display showing statistics that characterize the performance of a digital audio system. (*Source:* Tektronix, Beaverton, OR.)

Defining Terms

Clipping: A distortion mechanism in an audio system in which the input level to one or more devices or circuits is sufficiently high in amplitude that the maximum input level of the device or circuit is exceeded, resulting in significant distortion of the output waveform.

Crest factor: In an audio system, the ratio of the true peak to the RMS value. For any signal but an ideal square wave the crest factor will be greater than one.

Group delay: The relative delay of the spectral components of a complex waveform. If the group delay is flat, all components arrive together.

Multitone testing: A measurement technique whereby an audio system is characterized by the simultaneous application of a combination of sine waves to a device under test.

Peak-equivalent sine: A method of specifying signal amplitude in an audio system, equal to the RMS level of a sine wave having the same peak-to-peak amplitude as the signal under consideration.

SMPTE IM: A method of measuring intermodulation distortion in an audio system standardized by the Society of Motion Picture and Television Engineers.

Weighting filter: A standardized filter used to impart predetermined characteristics to noise measurements in an audio system.

References

Benson, K.B., ed. 1988. *Audio Engineering Handbook*. McGraw–Hill, New York.

Benson, K.B. and Whitaker, J.C. 1990. *Television and Audio Handbook for Technicians and Engineers*. McGraw–Hill, New York.

Fink, D.G. and Christiansen, D., eds. 1989. *Electronics Engineers' Handbook*, 3rd ed. McGraw–Hill, New York.

Tektronix, 1995. 764 Digital Audio Monitor. Tektronix Application Note 21W7269, Tektronix, Beaverton, OR.

Thompson, B. 1992. Multitone Audio Testing. Tektronix Application Note 21W-7182, Tektronix, Beaverton, OR.

Whitaker, J.C. 1991. *Maintaining Electronic Systems*. CRC Press, Boca Raton, FL.

Whitaker, J.C. 1992. *Interconnecting Electronic Systems*. CRC Press, Boca Raton, FL.

Further Information

Most of the information presented in this chapter focuses on classic audio distortion mechanisms and analysis tools. There is also, however, an emerging discipline of digital audio analysis, which was only touched on briefly in this chapter. The best source for information on developing measurement techniques is the Audio Engineering Society (New York), which conducts trade shows and highly respected technical seminars annually in the U.S. and abroad. The society also publishes a monthly journal (*Journal of the Audio Engineering Society*) and numerous specialized books on the topic of audio technology in general, and digital audio in particular.

Readers are also directed to a well-respected book in the audio field, *Audio Engineering Handbook*, edited by K.B. Benson and published by McGraw–Hill. The publication date of the Benson work is 1988, and its coverage of digital audio is minimal. However, a second edition of the work, with extensive coverage of digital audio is underway. Another title of interest is *Principles of Digital Audio*, 3rd edition, written by Ken Pohlmann and published by McGraw–Hill (1996).

3
Video Display Distortion Mechanisms and Analysis

Jerry C. Whitaker
Editor-in-Chief

3.1 Introduction 19
3.2 Video Signal Spectra 19
   Minimum Video Frequency • Maximum Video Frequency • Horizontal Resolution • Video Frequencies Arising from Scanning
3.3 Measurement of Color Displays 23
3.4 Assessment of Color Reproduction 24
   Chromatic Adaptation and White Balance • Overall Gamma Requirements • Perception of Color Differences
3.5 Display Resolution and Pixel Format 26
   Contrast Ratio
3.6 Applications of the Zone Plate Signal 27
   Simple Zone Plate Patterns • Producing the Zone Plate Signal • The Time (Motion) Dimension
3.7 CRT Measurement Techniques 32
   Subjective CRT Measurements • Objective CRT Measurements

3.1 Introduction

New applications for electronic displays are pushing design engineers to produce systems that provide higher resolution, larger screen size, and better accuracy. More stringent performance demands improved quality assessment techniques in the manufacture, installation, and maintenance of display systems.[1]

3.2 Video Signal Spectra

The spectrum of the video signal arising from the scanning process in a television camera extends from a lower limit determined by the time rate of change of the average luminance of the scene to an upper limit determined by the time during which the scanning spots cross the sharpest vertical boundary in the scene as focused within the camera. This concept is illustrated in Fig. 3.1.

The distribution of spectrum components within these limits is determined by the following:

- The distribution of energy in the camera scanning spots
- Number of lines scanned per second
- Percentage of line-scan time consumed by horizontal blanking
- Number of fields scanned per second
- Rates at which the luminances and chrominances of the scene change in size, position, and boundary sharpness

[1] Portions of this chapter were adapted from: Whitaker J.C. 1993. *Electronic Displays: Technology, Design and Applications*. McGraw–Hill, New York. Used with permission.

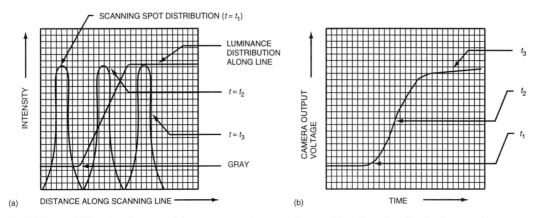

FIGURE 3.1 Video signal spectra: (a) camera scanning spot, shown with a Gaussian distribution, passing over a luminance boundary on a scanning line, (b) corresponding camera output signal resulting from convolution of the spot and luminance distributions.

To the extent that the contents and dynamic properties of the scene cannot be predicted, the spectrum limits and energy distribution are not defined. However, the spectra associated with certain static and dynamic test charts and tapes may be used as the basis for video system design. Among the configurations of interest are:

- Flat fields of uniform luminance and/or chrominance
- Fields divided into two or more segments of different luminance by sharp vertical, horizontal, or oblique boundaries

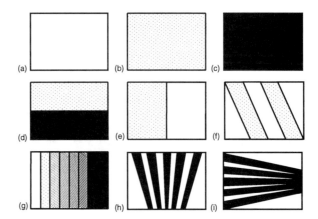

The latter case includes the horizontal and vertical wedges of test charts, as illustrated in Fig. 3.2, and the concentric circles of **zone plate** charts. The reproductions of such patterns typically display diffuse boundaries and other degradations that may be introduced by the camera scanning process, the amplitude and phase responses of the transmission system, the receiver scanning spots, and other artifacts associated with scanning.

FIGURE 3.2 Scanning patterns of interest in analyzing video signals: (a), (b), (c) flat fields useful for determining color purity and transfer gradient (gamma); (d) horizontal half-field pattern for measuring low-frequency performance; (e) vertical half-field for examining high-frequency transient performance; (f) display of oblique bars; (g) in monochrome, a tonal wedge for determining contrast and luminance transfer characteristics; in color, a display used for hue measurements and adjustments; (h) wedge for measuring horizontal resolution; (i) wedge for measuring vertical resolution.

The upper limit of the video spectrum actually employed in reproducing a particular image is most often determined by the amplitude-vs-frequency and phase-vs-frequency responses of the receiving system. These responses are selected as a compromise between the image sharpness demanded by typical viewers and the deleterious effects of noise, interference, and incomplete separation of the luminance and chrominance signals in the receiver.

Minimum Video Frequency

To reproduce a uniform value of luminance from top to bottom of an image scanned in the conventional interlaced fashion, the video signal spectrum must extend downward to include the field-scanning frequency. This frequency represents the lower limit of the spectrum arising from scanning an image whose luminance does not change. Changes in the average luminance are reproduced by extending the video spectrum to a lower frequency equal to the reciprocal of the duration of the luminance change. Because a given average luminance may persist for many minutes, the spectrum extends sensibly to zero frequency (DC). Various techniques of preserving or restoring the DC component are employed to extend the spectrum from the field frequency down to zero frequency.

Maximum Video Frequency

In the analysis of maximum operating frequency for a video system, three values must be distinguished:

- Maximum output signal frequency generated by the camera or other pickup/generating device
- Maximum modulating frequency corresponding to (1) the fully transmitted radiated sideband, or (2) the system used to convey the video signal from the source to the display
- Maximum video frequency present at the picture tube (display) control electrodes

The maximum camera frequency is determined by the design and implementation of the imaging element. The maximum modulating frequency is determined by the extent of the video channel reserved for the fully transmitted sideband. The channel width, in turn, is chosen to provide a value of horizontal resolution approximately equal to the vertical resolution implicit in the scanning pattern. The maximum video frequency at the display is determined by the device and support circuitry of the display system.

Horizontal Resolution

The horizontal resolution factor is the proportionality factor between horizontal resolution and video frequency. It may be expressed as

$$H_r = \frac{R_h}{\alpha} \times \iota$$

where:

H_r = horizontal resolution factor, lines/MHz
R_h = lines of horizontal resolution per hertz of the video waveform
α = aspect ratio of the display
ι = active line period, μs

For the National Television Systems Committee (NTSC), the horizontal resolution factor is

$$78.8 = \frac{2}{4/3} \times 52.5$$

Video Frequencies Arising from Scanning

The signal spectrum arising from scanning comprises a number of discrete components at multiples of the scanning frequencies. Each spectrum component is identified by two numbers, m and n, which describe the pattern that would be produced if that component alone were present in the signal. The value of m represents the number of sinusoidal cycles of brightness measured horizontally (in the width of the picture) and n the number of cycles measured vertically (in the picture height). The 0, 0 pattern is the DC component of the signal, the 0, 1 pattern is produced by the field-scanning frequency, and the 1, 0 pattern by the line scanning frequency. Typical patterns for various values of m and n are shown in Fig. 3.3. By combining a number of such patterns (including m and n values up to several hundred), in the appropriate amplitudes and phases, any image capable of being represented by the scanning pattern may be built up. This is a two-dimensional form of the Fourier series.

The amplitudes of the spectrum components decrease as the values of m and n increase. Because m represents the order of the harmonic of the line-scanning frequency, the corresponding amplitudes are those of the left-to-right variations in brightness. A typical spectrum resulting from scanning a static scene is shown in Fig. 3.4. The components of major magnitude include:

- The DC component
- Field-frequency component
- Components of the line frequency and its harmonics

Surrounding each line-frequency harmonic is a cluster of components, each separated from the next by an interval equal to the field-scanning frequency.

It is possible for the clusters surrounding adjacent line-frequency harmonics to overlap one another. This corresponds to two patterns in Fig. 3.4 situated on adjacent vertical columns, which produce the same value of video frequency when scanned. Such intercomponent confusion of spectral energy is fundamental to the scanning process. Its effects are visible when a heavily striated pattern (such as that of a fabric having an accented weave) is scanned with the striations approximately parallel to the scanning lines. In the NTSC and PAL color systems, in which the luminance and chrominance signals occupy the same spectral region (one being interlaced in frequency with the other), such intercomponent confusion may produce prominent color fringes. Precise filters, which sharply separate the luminance and chrominance signals (comb filters), can remove this effect, except in the diagonal direction.

In static and slowly moving scenes, the clusters surrounding each line-frequency harmonic are compact, seldom extending further than 1 or 2 kHz on either side of the line-harmonic frequency. The space remaining in the signal spectrum is unoccupied and

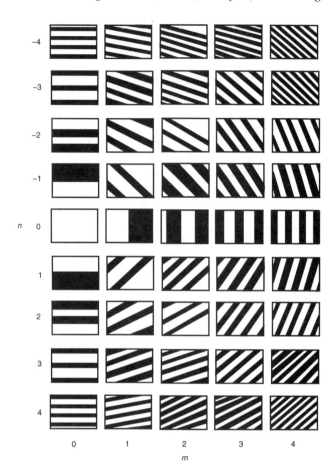

FIGURE 3.3 An array of image patterns corresponding to indicated values of m and n.

may be used to accommodate the spectral components of another signal having the same structure and frequency spacing. For scenes in which the motion is sufficiently slow for the eye to perceive the detail of the moving objects, it may be safely assumed that less than half the spectral space between line-frequency harmonics is occupied by energy of significant magnitude. It is on this principle that the NTSC and PAL compatible color television systems are based. The SECAM system uses frequency-modulated chrominance signals, which are not frequency interlaced with the luminance signal.

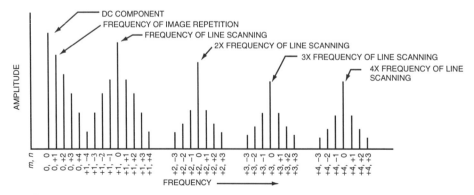

FIGURE 3.4 The typical spectrum of a video signal, showing the harmonics of the line-scanning frequency surrounded by clusters of components separated at intervals equal to the field-scanning frequency.

3.3 Measurement of Color Displays

The chromaticity and luminance of a portion of a color display device may be measured in several ways. The most fundamental approach involves a complete spectroradiometric measurement followed by computation using tables of color matching functions. Portable spectroradiometers with builtin computers are available for this purpose. Another method, somewhat faster but less accurate, involves the use a photoelectric colorimeter. These devices have spectral sensitivities approximately equal to the CIE color-matching functions and thus provide direct readings of tristimulus values.

For setting up the reference white, it is often simplest to use a split-field visual comparator and to adjust the display device until it matches the reference field (usually D65) of the comparator. However, because there is usually a large spectral difference (large metamerism) between the display and the reference, different observers

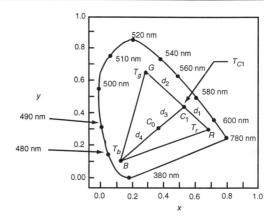

FIGURE 3.5 The CIE 1931 chromaticity diagram illustrating use of the center of gravity law ($T_r d_1 = T_g d_2$, $T_{c1} = T_r + T_g$, $T_{c1} d_3 = T_b d_4$).

will often make different settings by this method. Thus settings by one observerÑor a group of observersÑwith normal color vision are often used simply to provide a reference point for subsequent photoelectric measurements.

An alternative method of determining the luminance and chromaticity coordinates of any area of a display involves measurement of the output of each phosphor separately and combining them using the center of gravity law in which the total tristimulus output of each phosphor is considered as an equivalent weight located at the chromaticity coordinates of the phosphor.

Consider the CIE chromaticity diagram shown in Fig. 3.5 to be a uniform flat surface positioned in a horizontal plane. For the case illustrated, the center of gravity of the three weights (T_r, T_g, T_b), or the balance point, will be at the point Co. This point determines the chromaticity of the mixture color. The luminance of the color Co will be the linear sum of the luminance outputs of the red, green, and blue phosphors. The chromaticity coordinates of the display primaries may be obtained from the manufacturer. The total tristimulus output of one phosphor may be determined by turning off the other two cathode ray tube (CRT) guns, measuring the luminance of the specified area, and dividing this value by the y chromaticity coordinate of the energized phosphor. This procedure is then repeated for the other two phosphors. From these data the color resulting from given excitations of the three phosphors may be calculated as follows:

- Chromaticity coordinates of red phosphor = xr, yr
- Chromaticity coordinates of green phosphor = xg, yg
- Chromaticity coordinates of blue phosphor = xb, yb
- Luminance of red phosphor = Yr
- Luminance of green phosphor = Yg
- Luminance of blue phosphor = Yb

Total tristimulus value of red phosphor = $X_r + Y_r + Z_r = Y_r/y_r = T_r$. Total tristimulus value of green phosphor = $X_g + Y_g + Z_g = Y_g/y_g = T_g$. Total tristimulus value of blue phosphor = $X_b + Y_b + Z_b = Y_b/y_b = T_b$.

Consider T_r as a weight located at the chromaticity coordinates of the red phosphor and T_g as a weight located at the chromaticity coordinates of the green phosphor. The location of the chromaticity coordinates of color $C1$ (blue gun of color CRT turned off) can be determined by taking moments along line RG to determine the center of gravity of weights T_r and T_g,

$$T_r \times d_1 = T_g \times d_2$$

The total tristimulus value of $C1$ is equal to $T_r + T_g = T_{C1}$. Taking moments along line C_1B will locate the chromaticity coordinates of the mixture color Co,

$$T_{C1} \times d_3 = T_b \times d_4$$

The luminance of the color Co is equal to $Y_r + Y_g + Y_b$.

3.4 Assessment of Color Reproduction

A number of factors may contribute to poor color rendition in a display system. To assess the effect of these factors, it is necessary to define system objectives, and then establish a method of measuring departures from the objectives. Visual image display may be categorized as follows:

- **Spectral color reproduction**. The exact reproduction of the spectral power distributions of the original stimuli. Clearly this is not possible in a video system with three primaries.
- Exact color reproduction. The exact reproduction of tristimulus values. The reproduction is then a metameric match to the original. Exact color reproduction will result in equality of appearance only if the viewing conditions for the picture and the original scene are identical. These conditions include the angular subtense of the picture, the luminance and chromaticity of the surround, and glare. In practice, exact color reproduction often cannot be achieved because of limitations on the maximum luminance that can be produced on a color monitor.
- Colorimetric color reproduction. A variant of exact color reproduction in which the tristimulus values are proportional to those in the original scene. In other words, the chromaticity coordinates are reproduced exactly, but the luminances are all reduced by a constant factor. Traditionally, color video systems have been designed and evaluated for colorimetric color reproduction. If the original and the reproduced reference whites have the same chromaticity, if the viewing conditions are the same, and if the system has an overall gamma of unity, colorimetric color reproduction is indeed a useful criterion. However, these conditions often do not hold and then colorimetric color reproduction is inadequate.
- Equivalent color reproduction. The reproduction of the original color appearance. This might be considered as the ultimate objective but cannot be achieved because of the limited luminance that can be generated in a display system.
- **Corresponding color reproduction**. A compromise in which colors in the reproduction have the same appearance as the colors in the original would have had if they had been illuminated to produce the same average luminance level and the same reference white chromaticity as that of the reproduction. For most purposes, corresponding color reproduction is the most suitable objective of a color video system.
- Preferred color reproduction. A departure from the preceding categories that recognizes the preferences of the viewer. It is sometimes argued that corresponding color reproduction is not the ultimate aim for some display systems, such as color television, but that account should be taken of the fact that people prefer some colors to be different from their actual appearance. For example, sun-tanned skin color is preferred to average real skin color, and sky is preferred bluer and foliage greener than they really are.

Even if corresponding color reproduction is accepted as the target, it is important to remember that some colors are more important than others. For example, flesh tones must be acceptableÑnot obviously reddish, greenish, purplish, or otherwise incorrectly rendered. Similarly the sky must be blue and the clouds white, within the viewer's range of acceptance. Similar conditions apply to other well-known colors of common experience.

Chromatic Adaptation and White Balance

With properly adjusted cameras and displays, whites and neutral grays are reproduced with the chromaticity of D65. Tests have shown that such whites (and grays) appear satisfactory in home viewing situations even if the ambient light is of quite different color temperature. Problems occur, however, when the white balance is slightly different from one camera to the next or when the scene shifts from studio to daylight or vice versa. In the first case, unwanted shifts of the displayed white occur; whereas in the other, no shift occurs even though the viewer subconsciously expects a shift.

By always reproducing a white surface with the same chromaticity, the system is mimicking the human visual system, which adapts so that white surfaces always appear the same whatever the chromaticity of the illuminant (at least within the range of common light sources). The effect on other colors, however, is more complicated. In video cameras the white balance adjustment is usually made by gain controls on the R, G, and B channels. This is similar to the von Kries model of human chromatic adaptation, although the R, G, and B primaries of the model are not the same as the video primaries. It is known that the von Kries model does not accurately account for the appearance of colors after chromatic adaptation, and so it follows that simple gain changes in a video camera is not the ideal approach. Nevertheless, this approach seems to work well in practice, and the viewer does not object to the fact, for example, that the relative increase in the luminances of reddish objects in tungsten light is lost.

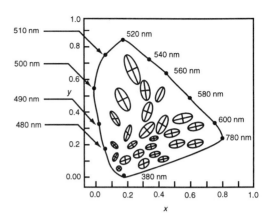

FIGURE 3.6 Ellipses of equally perceptible color differences.

Overall Gamma Requirements

Colorimetric color reproduction requires that the overall gamma of the system, including the camera, the display, and any gamma-adjusting electronics, be unity. This simple criterion is the one most often used in the design of a video color rendition system. However, the more sophisticated criterion of corresponding color reproduction takes into account the effect of the viewing conditions. In particular, several studies have shown that the luminance of the surround is important. For example, a dim surround requires a gamma of about 1.2, and a dark surround requires a gamma of about 1.5 for optimum color reproduction.

Perception of Color Differences

The CIE 1931 chromaticity diagram does not map chromaticity on a uniform-perceptibility basis. A just-perceptible change of chromaticity is not represented by the same distance in different parts of the diagram. Many investigators have explored the manner in which perceptibility varies over the diagram. The most often quoted study is that of MacAdam [1942] who identified a set of ellipses that are contours of equal perceptibility about a given color, as shown in Fig. 3.6.

From this and similar studies it is apparent, for example, that large distances represent relatively small perceptible changes in the green sector of the diagram. In the blue region, much smaller changes in the chromaticity coordinates are readily perceived.

Further, viewing identical images on dissimilar displays can result in observed differences in the appearance of the image [Bender and Blount 1992]. There are several factors that affect the appearance of the image:

- Physical factors, including display gamut, illumination level, and black point
- Psychophysical factors, including chromatic induction and color constancy

Each of these factors interact in such a way that prediction of the appearance of an image on a given display becomes difficult. System designers have experimented with colorimetry to facilitate the sharing of image data among display devices that vary in manufacture, calibration, and location. Of particular interest is the application of colorimetry to imaging in a networked window system environment, where it is often necessary to assure that an image displayed remotely looks like the image displayed locally.

Studies have indicated that image context and image content are also factors that affect color appearance. The use of highly chromatic backgrounds in a windowed display system is popular, but will impact the appearance of the colors in the foreground.

3.5 Display Resolution and Pixel Format

The **pixel** represents the smallest resolvable element of a display. The size of the pixel varies from one type of display to another. In a monochrome CRT, pixel size is determined primarily by the following factors:

- Spot size of the electron beam (the current density distribution)
- Phosphor particle size
- Thickness of the phosphor layer

The term *pixel* was developed in the era of monochrome television, and the definition was—at that time—straightforward. With the advent of color triad-based CRTs and solid state display systems, the definition is not nearly so clear.

For a color CRT, a single triad of red, green, and blue phosphor dots constitutes a single pixel. This definition assumes that the mechanical and electrical parameters of the CRT will permit each triad to be addressed without illuminating other elements on the face of the tube. Most display systems, however, will not meet this criteria. Depending on the design, a number of triads may constitute a single pixel in a CRT display. A more all-inclusive definition for the pixel is the smallest spatial-information element as seen by the viewer [Tannas 1985].

Dot pitch is one of the principle mechanical criteria of a CRT that determines—to a large extent—the resolution of the display. Dot pitch is defined as the center-to-center distance between adjacent green phosphor dots of the red, green, blue triad.

The **pixel format** is the arrangement of pixels into horizontal rows and vertical columns. For example, an arrangement of 640 horizontal pixels by 480 vertical pixels results in a 640 × 480 pixel format. This description is not a resolution parameter by itself, simply the arrangement of pixel elements on the screen. *Resolution* is the measure of the ability to delineate picture detail; it is the smallest discernible and measurable detail in a visual presentation (standards and definitions committee, Society for Information Display).

Pixel density is a parameter closely related to resolution, stated in terms of pixels per linear distance. Pixel density specifies how closely the pixel elements are spaced on a given display. It follows that a display with a given pixel format will not provide the same pixel density (resolution) on a large size screen, such as 19-in diagonal, as on a small size screen, such as 12-in diagonal.

Television lines is another term used to describe resolution. The term refers to the number of discernible lines on a standard test chart. As before, the specification of television lines is not by itself a description of display resolution. A 525-line display on a 17-in monitor will appear to have greater resolution to a viewer than a 525-line display on a 30-in monitor. Pixel density is the preferred resolution parameter.

Contrast Ratio

The purpose of a video display is to convey information by controlling the illumination of phosphor dots on a screen, or by controlling the reflectance or transmittance of a light source. The contrast ratio specifies the observable difference between a pixel that is switched on and its corresponding off state,

$$C_r = \frac{L_{\text{on}}}{L_{\text{off}}}$$

where:

C_r = contrast ratio of the display
L_{on} = luminance of a pixel in the on state
L_{off} = luminance of a pixel in the off state

The area encompassed by the contrast ratio is an important parameter when considering the performance of a display. Two contrast ratio divisions are typically specified:

- *Small area*, comparison of the on and off states of a pixel-sized area
- *Large area*, comparison of the on and off states of a group of pixels

For most display applications, the small area contrast ratio is the more critical parameter.

3.6 Applications of the Zone Plate Signal

The increased information content of advanced, high-definition display systems requires sophisticated processing to make recording and transmission practical.[2] This processing uses various forms of bandwidth compression, scan rate changes, motion detection and compensation algorithms, and other techniques. Zone plate patterns are well suited to exercising a complex video system in the three dimensions of its signal spectrum: horizontal, vertical, and temporal. Zone plate signals, unlike most conventional test signals, can be complex and dynamic. Because of this, they are capable of simulating much of the detail and movement of actual picture video, exercising the system under test with signals representative of the intended application. These digitally generated and controlled signals also have other important characteristics needed in test signals.

A signal intended for meaningful testing of a video system must be carefully controlled, so that any departure from a known parameter of the signal is attributable to a distortion or other change in the system under test. The test signal must also be predictable, so it can be accurately reproduced at other times or places. These constraints have usually led to test signals that are electronically generated. In a few special cases, a standardized picture has been televised by a camera or monoscopeÑusually for a subjective, but more detailed evaluation of system performance. A zone plate is a physical optical pattern, which was first used by televising it in this way. Now that electronic generators are capable of producing similar patterns, the label zone plate is applied to the wide variety of patterns created by video test instruments.

Conventional test signals, for the most part limited by the practical considerations of electronic generation, have represented relatively simple images. Each signal is capable of testing a narrow range of possible distortions; several test signals are needed for a more complete evaluation. Even with several signals, this method may not reveal all possible distortions, or allow study of all pertinent characteristics. This is especially true in video systems employing sophisticated signal processing.

Simple Zone Plate Patterns

The basic testing of a video communication channel historically has involved the application of several single frequenciesÑin effect, spot checking the spectrum of interest. A well known and quite practical adaptation of this idea is the multiburst signal, as shown in Fig. 3.7. This test waveform has been in use since the earliest days of video. The multiburst signal provides several discrete frequencies along a TV line.

The frequency sweep signal is an improvement on multiburst. Although harder to implement in earlier generators, it was easier to use. The frequency sweep signal, illustrated in Fig. 3.8, varies the applied signal

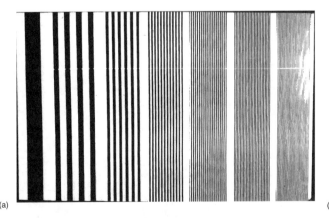
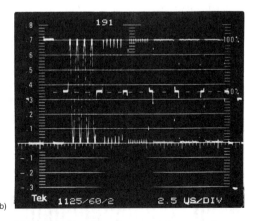

FIGURE 3.7 Multiburst video test waveform: (a) picture display, (b) multiburst signal as viewed on a waveform monitor (1H). (*Source:* Tektronix, Beaverton, OR.)

---

[2]Portions of this section were adapted from: Tektronix. 1992. Broadening the applications of zone polate generators. Application Note 20W7056, Tektronix, Beaverton, OR.

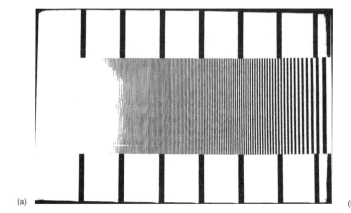 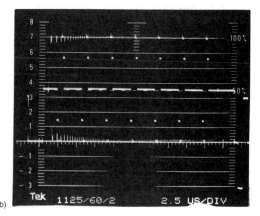

FIGURE 3.8 Conventional sweep frequency test waveform: (a) picture display, (b) waveform monitor display, with markers (1H). (*Source:* Tektronix, Beaverton, OR.)

frequency continuously along the TV line. [Figure 3.8 and other photographs in this section show the beat effects introduced by the screening process used for photographic printing. This is largely unavoidable. The screening process is quite similar to the scanning or sampling of a television imageÑthe patterns are designed to identify this type of problem.] In some cases, the signal is swept as a function of the vertical position (field time). Even in these cases, the signal being swept is appropriate for testing the spectrum of the horizontal dimension of the picture.

Figure 3.9 shows the output of a zone plate generator configured to produce a horizontal single-frequency output. Figure 3.10 shows a zone plate generator configured to produce a frequency sweep signal. Electronic test patterns, such as these, may be used to evaluate the following system characteristics:

- Channel frequency response
- Horizontal resolution
- Moirâ effects in recorders and displays
- Other impairments

Patterns that test vertical (field) response have, traditionally, been less frequently used. As new technologies implement conversion from interlaced to progressive scan, line doubling display techniques, vertical anti-aliasing filters, scan conversion, motion detection, or other processes that combine information from line to line, vertical testing patterns will be more in demand.

In the vertical dimension, as well as the horizontal, tests may be done at a single frequency or with a frequency sweep signal. Figure 3.11 illustrates a magnified vertical rate waveform display. Each dash in the photograph

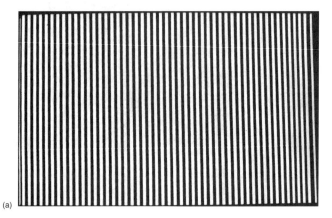 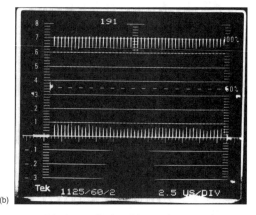

FIGURE 3.9 Single horizontal frequency test signal from a zone plate generator: (a) picture display, (b) waveform monitor display (1H). (*Source:* Tektronix, Beaverton, OR.)

Video Display Distortion Mechanisms and Analysis

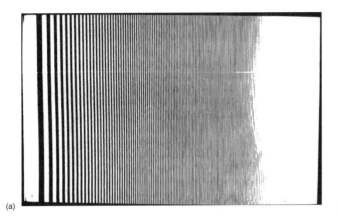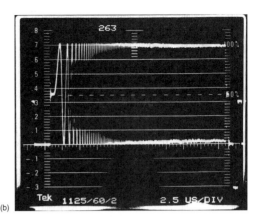

FIGURE 3.10 Horizontal frequency sweep test signal from a zone plate generator: (a) picture display, (b) waveform monitor display (1H). (*Source:* Tektronix, Beaverton, OR.)

represents one horizontal scan line. Sampling of vertical frequencies is inherent in the scanning process, and the photo shows the effects on the signal waveform. Note also that the signal voltage remains constant during each line, and changes only from line to line in accord with the vertical dimension sine function of the signal. Figure 3.12 shows a vertical frequency sweep picture display.

The horizontal and vertical sinewaves and sweeps are quite useful, but they do not use the full potential of a zone plate signal source.

Producing the Zone Plate Signal

A zone plate signal is created in real time by the test signal generator. The value of the signal at any instant is represented by a number in the digital hardware. This number is incremented as the scan progresses through the three dimensions that define a point in the video image; horizontal position, vertical position, and time.

The exact method in which these dimensions alter the number is controlled by a set of coefficients. These coefficients determine the initial value of the number and control the size of the increments as the scan progresses along each horizontal line, from line to line vertically, and from field to field temporally. A set of coefficients uniquely determines a pattern, or a sequence of patterns when the time dimension is active.

This process produces a sawtooth waveform; overflow in the accumulator holding the signal number effectively resets the value to zero at the end of each cycle of the waveform. Usually it is desirable to minimize the harmonic energy content of the output signal; in this case, the actual output is a sine function of the number generated by the incrementing process.

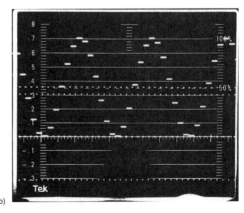

FIGURE 3.11 Single vertical frequency test signal: (a) picture display, (b) magnified vertical rate waveform, showing the effects of scan sampling. (*Source:* Tektronix, Beaverton, OR.)

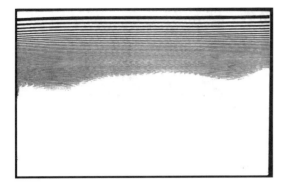

FIGURE 3.12 Vertical frequency sweep picture display. (*Source:* Tektronix, Beaverton, OR.)

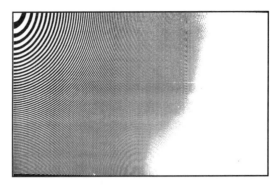

FIGURE 3.13 Combined horizontal and vertical frequency sweep picture display. (*Source:* Tektronix, Beaverton, OR.)

Complex Patterns

A pattern of sinewaves or sweeps in multiple dimensions may be produced, using the unique architecture of the zone plate generator. The pattern shown in Fig. 3.13, for example is a single signal sweeping both horizontally and vertically. Figure 3.14 shows the waveform of a single selected line (line 263 in the 1125/60/2 HDTV system). Note that the horizontal waveform is identical to Fig. 3.8(b), even though the vertical dimension sweep is now also active. Actually, different lines will give slightly different waveforms. The horizontal frequency and sweep characteristics will be identical, but the starting phase must be different from line to line to construct the vertical signal.

Figure 3.15 shows a two-axis sweep pattern that is most often identified with zone plate generators; perhaps because it quite closely resembles the original optical pattern. In this circle pattern, both horizontal and vertical frequencies start high, sweep to zero (in the center of the screen), and sweep up again to the end of their respective scans. The concept of two-axis sweeps is actually more powerful than it might, at first, appear. In addition to purely horizontal or vertical effects, there are possible distortions or artifacts that are only apparent with simultaneous excitation in both axes. In other words, the response of a system to diagonal detail may not be predictable from information taken from the horizontal and vertical responses.

Consider an example from NTSC. A comb filter will suppress crosscolor effects from a horizontal luminance frequency signal near the subcarrier frequency (such as the higher frequency packets in a NTSC multiburst signal). If, however, the right vertical component is added, creating a specific diagonal luminance pattern, even the most complex decoders will interpret the pattern as colored. In this case, the two-axis sweep shows very different effects than the same sweeps shown individually. A multidimensional sweep is a powerful tool for identifying analogous effects in other complex signal processing systems.

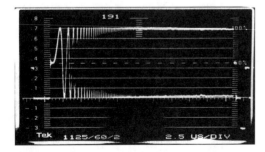

FIGURE 3.14 Combined horizontal and vertical frequency sweeps, selected line waveform display (1H). This figure shows the maintenance of horizontal structure in the presence of vertical sweep. (*Source:* Tektronix, Beaverton, OR.)

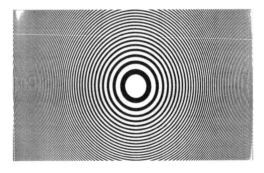

FIGURE 3.15 The best known zone plate pattern, combined horizontal and vertical frequency sweeps with zero frequency in the center screen. (*Source:* Tektronix, Beaverton, OR.)

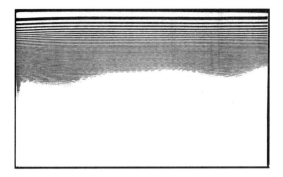

FIGURE 3.16 Vertical frequency sweep picture display. (*Source:* Tektronix, Beaverton, OR.)

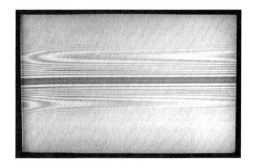

FIGURE 3.17 The same vertical sweep as Fig. 3.16, except that appropriate pattern motion has been added to freeze the beat pattern in the center screen for photography or other analysis. (*Source:* Tektronix, Beaverton, OR.)

The Time (Motion) Dimension

Incrementing the number in the accumulator of the zone plate generator from frame to frame (or field to field in an interlaced system) creates a predictably different pattern for each vertical scan. This, in turn, creates apparent motion and exercises the signal spectrum in the temporal dimension. Analogous to the single-frequency and frequency sweep examples given previously, appropriate setting of the time-related coefficients will create constant motion or motion sweep (acceleration).

Specific motion detection and interpolation algorithms in a system under test may be exercised by determining the coefficients of a critical sequence of patterns. These patterns may then be saved for subsequent testing during development or adjustment. In an operational environment, appropriate response to a critical sequence could ensure expected operation of the equipment or facilitate fault detection.

Although motion artifacts are difficult to portray in the still image constraints of a printed book, the following example gives some idea of the potential of a versatile generator. In Fig. 3.16 the vertical sweep maximum frequency has been increased to the point where it is zero-beating with the scan at the bottom of the screen. (The cycles per picture height of the pattern matches the lines per picture height per field of the scan.) Actually, in direct viewing, there is another noticeable artifact in the vertical center of the screen; it is an harmonic beat related to the gamma of the display CRT. Because of interlace, this beat flickers at the field rate. The photograph integrates the interfield flicker and thereby hides the artifact, which is readily apparent when viewed in real time.

Figure 3.17 is identical to the previous photograph, except for one important difference—upward motion of $\frac{1}{2}$ cycle per field has been added to the pattern. Now the sweep pattern itself is integrated out, as is the first-order beat at the bottom. The harmonic effects in center screen no longer flicker, because the change of scan vertical position from field to field is compensated by a change in position of the image. The resulting beat pattern does not flicker and is easily photographed or, perhaps, scanned to determine depth of modulation.

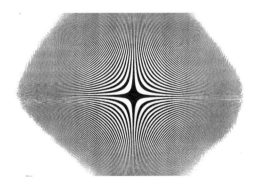

FIGURE 3.18 A hyperbolic variation of the two-axis zone plate frequency sweep. (*Source:* Tektronix, Beaverton, OR.)

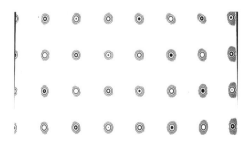

FIGURE 3.19 A two-axis frequency sweep in which the range of frequencies is swept several times in each axis. Complex patterns such as this may be created for specific test requirements. (*Source:* Tektronix, Beaverton, OR.)

A change in coefficients produces hyperbolic, rather than circular two-axis patterns, as shown in Fig. 3.18. Another interesting pattern, which has been suggested for checking complex codecs, is shown in Fig. 3.19. This is also a moving pattern, which was slightly altered to freeze some aspects of the movement to take the photograph.

3.7 CRT Measurement Techniques

A number of different techniques have evolved for measuring the static performance of CRT display devices and systems.[3] Most express the measured device performance in a unique figure of merit or metric. Whereas each approach provides useful information, the lack of standardization in measurement techniques makes it difficult or impossible to directly compare the performance of a given class of CRT.

Regardless of the method used to measure performance, the operating parameters must be set for the anticipated operating environment. Key parameters include:

- Input signal level
- System/display line rate
- Luminance (brightness)
- Contrast
- Image size
- Aspect ratio

If the display is used in more than one environmental condition, such as under day and night conditions, a set of measurements is appropriate for each application.

Subjective CRT Measurements

Three common subjective measurement techniques are used to assess the performance of a CRT:

- *Shrinking raster*
- *Line width*
- *TV limiting resolution*

Predictably, subjective measurements tend to exhibit more variability than objective measurements. Although not generally used for acceptance testing or quality control, subjective CRT measurements provide a fast and relatively simple means of performance assessment. Results are usually consistent when performed by the same observer. However, results for different observers are often not consistent because different observers use different visual criteria to make their judgments.

The shrinking raster and line width techniques are used to estimate the vertical dimension of the CRT beam spot size (*footprint*). There are several underlying assumptions with this approach:

- The spot is assumed to be symmetrical and Gaussian in the vertical and horizontal planes.
- The display modulation transfer function (MTF) calculated from the spot size measurement results in the best performance envelope that can be expected from the CRT.
- The modulating electronics are designed with sufficient bandwidth so that spot size is the limiting performance parameter.
- The modulation contrast at low spatial frequencies approaches 100%.

Depending on the application, not all of these assumptions are valid:

- Assumption 1. Verona [1992] has reported that the symmetry assumption is generally not true. The vertical spot profile is only an approximation to the horizontal spot profile; most spot profiles exhibit some degree of astigmatism. However, significant deviations from the symmetry and Gaussian assumptions result in only minor deviations from the projected performance when the assumptions are correct.

---

[3]This section is based on: Verona, R. 1992. Comparison of CRT display measurement techniques. In *Helmet-Mounted Displays III*, ed. T.M. Lippert. *Proc. SPIE* 1695:117–127.

- Assumption 2. The optimum performance envelope assumption infers that other types of measurements will result in the same or lower modulation contrast at each spatial frequency. The MTF calculations based on a beam footprint in the vertical axis indicate the optimum performance that can be obtained from the display because finer detail (higher spatial frequency information) cannot be written onto the screen smaller than the spot size.
- Assumption 3. The modulation circuit bandwidth must be sufficient to pass the full incoming video signal. Typically, the video circuit bandwidth is not a problem with current technology circuits, which are usually designed to provide significantly more bandwidth than the CRT is capable of displaying. However, in cases where this assumption is not true, the calculated MTF based purely on the vertical beam profile will be incorrect. The calculated performance will be better than the actual performance of the display.
- Assumption 4. The calculated MTF is normalized to 100% modulation contrast at zero spatial frequency and ignores the light scatter and other factors that degrade the actual measured MTF. Independent modulation contrast measurements at a low spatial frequency can be used to adjust the MTF curve to correct for the normalization effects.

Shrinking Raster Method

The shrinking raster measurement procedure is relatively simple. Steps include the following:

- The brightness and contrast controls are set for the desired peak luminance with an active raster background luminance (1% of peak luminance) using a stair-step video signal.
- While displaying a flat field video signal input corresponding to the peak luminance, the vertical gain/size is reduced until the raster lines are barely distinguishable.
- The raster height is measured and divided by the number of active scan lines to estimate the average height of each scan line. The number of active scan lines is typically 92% of the line rate. (For example, a 525 line display has 480 active lines, an 875 line display has 817, and a 1025 line display has 957 active lines.)

The calculated average line height is typically used as a stand-alone metric of display performance.

The most significant shortcoming of the shrinking raster method is the variability introduced through the determination of when the scan lines are *barely distinct* to the observer. Blinking and eye movements often enhance the distinctness of the scan lines; lines that were indistinct become distinct again. Further, although the shrinking raster procedure can produce acceptable results on large format CRT displays, it is less accurate for miniature devices (1-in diameter and less). The nominal line spacing for a 1-in CRT operating with full raster (at 875 line-rate video) is approximately 15.6 to 15.8 μm. Because the half-intensity spot width is about 20–22 μm, there is a significant amount of spot overlap between the scan lines before the raster height is reduced.

Line Width Method

The line-width measurement technique requires a microscope with a calibrated graticule. The focused raster is set to a 4:3 aspect ratio, and the brightness and contrast controls are set for the desired peak luminance with an active raster background luminance (1% of peak luminance) using a stair-step video signal. A single horizontal line at the anticipated peak operating luminance is presented in the center of the display. The spot is measured by comparing its luminous profile with the graticule markings. As with the shrinking raster technique, determination of the line edge is subjective.

TV Limiting Resolution Method

This technique involves the display of two-dimensional, high-contrast bar patterns or lines of various size, spacing, and angular orientation. The observer subjectively determines the limiting resolution of the image by the smallest set of bars that can be resolved. There are several potential errors with this technique, including:

- A phenomenon called *spurious resolution* can occur that leads the observer to overestimate the limiting resolution. Spurious resolution occurs beyond the actual resolution limits of the display. It appears as fine structures that can be perceived as line spacings closer than the spacing at which the bar pattern first completely blurs. This situation arises when the frequency response characteristics fall to zero, then go negative, and perhaps oscillate as they die out. At the bottom of the negative trough, contrast is restored, but in reverse phase (white becomes black and black becomes white).

- The use of test charts imaged with a video camera can lead to incorrect results because of the addition of camera resolution considerations to the measurement. Electronically generated test patterns are more reliable image sources.

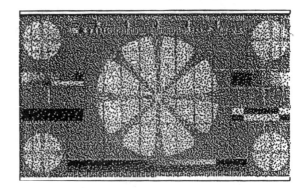

FIGURE 3.20 Wide aspect ratio resolution test chart produced by an electronic signal generator. (*Source:* Tektronix, Beaverton, OR.)

The proper setting of brightness and contrast is required for this measurement. Brightness and contrast controls are adjusted for the desired peak luminance with an active raster background luminance (1% of peak luminance) using a stair-step video signal. Too much contrast will result in an inflated limiting resolution measurement; too little contrast will result in a degraded limiting resolution measurement.

Electronic resolution pattern generators typically provide a variety of resolution signals from 100 to 1000 *TV lines/picture height* (TVL/PH) or more in a given multiple (such as 100). Figure 3.20 illustrates an electronically-generated resolution test pattern for high- definition video applications.

Application Considerations

The subjective techniques discussed in this section, with the exception of TV limiting resolution, measure the resolution of the *display*. The TV pattern test measures *image* resolution, which is quite different.

Consider as an example a video display in which the scan lines can just be perceivedÑabout 480 scan lines per picture height. This indicates a *display* resolution of at least 960 TV linesÑcounting light *and* dark lines, per the convention. If a pattern from an electronic generator is displayed, observation will show the image beginning to deteriorate at about 340 TV lines. This characteristic is the result of beats between the image pattern and the raster, with the beat frequency decreasing as the pattern spatial frequency approaches the raster spatial frequency. This ratio of 340/480 = 0.7 (approximately) is known as the *Kell factor*. Although debated at length, the factor does not change appreciably in subjective observations.

Objective CRT Measurements

Four common types of objective measurements may be performed to assess the capabilities of a CRT:

- *Half-power width*
- *Impulse Fourier transform*
- *Knife edge Fourier transform*
- *Discrete frequency*

Although more difficult to perform than the subjective measurements discussed so far, objective CRT measurement techniques offer greater accuracy and better repeatability. Some of the procedures require specialized hardware and/or software.

Half-Power Width Method

The half-power width technique is appropriate for large displays (9 in and larger), but is not reliable when used to measure line width on a miniature CRT. A single horizontal line is activated with the brightness and contrast controls set to a typical operating level (as discussed previously). The line luminance is equivalent to the highlight luminance (maximum signal level). The central portion of the line is imaged with a microscope in the plane of a variable width slit. The open slit allows all of the light from the line to pass through to a photodetector. The output of the photodetector is displayed on an oscilloscope. As the slit is gradually closed, the peak amplitude of the photodetector signal decreases. When the signal drops to 50% of its initial value, the slit width is recorded. The width measurement divided by the microscope magnification represents the half-power width of the horizontal scan line. The half-power width of a miniature CRT may be measured using a scanning microphotometer and software to perform numerical integration on the luminance profile.

The half-power width is defined as the distance between symmetrical integration limits, centered about the maximum intensity point, which encompasses half of the total power under the intensity curve. The half-power width is not the same as the half-intensity width measured between the half-intensity points. The half-intensity width is theoretically 1.75 times greater than the half-power width for a Gaussian spot luminance distribution.

It should be noted that the half-power line width technique relies on line width to predict the performance of the CRT. Many of the precautions outlined in the previous section apply here also. The primary difference, however, is that line width is measured under this technique objectively, rather than subjectively.

Fourier Transform Methods

The impulse Fourier transform technique involves measuring the luminance profile of the spot and then taking the Fourier transform of the distribution to obtain the MTF. The MTF, by definition, is the Fourier transform of the line spread function. Commercially available software may be used to perform these measurements using either an impulse or knife edge as the input waveform. Using the vertical spot profile as an approximation to the horizontal spot profile is not always appropriate, and the same reservations expressed in the previous section apply in this case.

The measurement is made by generating a single horizontal line on the display with the brightness and contrast set as discussed previously. A microphotometer with an effective slit aperture width approximately 1/10 the width of the scan line is moved across the scan line (the long slit axis parallel to the scan line). The data taken is stored in an array, which represents the luminance profile of the CRT spot, distance vs luminance. The microphotometer is calibrated for luminance measures and for distance measures in the object plane. Each micron step of the microphotometer represents a known increment in the object plane. The software then calculates the MTF of the CRT based on its line spread from the calibrated luminance and distance measurements. Finite slit width corrections may also be made to the MTF curve by dividing it by a measurement system MTF curve obtained from the luminance profile of an ideal knife edge aperture or a standard source.

The knife edge Fourier transform measurement may be conducted using a low spatial frequency vertical bar pattern (5–10 cycles) across the display with the brightness and contrast controls set as discussed previously. The frequency response of the square wave pattern generator and video pattern generator should be greater than the frequency response of the display system (100 MHz is typical). The microphotometer scans from the center of a bright bar to the center of a dark bar (left to right), measuring the width of the boundary and comparing it to a knife edge. The microphotometer slit is oriented vertically, with its long axis parallel to the bars. The scan is usually made from a light bar to a dark bar in the direction of spot movement. This procedure is preferred because waveforms from scans in the opposite direction may contain anomalies. When the beam is turned on in a square wave pattern, it tends to overshoot and oscillate. This behavior produces artifacts in the luminance profile of the bar edge as the beam moves from an off to an on state. In the on-to-off direction, however, the effects are minimal and the measured waveform does not exhibit the same anomalies that can corrupt the MTF calculations.

The bar edge (knife edge) measurement, unlike the other techniques discussed so far, uses the horizontal spot profile to predict display performance. All of the other techniques use the vertical profile as an approximation of the more critical horizontal spot profile. The bar edge measurement will yield a more accurate assessment of display performance because the displayed image is being generated with a spot scanned in the horizontal direction.

Discrete Frequency Method

The discrete sine wave frequency response measurement technique provides the most accurate representation of display performance. With this approach there are no assumptions implied about the shape of the spot, the electronics bandwidth, or low-frequency light scatter. Discrete spatial frequency sine wave patterns are used to obtain a discrete spatial frequency MTF curve that represents the signal-in to luminance-out performance of the display.

The measurement begins by setting the brightness and contrast as discussed previously, with black level luminance at 1% of the highlight luminance. A sine wave signal is produced by a function generator and fed to a pedestal generator where it is converted to an RS-170A or RS-343 signal, which is then applied to the CRT. The modulation and resulting spatial frequency pattern is measured with a scanning microphotometer. The highlight and black level measurements are used to calculate the modulation constant for each spatial frequency from *5 cycles/display width* to the point that the modulation constant falls to less than 1%. The modulation constant values are then plotted as a function of spatial frequency, generating a discrete spatial frequency MTF curve.

Defining Terms

Corresponding color reproduction: The characteristics of a system in which colors in the reproduction have the same appearance that colors in the original would have if they had been illuminated to produce the same average luminance level and the same reference white chromaticity as that of the reproduction. For most purposes, *corresponding color reproduction* is the most suitable objective of a color video system.

Dot pitch: The center-to-center distance between adjacent green phosphor dots of the red, green, blue triad in a color display.

Pixel: The smallest spatial-information element as seen by the viewer in an imaging system.

Pixel density: A parameter that specifies how closely the pixel elements are spaced on a given display, usually stated in terms of pixels per linear distance.

Pixel format: The arrangement of pixels into horizontal rows and vertical columns.

Spectral color reproduction: The exact reproduction of the spectral power distributions of the original stimuli. Although this is the ultimate goal of any color system, it is not possible in a video system with three primaries.

Zone plate: A complement of test signals for video systems that permit the engineer to exercise a complex video system in the three dimensions of its signal spectrum: horizontal, vertical, and temporal.

References

Anstey, G. and Dore, M.J. 1980. Automatic measurement of cathode ray tube MTFs. Royal Signals and Radar Establishment.

Baldwin, M.W. Jr. 1940. The subjective sharpness of simulated television images. *Proc. IRE.* 28(Oct.):458.

Baldwin, M.W. Jr.1951. Subjective sharpness of additive color pictures. *Proc. IRE* 39(Oct.):1173–1176.

Bartleson, C.J. and Breneman, E.J. 1967. Brightness reproduction in the photographic process. *Photog. Sci. Eng.* 11:254–262.

Bender, W. and Blount, A. 1992. The role of colorimetry and context in color displays. In *Human Vision, Visual Processing, and Digital Display III*, ed. B.E. Rogowitz. *Proc. SPIE* 1666:343–348.

Benson, K.B. 1973. Report on sources of variability in color reproduction as viewed on the home television receiver. *IEEE Trans. BTR* 19:269–275.

Benson, K.B. and Whitaker, J.C., eds. 1991. *Television Engineering Handbook*, revised ed. McGraw–Hill, New York.

Benson, K.B. and Whitaker, J.C. 1989. *Television and Audio Handbook for Engineers and Technicians.* McGraw–Hill, New York.

Bingley, F.J. 1948. The application of projective geometry to the theory of color mixture. *Proc. IRE* 36:709–723.

Brodeur, R., Field, K.R., and McRae, D.H. 1971. Measurement of color rendition in color television. In *Proceedings of the ISCC Conference on Optimum Reproduction of Color* (Williamsburg, VA) ed. M. Pearson. Graphic Arts Research Center, Rochester, NY.

Castellano, J.A. 1992. *Handbook of Display Technology.* Academic, New York.

Crost, M.E. 1967. Display devices and the human observer. *Proc. Interlab. Sem. Component Technol.* Pt. 1, R&D Tech. Rept. ECOM-2865. U.S. Army Electronics Command, Fort Monmouth, NJ. Aug.

DeMarsh, L.E. 1974. Colorimetric standards in US color television. *J. SMPTE* 83:1–5.

DeMarsh, L.E. 1977. Color rendition in television. *IEEE Trans. CE* 23:149–157.

Donofrio, R. 1972. Image sharpness of a color picture tube by modulation transfer techniques. *IEEE Trans. Broadcast Television Receivers* BTR-18(1):16.

Epstein, D.W. 1953. Colorimetric analysis of RCA color television system. *RCA Review* 14:227–258.

Eshbach, O.W. 1936. *Handbook of Engineering Fundamentals*, 2nd ed. Wiley, New York.

Fink, D.G. 1955. *Color Television Standards.* McGraw–Hill, New York.

Fink, D.G. and Christiansen, D., eds. 1982. *Electronics Engineers' Handbook*, 2nd eds. McGraw–Hill, New York.

Fink, D.G. and Wayne Beaty H., eds. 1978. *Standard Handbook for Electrical Engineers*, 11th ed. McGraw–Hill, New York.

Green, M. 1992. Temporal sampling requirements for stereoscopic displays. In *Human Vision, Visual Processing, and Digital Display III*, ed. B.E. Rogowitz. *Proc. SPIE* 1666:101–111.

Herman, S. 1975. The design of television color rendition. *J. SMPTE* 84:267–273.

Hunt, R.W.G. 1975. *The Reproduction of Colour*, 3rd ed. Fountain, England.

Hunter, R. 1975. *The Measurement of Appearance*. Wiley, New York.

Jenkins, A.J. 1981. Modulation transfer function (MTF) measurements on phosphor screens. In *Assessment of Imaging Systems: Visible and Infrared* (Sira), Vol. 274, pp. 154–158, SPIE. Bellingham, WA.

Kucherrov, G.V. et al. 1974. Application of the modulation transfer function method to the analysis of cathode-ray tubes. *Radio Eng. Elec. Phys.*, 19(Feb.):150–152.

MacAdam, D.L. 1942. Visual sensitivities to color differences in daylight. *J. Opt. Soc. Am.* 32:247–274.

Middlebrook, B. and Day, M. 1975. Measure CRT spot size to pack more information into high-speed graphic displays: You can do it with the vernier line method. *Elec. Design* 15(July 19):58–60.

Naiman, A.C. and Makous, W. 1992. Spatial non-linearities of grayscale CRT pixels. In *Human Vision, Visual Processing, and Digital Display III*, ed. B.E. Rogowitz. *Proc. SPIE* 1666:41–56.

Neal, C.B. 1973. Television colorimetry for receiver engineers. *IEEE Trans. BTR* 19:149–162.

Pearson, M., ed. 1971. *Proceedings of the ISCC Conference on Optimum Reproduction of Color* (Williamsburg, VA). Graphic Arts Research Center, Rochester, NY.

Pritchard, D.H. 1977. US color television fundamentals—A review. *IEEE Trans. CE* 23:467–478.

Rash, C.E. and Verona, R.W. Temporal aspects of electro-optical imaging systems. In *Imaging Sensors and Displays*, Vol. 765, pp. 22–25, SPIE. Bellingham, WA.

Reinhart, W.F. 1992. Gray-scale requirements for anti-aliasing of stereoscopic graphic imagery. In *Human Vision, Visual Processing, and Digital Display III*, ed. B.E. Rogowitz. *Proc. SPIE* 1666:90–100.

SMPTE. 1977. Setting chromaticity and luminance of white for color television monitors using shadow mask picture tubes. SMPTE Recommended Practice 71-1977, Society of Motion Picture and Television Engineers.

Sproson, W.N. 1983. *Colour Science in Television and Display Systems*. Adam Hilger, Bristol, England.

Sullivan, J.R. and Ray, L.A. 1992. Secondary quantization of gray-level images for minimum visual distortion. In *Human Vision, Visual Processing, and Digital Display III*, ed. B.E. Rogowitz. *Proc. SPIE* 1666:27–40.

Tannas, L.E. Jr. 1985. Flat Panel Display and CRTs, p.18. Van Nostrand Reinhold, New York.

Tektronix. 1992. Broadening the application of zone plate generatings. Aplication Note 20W7056, Tektronix, Beaverton, OR.

Thomas, G.A. 1986. An improved zone plate test signal generator. *Proceedings IBC*, 11th International Broadcasting Conference, Brighton, UK, pp. 358–361.

Thomas, W. Jr., ed. 1973. *SPSE Handbook for Photographic Science and Engineering*. Wiley, New York.

Verona, R.W. 1992. Comparison of CRT display measurement techniques. In *Helmet-Mounted Displays III*, ed. T.M. Lippert. *Proc. SPIE* 1695:117–127.

Wentworth, J.W. 1955. *Color Television Engineering*. McGraw–Hill, New York.

Weston, M. 1982. The zone plate: Its principles and applications. *EBU Review—Technical* (195, Oct.).

Whitaker, J.C. 1993. *Electronic Displays: Technology, Design and Applications*. McGraw–Hill, New York.

Wintringham, W.T. 1951. Color television and colorimetry. *Proc. IRE* 39:1135–1172.

Further Information

The SPIE (the International Society for Optical Engineering) offers a wide variety of publications on the topic of display systems engineering, measurement, and quality control. The organization also holds technical seminars on this and other topics several times each year. SPIE is headquartered in Bellingham, WA. Two books on the topic of display technology are also recommended:

Sherr, S. 1994. *Fundamentals of Display System Design*, 2nd ed. Wiley-Interscience, New York.

Whitaker, J.C. 1994. *Electronic Displays: Technology, Design and Applications*. McGraw–Hill, New York.

4
Radio Frequency Distortion Mechanisms and Analysis

Samuel O. Agbo
California Polytechnic State University, San Luis Obispo

4.1 Introduction 38
4.2 Types of Distortion 39
 Linear Distortion • Nonlinear Distortion • Distortion Due to Time-Variant Multipath Channels
4.3 The Wireless Radio Channel 42
 Ground Waves • Tropospheric Waves • Sky Waves
4.4 Effects of Phase and Frequency Errors in Coherent Demodulation of AM Signals 45
 Double-Sideband Suppressed-Carrier (DSB-SC) Demodulation Errors • Single-Sideband Suppressed-Carrier (SSB-SC) Demodulation Errors
4.5 Effects of Linear and Nonlinear Distortion in Demodulation of Angle Modulated Waves 49
 Amplitude Modulation Effects of Linear Distortion • Effects of Amplitude Non-linearities on Angle Modulated Waves • Phase Distortion in Angle Modulation
4.6 Interference as a Radio Frequency Distortion Mechanism 53
 Interference in Amplitude Modulation • Interference in DSB-SC AM • Interference in SSB-SC • Interference in Angle Modulation

4.1 Introduction

Radio frequency (RF) communication is usually understood to mean radio and television broadcasting, cellular radio communication, point-to-point radio communication, microwave radio, and other wireless radio communication. For such wireless radio communication, the communication channel is the atmosphere or free space. It is in this sense that RF distortion is treated in this section, although RF signals are also transmitted through other media such as coaxial cables.

Distortion refers to the corruption encountered by a signal during transmission or processing that mutilates the signal waveform. A simple model of the atmospheric radio channel, as shown in Fig. 4.1, will help to illustrate the sources of radio frequency distortion. The channel is modeled as a linear time-variant system with additive noise and additive interference. In the figure, $s(t)$ is the RF signal to be transmitted through the channel, $h(t, \tau)$ is the channel impulse response, $n(t)$ is the additive noise, $i(t)$ is the interference, and $r(t)$ is the received signal. Sources of signal corruption, such as signal attenuation, amplitude distortion, phase distortion, multipath effects and time-varying channel characteristics, constitute the linear time-variant channel impulse response. Noise and interference are additive sources of signal corruption. Corruption by noise may result in a roughness of the output waveform, but the original signal waveform is usually discernible. This is one sense in which noise differs from the distortion mechanisms discussed in this section.

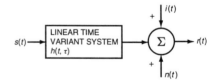

FIGURE 4.1 Model of the atmospheric radio channel as a linear time-variant system with additive noise and additive interference.

Radio Frequency Distortion Mechanisms and Analysis

The impulse response $h(t, \tau)$ is the response of the channel at time t due to an impulse $\delta(t - \tau)$ applied at time $t - \tau$. Let $\alpha_k(t)$ represent the time-dependent attenuation factors for the m multipath components of an atmospheric radio channel. Then a simple representation of the impulse response is

$$h(t, \tau) = \sum_{k=1}^{m} \alpha_k(t)\delta(t - \tau_k) \tag{4.1}$$

The received signal, which is the convolution of this impulse response with the signal $s(t)$, plus the additive noise and interference acquired in the channel, is given by

$$r(t) = \sum_{k=1}^{m} \alpha_k s(t - \tau_k) + i(t) + n(t) \tag{4.2}$$

Radio frequency signals are processed in the transmitter, prior to transmission, and in the receiver, prior to demodulation, by electronic circuits. Such processing can be regarded in a generalized way as filtering. Distortion introduced by nonideal filters and the atmospheric radio channel will be discussed. Following this, the effects on demodulation outputs of frequency and phase distortion and interference on radio frequency signals will be discussed.

4.2 Types of Distortion

Linear Distortion

The processing of radio frequency signals in transmitter or receiver by electronic circuits can be viewed as forms of filtering. Such processing will introduce distortion unless the filter is ideal or distortionless within a band of frequencies equal to or exceeding the bandwidth of the input signal. A filter is distortionless if its input and output signals have identical waveforms, save for a constant amplitude gain or attenuation and a constant time delay for all of the frequency components. Thus, if $g(t)$ with a bandwidth W is the input signal into an ideal filter of bandwidth B, which is greater than W, the output signal $y(t)$ is given by

$$y(t) = \alpha g(t - \tau) \tag{4.3}$$

In Eq. (4.3) α is the constant gain or attenuation and τ is the time delay in passing through the system. With this definition, $H(f)$, the Fourier transforms of the ideal filter impulse response and $Y(f)$, the filter output are given by, respectively,

$$H(f) = \alpha e^{-j2\pi f \tau} \tag{4.4}$$

$$Y(f) = G(f)H(f)$$
$$= \alpha G(f) e^{-j2\pi f \tau} \tag{4.5}$$

Figure 4.2 shows the frequency-domain representation of an ideal and a nonideal filter and the time-domain representation of the input and output signals. Figures 4.2(a), 4.2(b), and 4.2(c) show, respectively, the frequency response of the ideal filter, the input signal waveform, and the output signal waveform, which is a time delayed version of the input waveform. Both the filter response and the input signal are bandlimited to B Hz.

Figure 4.2(d) shows a nonideal filter frequency response with ideal phase but nonideal amplitude response, which is given by

$$|H(f)| = \begin{cases} \alpha \cos 2\pi fT, & |f| < B \\ 0; & |f| > B \end{cases} \tag{4.6}$$

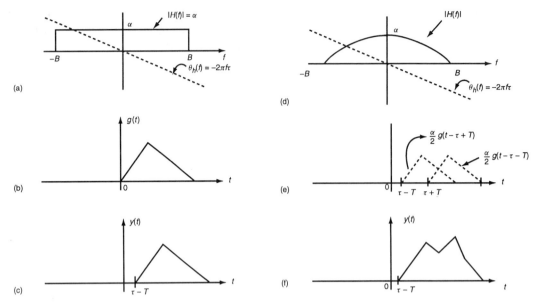

FIGURE 4.2 (a) Ideal filter frequency response, (b) waveform of input signal, (c) waveform of output signal from ideal filter, (d) nonideal filter frequency response with ideal phase but nonideal amplitude response, (e) output waveform of nonideal filter showing its components as time shifted versions of the input signal, (f) output waveform of nonideal filter: a dispersed version of the input signal.

By employing the time-shifting property of the Fourier transform, the time-domain output signal is obtained as

$$y(t) = \frac{\alpha}{2}[g(t - \tau - T) + g(t - \tau + T)] \qquad (4.7)$$

From Eq. 4.7, it is clear that the output signal consists of two scaled and time shifted versions of the input signal as shown in Fig. 4.2(e). The sum of these shifted versions shown in Fig. 4.2(f) indicates that linear distortion results in **dispersion** of the output signal. It is easy to see that nonideal phase response, which implies different delays at different frequencies, will also lead to dispersion. Thus, linear distortion causes pulse spreading and interference with neighboring pulses. Consequently, linear distortion should be avoided in time-division multiplexing (TDM) systems where it causes **crosstalk** with neighboring channels. In frequency-division multiplexing (FDM), linear distortion corrupts the signal spectrum, but the pulse spreading does not result in crosstalk because the channels are adjacent in frequency, not in time.

The ideal filter of Fig. 4.2 is a low-pass filter. Bandpass and high-pass filters are more relevant for the processing of RF signals in the transmitter, prior to transmission, and in the receiver, prior to demodulation. For bandpass and high-pass filters, the condition for distortionless filtering is the same as previously indicated: the input signal bandwidth should not exceed the filter bandwidth, and the filter should have a constant gain and the same time delay for all frequencies within its bandwidth. Figure 4.3 illustrates the frequency responses for ideal bandpass and ideal high-pass filters.

Nonlinear Distortion

Nonlinearity in system response usually arises in cases involving large signal amplitudes. For a simple case of memoryless, nonlinear system or channel, with a low-pass input signal $g(t)$, of bandwidth W Hz, the output signal $y(t)$ is given by

$$y(t) = a_0 + a_1 g(t) + a_2 g^2(t) + a_3 g^3(t) + \cdots + a_k g^k(t) + \cdots \qquad (4.8)$$

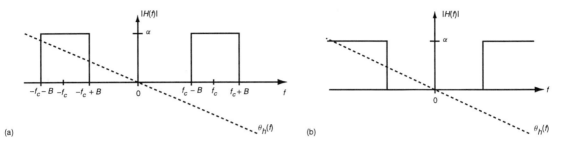

FIGURE 4.3 (a) Ideal bandpass filter, (b) high-pass filters.

Consider the term $a_k g^k(t)$ in the equation. Let $G(f)$ be the Fourier transform of $g(t)$. Then the Fourier transform of $g^k(t)$ is a $k-1$ fold convolution of $G(f)$. Thus, the bandwidth of $g^k(t)$ is $k-1$ times the bandwidth of $g(t)$. Suppose m is the last significant term of the power series of the equation, then the bandwidth of the output signal is $(m-1)W$ Hz. Thus, the bandwidth of the output signal will far exceed the bandwidth of the input signal.

Figure 4.4 illustrates the spectrum of the input and the output signals. If the input signal is a bandpass signal of bandwidth W and carrier frequency f_c, which would be appropriate for a radio frequency signal, the spectrum of the output signal would be much more complex. In this case, the kth term of the power series of the output signal $y(t)$ will, in general, contribute a component of bandwidth $(k-1)W$ centered at kf_c, in addition to smaller bandwidth components centered at lower harmonics of f_c.

The preceding discussion shows that nonlinear distortion causes spreading in the frequency domain. Such spreading of the spectrum causes interference in FDM systems because it results in frequency-domain overlap of neighboring channels. However, nonlinear distortion does not cause adjacent channel interference in TDM as the channels in that case are not adjacent in frequency, but in time.

Distortion Due to Time-Variant Multipath Channels

Radio waves from a transmitting antenna often reach the receiving antenna through many different paths. Examples include different reflecting layers in the **ionosphere**, numerous scattering points in the **troposphere**, reflections from the Earth's surface, and the direct line-of-sight path. Each of these paths will contribute a different attenuation and time delay. Shown in Fig. 4.5 is a model of the time-variant multipath radio channel. For the sake of simplicity, the effect of noise has been ignored.

This channel is time dispersive because the different time delays entail a spread in time of the composite received signal relative to the transmitted signal. In addition, because of variations with time and locality in weather, temperature and other atmospheric conditions, the time delay and attenuation of these paths vary with time. Because these variations in channel characteristics are generally unpredictable, a statisticall characterization of the time-variant multipath channel is appropriate.

To simplify the analysis, consider the transmission of an unmodulated carrier of unit amplitude, $s(t) = \cos 2\pi f_c t$, through an m multipath channel in the absence of noise. Let $\alpha_k(t)$ be the attenuation and $\tau_k(t)$ be

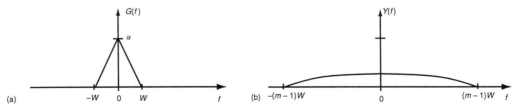

FIGURE 4.4 Input and output spectrum for a channel with nonlinear distortion: (a) amplitude spectrum of output signal, (b) amplitude spectrum of input signal.

time delay contributed by the kth path at the receiver. The received signal is given by

$$r(t) = \sum_{k=1}^{m} \alpha_k(t) \cos[2\pi f_c(t - \tau_k(t))]$$

$$= Re\left[\sum_{k=1}^{m} \alpha_k(t) e^{j2\pi f_c t} e^{-j2\pi f_c \tau_k(t)}\right] \quad (4.9)$$

$$= [\cos 2\pi f_c t] Re\left[\sum_{k=1}^{m} \alpha_k e^{-j2\pi f_c \tau_k(t)}\right]$$

It is easy to see that the response of the channel to the unmodulated carrier is

$$h(t, \tau) = Re\left[\sum_{k=1}^{m} \alpha_k(t) e^{-j2\pi f_c \tau_k(t)}\right] \quad (4.10)$$

Note that although the channel input was an unmodulated carrier, the channel output contains a frequency spread around the carrier, resulting from the time-varying delays of the various paths. The bandwidth of this output signal is a measure of how rapidly the channel characteristics are changing. A large change in channel characteristics is required to produce a significant change in the time-varying attenuation $\alpha_k(t)$. Because f_c is a large number, a change of 2π in the phase of the output signal $-2\pi f_c \tau_k(t)$ corresponds to only a change of $1/f_c$ in the time delay $\tau_k(t)$. Thus, the attenuation changes slowly, while the phase (or frequency deviation) changes rapidly.

Because these changes are random in nature, the channel impulse response can be modeled as a random process. Constructive addition of the phases of the various paths result in a strong signal, whereas destructive addition of the phases results in a weak signal. Thus, the multipath propagation results in signal fading, primarily due to the time-variant phase changes. When m, the number of multiple propagation paths, is large, the central limit theorem applies and the channel response will approximate a Gaussian random process.

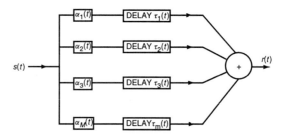

FIGURE 4.5 Model of a noiseless, time-variant, multipath channel.

4.3 The Wireless Radio Channel

In wireless radio communication, electromagnetic waves are radiated into the atmosphere or free space via a transmitting antenna. For efficient radiation, the antenna length should exceed one-tenth of the wavelength of the electromagnetic wave. The wavelength λ is given by

$$\lambda = \frac{c}{f} \quad (4.11)$$

In Eq. (4.11) $c = 3 * 10^8$ m/s, is the free-space velocity of electromagnetic waves and f is the frequency in Hz. Wireless radio communicating covers a wide range of frequencies from about 10 kHz in the very low frequency (VLF) band to above 300 GHz in the extra high frequency (EHF) band. Depending on the frequency band and the distance between the transmitter and the receiver, the transmitted wave may reach the receiving antenna by one or a combination of several propagation paths, as shown in Fig. 4.6.

Sky waves are those that reach the receiver after being reflected from the ionosphere, a band of reflecting layers ranging in altitude from 100 to 300 km. Tropospheric waves are those that reach the receiver after being reflected (scattered) by inhomogeneities within the troposphere, the region within 10 km of the Earth's surface. Other waves

Radio Frequency Distortion Mechanisms and Analysis

TABLE 4.1 Classification of Wireless Radio Frequency Channels by Frequency Bands, Typical Uses, and Wave Propagation Modes

Frequency Band	Name	Typical Uses	Propagation Mode
3–30 kHz	Very low frequency (VLF)	Navigation, sonar	Ground waves
30–300 kHz	Low frequency (LF)	Navigation, telephony, telegraphy	Ground waves
0.3–3 MHz	Medium frequency (MF)	AM Broadcasting, amateur, and CB radio	Sky waves and ground waves
3–30 MHz	High frequency (HF)	Mobile, amateur, CB, and military radio	Sky waves
30–300 MHz	Very high frequency (VHF)	VHF TV, FM Broadcasting Police, air traffic control	Sky waves, tropospheric waves
0.3–3 GHz	Ultra high frequency (UHF)	UHF TV, radar, satellite communication	Tropospheric waves, space waves
3–30 GHz	Super high frequency (SHF)	Space and satellite, radar microwave relay	Direct waves, ionospheric penetration waves
30–300 GHz	Extra high frequency (EHF)	Experimental, radar radio astronomy	Direct waves, ionospheric penetration waves

reach the receiving antenna via ground waves, which consists of space waves and surface waves. Surface waves (in the LF and VLF ranges) have wavelengths comparable to the altitude of the ionosphere. The ionosphere and the ground surface act as guiding planes, so that surface waves follow the curvature of the Earth, permitting transmission over great distances. Space waves consist of direct (line-of-sight) waves and ground reflected waves, which arrive at the receiver after being reflected from the ground. A classification of radio frequency communication by frequency bands, typical uses, and electromagnetic wave propagation modes is given in Table 4.1.

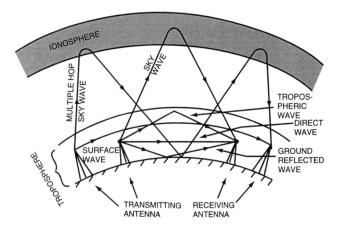

FIGURE 4.6 Wireless radio wave propagation paths.

Ground Waves

In the region adjacent to the Earth's surface, free-space conditions are modified by the surface topography, physical structures such as buildings, and the atmosphere. One effect of the atmosphere is to casuse a refraction of the wave, which carries it beyond the optical horizon. Consequently, the **radio horizon** is longer than the optical horizon. Hills and building structures cause multiple reflections and shieldings, which contribute both attenuation and phase change to the received signal. Ground reflected waves travel a different path length and undergo different attenuation and time delay relative to direct waves and depending on the point of reflection.

In addition, at each point of reflection on the Earth's surface, there is a phase change, which depends on the condition of the surface, but is typically a 180° phase change.

Consider the simplified model of ground wave propagation, shown in Fig. 4.7, which involves only two transmission paths: a direct wave and a ground reflected wave. The curved Earth's surface is represented by a flat Earth model after correction has been made for the longer radio horizon. The height of the transmitting antenna and the receiving antenna are h_T and h_r, respectively.

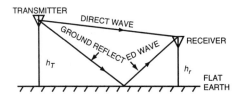

FIGURE 4.7 Simplified model for ground wave propagation showing the direct wave and the ground reflected wave.

The direct wave travels a path length x from the transmitting antenna to the receiving antenna. The attenuation factor for the atmospheric medium is α. Let the corresponding distance traveled by the ground reflected wave be $x + \Delta x$ and the transmitted signal be $s(t)$. Note that a path length equal to the wavelength λ corresponds to a phase change of 2π. The received direct wave $r_d(t)$, the received ground reflected wave $r_g(t)$, and the composite received wave $r(t)$ are given, respectively, by

$$r_d(t) = e^{-\alpha x} s(t) \cos\left(\frac{2\pi x}{\lambda}\right) \tag{4.12}$$

$$r_g(t) = e^{-\alpha(x+\Delta x)} s(t) \cos\left[\frac{2\pi}{\lambda}(x + \Delta x) - \pi\right] \tag{4.13}$$

$$r(t) = r_d(t) + r_g(t) \tag{4.14}$$

From the preceding equations, it is seen that even for small differences in path lengths, the received signals from the direct and the ground reflected waves tend to have large phase differences. Consequently, the composite received signal has significant attenuation and phase error relative to the transmitted signal.

Tropospheric Waves

Waves propagating through the troposphere are only seriously attenuated by adverse atmospheric conditions at frequencies above 10 GHz. Above 10 GHz, heavy rain causes severe attenuation. However, light rain, clouds, and fog cause serious attenuation only at frequencies above 30 GHz. Attenuation due to snow is negligible at all frequencies. On the other hand, the inhomogeneities in the atmosphere due to the weather and other conditions are favorable for tropospheric scatter propagation of radio waves within the 30 MHz–4 GHz frequency band. Within this range of frequencies, the inhomogeneities within the troposphere form effective point scatterers that deflect the transmitted waves downward toward the receiving antenna. By employing large transmitted power and highly directional antennas, long transmission ranges of up to 400 mi can be obtained.

The range of frequencies favored for tropospheric propagation is essentially the domain of microwave transmission, UHF, and VHF radio. Although atmospheric attenuation is higher for frequencies above 10 GHz, it is still quite significant within the 30 MHz–4 GHz range of frequencies. Hence, microwave repeater stations are spaced about 30 mi apart. This spacing is long compared to the spacing of about 3 mi for repeaters for coaxial cable systems. However, such spacing is relatively close, compared with the more recent development of optical fiber transmission in which repeater spacing of above 300 km has been achieved.

Sky Waves

The ionosphere consists of several layers containing ionized gases at altitudes of 100–300 km. In order of increasing height, these are the C, D, E, F_1, and F_2 layers. The ionization is caused by solar radiation, mainly due to ultraviolet and cosmic rays. Consequently, the electron density in the layers varies with the seasons and with the time of day and night. The gas molecules in the denser, lower layers cause more collisions and hence a higher recombination rate of electrons and ions. At night, the reduced level of radiation and the greater recombination rates cause the lower layers to disappear, so that only the F_2 layer remains.

Electromagnetic wave propagation through the ionosphere is by means of sky waves, mainly in the MF, HF, and VHF bands. The waves are successively refracted at the layers until they are eventually reflected back toward the Earth. At higher frequencies, the waves pass through the ionosphere without being reflected from it. Communication at these higher frequencies depends on a line-of-sight basis. During the daytime, ionospheric propagation is severely attenuated at frequencies below 2 MHz, due to severe absorption in the lower layers, especially the D layer. At night, after the disappearance of these lower layers, powerful AM transmission utilizing the F_2 layer is possible over the range of 100–250 mi. Of course, the maximum transmission range for sky waves depend on the frequency of transmission.

Figure 4.8 shows the geometry relevant to the determination of the range for ionospheric propagation. Because the altitudes involved are much greater than the antenna heights, the latter have been ignored. Rays from

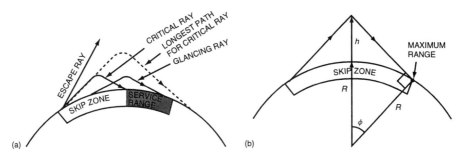

FIGURE 4.8 Geometry of ionospheric single hop: (a) raypaths, skip zone, and service range, (b) geometry for computation of maximum range.

the transmitting antenna at angles of elevation greater than the critical angle of elevation for the frequency of transmission result in escape rays, which are not reflected back from the ionosphere. The ray leaving the transmitter at the critical angle, the critical ray, which is reflected from the lowest possible layer, returns to the Earth closest to the transmitter than any other ray, giving the minimum range or the skip zone. The distance between this minimum range and the range of the glancing ray (which is tangential to the Earth at the receiver) is the service range.

The greatest transmission range is obtained when the critical ray is reflected from a layer at so high an altitude that it is also tangential to the Earth's surface at the receiver. The geometry for this case is shown in Fig. 4.8(b) in which the raypaths have been idealized as straight lines, ignoring the curvature of the paths due to refraction. The height of the virtual reflecting point above the Earth's surface is h, and R is the radius of the Earth. The angle ϕ and the maximum range d are given, respectively, by

$$\phi = \cos^{-1}\left(\frac{R}{R+h}\right) \tag{4.15}$$

$$d = 2R\cos^{-1}\left(\frac{R}{R+h}\right) \tag{4.16}$$

This computation is for a single-hop reflection. For multiple hops, the transmission range is considerably greater, although the signal strength would be correspondingly reduced. The condition of the ionosphere within a locality greatly affects the critical angle of elevation and the transmission path. Conditions vary considerably with time and locality within the ionosphere. Such irregularities within the ionosphere can be used for scatter propagation, much like tropospheric scatter propagation, thereby resulting in increased transmission range of up to 1000 mi. However, the undesirable effects of ionospheric inhomogeneities far exceed this favorable effect.

Ionospheric irregularities result from differences in electron densities in the various layers. The irregularities can be quite large in spatial dimensions, ranging from tens of meters to several hundreds of kilometers. Some of these irregularities travel through the ionosphere at speeds in excess of 3000 km/h. These time-varying irregularities are responsible for the time-variant multipath effects and the resultant fading that plagues ionospheric transmission.

4.4 Effects of Phase and Frequency Errors in Coherent Demodulation of AM Signals

RF signals acquire phase and amplitude distortion in the wireless radio channel and in processing circuits, through the many mechanisms previously discussed. For coherent (or synchronous) detection, a local carrier, which matches the transmitted signal in phase and frequency, is required at the receiver. In general, such a match is hard to obtain. These mismatch errors are additional to the phase and amplitude errors acquired in the channel.

Double-Sideband Suppressed-Carrier (DSB-SC) Demodulation Errors

The receiver for the synchronous demodulation of double-sideband suppressed-carrier (DSB-SC) waves is shown in Fig. 4.9. In general, it is easy to correct for the effect of attenuation or amplitude errors through amplification, filtering, or other processing in the receiver. It is more difficult to correct for the effect of phase or frequency errors.

The received signal at the input to the receiver and the local carrier are given, respectively, by

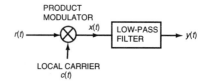

FIGURE 4.9 Receiver for synchronous demodulation of DSB-SC signals.

$$r(t) = m(t) \cos 2\pi f_c t \tag{4.17}$$

$$c(t) = \cos[2\pi(f_c + \Delta f)t + \theta] \tag{4.18}$$

In the last two equations, $m(t)$ is the message signal of bandwidth W, where f_c is the carrier frequency, Δf is the frequency error, and θ is the phase error. The output of the product modulator is $x(t)$. Thus, $x(t)$ and $y(t)$ are given, respectively, by

$$x(t) = m(t) \cos(2\pi f_c t) \cos[2\pi(f_c + \Delta f)t + \theta]$$
$$= \frac{1}{2} m(t) \{ \cos[2\pi(2f_c + \Delta f)t + \theta] + \cos(2\pi \Delta f t + \theta) \} \tag{4.19}$$

$$y(t) = \frac{1}{2} m(t) \cos(2\pi \Delta f t + \theta) \tag{4.20}$$

It is easy to see that when both the frequency and the phase errors are zero, the demodulated output is $1/2 m(t)$, and the recovery of the message signal is perfect. Consider the case of zero phase error, but finite frequency error. The demodulated signal and its Fourier transform are given, respectively, by

$$y(t) = \frac{1}{2} m(t) \cos 2\pi \Delta f t \tag{4.21}$$

$$Y(f) = \frac{1}{4} [M(f + \Delta f) + M(f - \Delta f)] \tag{4.22}$$

Thus, the demodulated output is a replica of the message signal multiplied by a slowly varying cosine wave of frequency Δf. This is a serious type of distortion, which produces a beating effect. The human ear can tolerate at most about 30 Hz of such frequency drift. Figure 4.10 shows the frequency-domain illustration of this type of distortion. Figure 4.10(a) shows the amplitude spectrum of the original message signal. Figure 4.10(b) shows that the demodulated signal can be viewed as a DSB-SC signal whose carrier is the frequency error Δf. This error is so small that the replicas of the message signal shifted by $+\Delta f$ and $-\Delta f$ overlap in frequency and, therefore, interfere, giving rise to the beating effect.

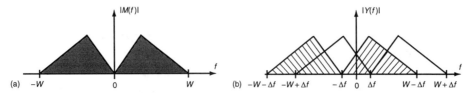

FIGURE 4.10 Frequency spectrum of message signal and demodulated output when there is a Δf frequency error between the local carrier and the carrier in the received DSB-SC signal: (a) message signal spectrum, (b) spectrum of demodulated output signal.

Next, consider the case in which the frequency error is zero, but the phase error is finite. The demodulated signal is given by

$$y(t) = \frac{1}{2}m(t)\cos\theta \qquad (4.23)$$

If θ, the phase error is constant, then it contributes only a constant attenuation, which is easily corrected for through amplification. However, in the extreme case in which $\theta = \pm\pi/2$, the demodulated signal amplitude is reduced to zero. On the other hand, if θ varies randomly with time, as is the case for multipath fading channels, the attenuation induced by the phase error varies randomly and represents a serious problem.

To avoid the undesirable effects of frequency and phase errors between the local and the received carriers, various schemes are employed. One such scheme employs a **pilot carrier** at the transmitter to provide a source of matching carrier at the receiver. Some other schemes employed for providing matching carriers are the *Costas receiver* and signal squaring.

Single-Sideband Suppressed-Carrier (SSB-SC) Demodulation Errors

The receiver for the demodulation of single-sideband suppressed-carrier (SSB-SC) signals is similar to the receiver for DSB-SC signals shown in Fig. 4.9. The local carrier is the same as for the DSB-SC signal, but the received signal is given by

$$r(t) = m(t)\cos 2\pi f_c t \mp m_h(t)\sin 2\pi f_c t \qquad (4.24)$$

In Eq. 4.24, the negative ($-$) sign applies to the upper-sideband (USB) case, whereas the positive sign ($+$) applies to the lower-sideband (LSB) case. The signal, $m_h(t)$, is the *Hilbert transform* of the message signal $m(t)$. The output of the product modulator is

$$\begin{aligned} x(t) &= [m(t)\cos 2\pi f_c t \mp m_h(t)\sin 2\pi f_c t]\cos[2\pi(f_c + \Delta f)t + \theta] \\ &= \frac{1}{2}m(t)\big\{\cos(2\pi\Delta ft + \theta) + \cos[2\pi(2f_c + \Delta f)t + \theta]\big\} \\ &\quad \pm \frac{1}{2}m_h(t)\big\{\sin(2\pi\Delta ft + \theta) - \sin[2\pi(2f_c + \Delta f)t + \theta]\big\} \end{aligned} \qquad (4.25)$$

The low-pass filter suppresses the bandpass components of the product modulator output. Hence, the demodulated output $y(t)$ is given by

$$y(t) = \frac{1}{2}m(t)\cos(2\pi\Delta ft + \theta) \pm \frac{1}{2}m_h(t)\sin[2\pi\Delta ft + \theta) \qquad (4.26)$$

Note that when both the frequency and the phase errors are set to zero in the equation, the demodulator output signal is $1/2 m(t)$, an exact replica of the desired message signal. Consider the case of zero phase error, but nonzero frequency error. In this case, the demodulated output signal $y(t)$ can be expressed as

$$y(t) = \frac{1}{2}m(t)\cos 2\pi\Delta ft \pm \frac{1}{2}m_h(t)\sin 2\pi\Delta ft \qquad (4.27)$$

This equation shows that when only frequency error is present between the received and local carriers, the demodulated signal is essentially an SSB-SC signal in which the small frequency error Δf plays the role of the carrier. A comparison of this equation with the expression for the SSB-SC signal shows that if the modulated signal is a USB signal, the demodulated output is and LSB signal, and vice versa.

Figure 4.11 shows the spectra of the message signal, the SSB-SC modulated signal (USB case), and two cases of the demodulated signal when only frequency errors are present. Figure 4.11(c) shows the spectrum of the demodulated output of the USB signal when the frequency error is negative. Figure 4.11(e) shows the spectrum of the demodulated output of the USB signal when the frequency error is positive.

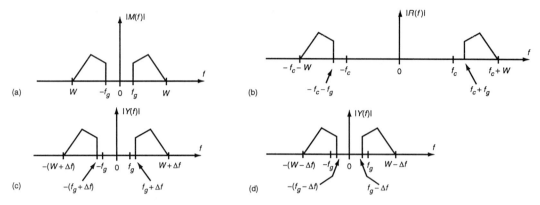

FIGURE 4.11 An illustration of the effect of frequency error in synchronous demodulation of SSB-SC signals: (a) amplitude spectrum of message signal, (b) amplitude spectrum of SSB (USB) signal, (c) amplitude spectrum of demodulated USB signal when $\Delta f < 0$, (d) amplitude spectrum of demodulated USB signal when $\Delta f > 0$.

By comparing the demodulated output of an SSB-SC signal with the message signal, it is easy to see that in each case, each frequency component in the demodulated output is shifted by Δf. This does not introduce as serious a distortion as in the DSB-SC carrier case in which each frequency component is shifted by $-\Delta f$ and $+\Delta f$. As shown in Fig. 4.11, $M(f)$, the spectrum of the message signal is zero in the frequency gap from zero to f_g, as should be the case for a well-processed SSB message. By ensuring that f_g exceeds the maximum possible value of Δf, the sidebands for the demodulated output of an SSB-SC wave will never overlap in frequency or interfere with each other as in the DSB-SC case.

For the case of USB modulation in which the frequency error is zero but the phase error is nonzero, the demodulated output is obtained by setting Δf equal to zero in Eq. (4.26). This gives $y(t)$, the demodulated output signal and its frequency domain representation $Y(f)$, respectively, as

$$y(t) = \frac{1}{2}[m(t)\cos\theta + m_h(t)\sin\theta] \tag{4.28}$$

$$Y(f) = \frac{1}{2}[M(f)\cos\theta + M(f)\sin\theta] \tag{4.29}$$

But the frequency domain representation of the Hilbert transform of the message signal, $m_h(f)$, can be expressed as

$$M_h(f) = -j\,\text{sgn}(f)M(f)$$
$$= \begin{cases} -jM(f); & f > 0 \\ jM(f); & f < 0 \end{cases} \tag{4.30}$$

Thus, $Y(f)$ can be expressed as

$$Y(f) = \begin{cases} \dfrac{1}{4}\left[M(f)e^{-j\theta} - jM(f)\left(\dfrac{-e^{-j\theta}}{j}\right)\right]; & f > 0 \\[6pt] \dfrac{1}{4}\left[M(f)e^{j\theta} + jM(f)\left(\dfrac{e^{j\theta}}{j}\right)\right]; & f < 0 \end{cases}$$

$$= \begin{cases} \dfrac{1}{2}M(f)e^{-j\theta}; & f > 0 \\[6pt] \dfrac{1}{2}M(f)e^{j\theta}; & f < 0 \end{cases} \tag{4.31}$$

From the preceding equation, it is clear that each positive frequency component in the demodulated output signal undergoes a phase shift of $-\theta$, while each negative frequency component undergoes a phase shift of $+\theta$. Thus, phase errors in SSB-SC signals results in phase distortion in the demodulated outputs. For voice signals, the human ear is relatively insensitive to phase distortion. However, this type of phase distortion, which produces a constant phase shift at all message signal frequencies, is quite objectionable in music signals and intolerable in video signals.

4.5 Effects of Linear and Nonlinear Distortion in Demodulation of Angle Modulated Waves

The amplitude of an angle modulated wave is supposed to be constant while the information being transmitted is contained in the phase. Linear distortion can be introduced by the channel, non-ideal filters and other sub-circuits. In general, linear distortion results in time-variant amplitude and phase errors in the angle modulated wave. In effect, the angle modulated wave now contains both amplitude and phase modulation. Nonlinear distortion usually results from channel or power amplifier nonlinearities. Amplitude nonlinearities result in undesirable (or residual) amplitude modulation (AM), whereas phase nonlinearities result in phase distortion of the angle modulated wave. The effect of undesirable amplitude modulation resulting from linear and nonlinear distortion are discussed under separate subheadings. The effects of the time-variant phase errors resulting from linear distortion and from phase nonlinearities are discussed together under phase distortion in angle modulation.

Amplitude Modulation Effects of Linear Distortion

In an angle modulated waveform, the information resides in the phase of the modulated wave. Such an angle modulated wave can be demodulated by first passing it through a circuit whose amplitude response is proportional to the frequency of the input signal. The resulting intermediate signal has both amplitude modulation and angle modulation, and the information also resides in its amplitude. The demodulation process is then completed by passing this intermediate signal through an envelope detector.

A circuit which can implement this process is shown in Fig. 4.12. The differentiator input signal is the angle modulated wave. The amplitude of its output is proportional to the signal frequency, as shown in Fig. 4.12(b). Thus, the differentiator output is an amplitude as well as a frequency modulated waveform. The envelope detector output contains a replica of the message signal. Consider the case in which the angle modulation involved is frequency modulation (FM). Suppose the FM signal contains no amplitude or phase errors, its amplitude will be constant. For simplicity, it can be assumed to be of unit amplitude. Let f_c be the carrier frequency, k_f be the frequency sensitivity of the modulator, and $m(t)$ be the message signal. Then the FM signal $s(t)$, the differentiator output $s'(t)$, and the envelop detector output $y(t)$, are given, respectively, by

$$s(t) = \cos\left[2\pi f_c t + 2\pi k_f \int_{-\infty}^{t} m(\lambda)d\lambda\right] \tag{4.32}$$

$$s'(t) = -[2\pi f_c + 2\pi k_f m(t)] \cos\left[2\pi f_c t + 2\pi k_f \int_{-\infty}^{t} m(\lambda)d\lambda\right] \tag{4.33}$$

$$y(t) = 2\pi f_c + 2\pi k_f m(t) \tag{4.34}$$

It is easy to see that an exact replica of the message signal can be recovered from the envelope detector output.

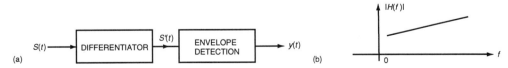

FIGURE 4.12 FM demodulator employing a differentiator and an envelope detector: (a) FM demodulator, (b) amplitude response of an ideal differentiator.

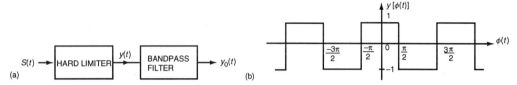

FIGURE 4.13 The bandpass limiter and an illustration of the output waveform of the hard limiter: (a) Bandpass limiter, (b) output waveform of the hard limiter.

Consider the case in which the FM signal contains a time variation in its amplitude due to the channel, nonideal filter, or other system effects. Let $a(t)$ represent the time variation in the amplitude relative to a the unit carrier amplitude. Then the FM signal and the output of the differentiator are given, respectively, by

$$s(t) = a(t) \cos\left[2\pi f_c t + 2\pi k_f \int_{-\infty}^{t} m(\lambda)d\lambda\right] \tag{4.35}$$

$$s'(t) = -a(t)[2\pi f_c + 2\pi k_f m(t)] \sin\left[2\pi f_c t + 2\pi k_f \int_{-\infty}^{t} m(\lambda)d\lambda\right]$$

$$+ a'(t) \cos\left[2\pi f_c t + 2\pi k_f \int_{-\infty}^{t} m(\lambda)d\lambda\right] \tag{4.36}$$

The output of the envelope detector, $y(t)$, is the envelope of the differentiator output given in the last equation. Thus, the envelope detector output is

$$y(t) = \left\{a^2(t)[2\pi f_c + 2\pi k_f m(t)]^2 + [a'(t)]^2\right\}^{\frac{1}{2}} \tag{4.37}$$

From the last equation it is easy to see that because of the time variation in the amplitude of the FM signal, it is not possible to extract an undistorted replica of the message signal from the envelope detector output. Even if the second term in Eq. (4.36) is ignored, the message signal is still multiplied by the time-variant amplitude. Thus, it is necessary to rid the FM wave of variations in its amplitude prior to demodulation. This is usually accomplished by passing the FM signal through a bandpass limiter, such as is illustrated in Fig. 4.13, prior to demodulation.

The FM wave containing time-variant amplitude errors can be expressed as

$$s(t) = a(t) \cos[\phi(t)] \tag{4.38}$$

Its phase $\phi(t)$ is given by

$$\phi(t) = 2\pi f_c t + 2\pi k_f \int_{-\infty}^{t} m(\lambda)d\lambda \tag{4.39}$$

The hard limiter is a device with a voltage transfer characteristic such that if its input is $s(\phi)$, then its output, $y(\phi)$ is given by

$$y(\phi) = \begin{cases} 1, & \phi > 0 \\ -1, & \phi < 0 \end{cases} \tag{4.40}$$

When the input to the hard limiter is the FM signal, $a(t)\cos[\phi(t)]$, the output of the hard limiter is the square waveform shown in Fig. 4.13(b). This periodic square waveform has a Fourier series expression given by

$$y[\phi(t)] = \frac{4}{\pi}\left[\cos[\phi(t)] - \frac{1}{3}\cos[3\phi(t)] + \frac{1}{5}\cos[5\phi(t)] + \cdots\right] \tag{4.41}$$

Note that in the last equation, the terms containing $\phi(t)$, $3\phi(t)$, $5\phi(t)$, ..., correspond to FM waves with carrier frequencies f_c, $3f_c$, $5f_c$, ... Hence, the bandpass filter centered at f_c selects only the desired FM signal with carrier frequency f_c. Note that the output of the bandpass limiter, which is the output of the bandpass filter, has a constant amplitude given by

$$y_0(t) = \frac{4}{\pi} \cos\left[2\pi f_c t + 2\pi k_f \int_{-\infty}^{t} m(\lambda) d\lambda\right] \quad (4.42)$$

Effects of Amplitude Nonlinearities on Angle Modulated Waves

Amplitude nonlinearities can arise from nonlinearities in the channel or from nonlinearities in processing circuits such as power amplifiers. Consider a channel or system with amplitude nonlinearities, whose input signal $x(t)$ and output signal $y(t)$ are related as in the following equation:

$$y(t) = a_0 + a_1 x(t) + a_2 x^2(t) + a_3 x^3(t) + \cdots \quad (4.43)$$

If the input signal $s(t)$ is the FM wave given by

$$s(t) = \cos[\phi(t)] = \cos\left[2\pi f_c t + 2\pi k_f \int_{-\infty}^{t} m(\lambda) d\lambda\right] \quad (4.44)$$

then the output signal $y(t)$ is given by

$$\begin{aligned}
y(t) &= a_0 + a_1 \cos[\phi(t)] + a_2 \cos^2[\phi(t)] + a_3 \cos^3[\phi(t)] + \cdots \\
&= a_0 + a_1 \cos[\phi(t)] + \frac{1}{2} a_2 [1 + \cos 2\phi(t)] + \frac{1}{2} a_3 \left\{ \cos[\phi(t)] + \frac{1}{2} \cos[\phi(t)] + \frac{1}{2} \cos[3\phi(t)] \right\} + \cdots \\
&= \left(a_0 + \frac{1}{2} a_2\right) + \left(a_1 + \frac{3}{4} a_3\right) \cos[\phi(t)] + \frac{1}{2} a_2 \cos[2\phi(t)] + \frac{1}{4} \cos[3\phi(t)] + \cdots \quad (4.45)
\end{aligned}$$

The phase angles $\phi(t)$, $2\phi(t)$, and $3\phi(t)$, correspond to FM waves with carrier frequencies f_c, $2f_c$, and $3f_c$. Thus if the nonlinear system or channel is followed by a bandpass filter centered at f_c, the filter output is the desired FM wave with a carrier f_c, given by

$$\begin{aligned}
y_0(t) &= \left(a_1 + \frac{3}{4} a_3\right) \cos[\phi(t)] \\
&= \alpha \cos\left[2\pi f_c t + 2\pi k_f \int_{-\infty}^{t} m(\lambda) d\lambda\right] \quad (4.46)
\end{aligned}$$

Let B_{FM} be the transmission bandwidth of the FM signal, W the bandwidth of the message signal and Δf the frequency deviation. The FM bandwidth is given for narrowband FM (NBFM) and for wideband FM (WBFM) by

$$B_{\text{FM}} = \begin{cases} 2(\Delta f + W), & \text{for NBFM} \\ 2(\Delta f + 2W), & \text{for WBFM} \end{cases} \quad (4.47)$$

To ensure that the FM wave with a carrier frequency of $2f_c$ does not interfere with the desired FM signal of carrier frequency f_c, the following condition must be satisfied:

$$f_c + \frac{1}{2} B_{\text{FM}} \leq 2\left(f_c - \frac{1}{2} B_{\text{FM}}\right) \quad (4.48)$$

This condition is equivalent to ensuring that $f_c \geq 1.5 B_{\text{FM}}$. By substituting the expressions for the bandwidths for NBFM and WBFM into the last equation, it is seen that the carrier frequency must satisfy the condition

$$f_c > \begin{cases} 3(\Delta f + W) & \text{for NBFM} \\ 3(\Delta f + 2W) & \text{for WBFM} \end{cases} \quad (4.49)$$

Note that in the case discussed earlier in which the nonlinearities are not time variant, the effect of passing an FM wave through a nonlinear channel or system, followed by a bandpass filter centered at the carrier frequency, is to produce a gain or attenuation in the constant amplitude. If the nonlinearities are time varying, so that the coefficients in the expression relating the channel input to the output are $a_0(t), a_1(t), a_2(t), \ldots$, a substitution of these time-varying coefficients into the preceding equation shows that the bandpass filter output will be a time-variant amplitude version of the FM signal at the correct carrier frequency. As before, such amplitude variations can be removed with a bandpass limiter, prior to demodulation. Thus, unlike amplitude modulation, FM is fairly tolerant of channel nonlinearities. Hence, FM is widely used in microwave and satellite communication systems where the nonlinearities of the power amplifiers need to be well tolerated.

Phase Distortion in Angle Modulation

Angle modulation systems are very sensitive to phase nonlinearities because in such systems the information resides in the phase of the modulated signal. Phase distortion of the modulated wave eventually results in distortion of the demodulated output. Phase distortion can result from phase nonlinearities of channel, repeater, filter, or amplifier responses. It can also result from nonideal linear filtering of an angle modulated wave with a time-variant amplitude.

Let the input signal to a nonideal linear filter be an FM signal $s_1(t)$ with envelope $E_1(t)$ and phase $\phi_1(t)$. Let the FM wave have a residual amplitude modulation due to distortion previously acquired in the transmission channel. Let the output $s_2(t)$ of the filter have an envelope $E_2(t)$ and a phase $\phi_2(t)$. Then the input and output signal can be expressed, respectively, as

$$s_1(t) = E_1(t) \cos[\phi_1(t)] \quad (4.50)$$

$$s_2(t) = E_2(t) \cos[\phi_2(t)] \quad (4.51)$$

Note that although $s_2(t)$ is a linear function of $s_1(t)$, both the phase and the envelope of the output signal are nonlinear functions of the output signal $s_2(t)$. Hence, they are both nonlinear functions of the input signal $s_1(t)$. Thus, in general, amplitude modulation at the input to a linear filter results in phase modulation at the output. This is known as AM–PM conversion. On the other hand, phase modulation at the input to a linear filter results in residual amplitude modulation at the filter output. This is known as PM–AM conversion. PM–AM conversion does not present much of a problem in angle modulated systems, since the resulting residual amplitude modulation is easily remedied with the aid of bandpass limiters. However, AM–PM conversion presents a great problem in angle modulated systems because it is usually not possible to separate the resulting phase distortion from the demodulated signal.

In addition to the mechanism just described, phase nonlinearities of amplifiers, filters, repeaters, etc., also cause AM–PM conversion. This type of distortion is common in microwave systems where it results from the dependence of the phase characteristic of the repeaters and amplifiers on the envelope of the input signal. Typically, the phase change $\Delta\phi_2(t)$ in the output signal is proportional to the corresponding change, in decibel, of the envelope of the input signal. Let $k°/\text{dB}$ be the constant of proportionality relating the change $\Delta E_1(t)$ in the envelope of the input signal $E_1(t)$ to the phase change in the output signal. Then the phase change is given by

$$\Delta\phi_2(t) = k 20 \log \frac{\Delta E_1(t)}{E_1(t)} \quad (4.52)$$

4.6 Interference as a Radio Frequency Distortion Mechanism

Interference here refers to the corruption of the received signal in the desired channel by adjacent or nearby channels. Although each channel has an assigned bandwidth, adjacent channel interference often occurs because practical transmission filters do not provide infinite signal suppression outside the desired or assigned bandwidth. Adjacent channel interference from a strong channel can be disastrous for a weak channel. The effect of interference will be discussed for AM, DSB-SC AM, SSB-SC AM, and angle modulation. For each case, f_c will represent the carrier frequency of the desired channel, whereas $f_c + \Delta f$ will represent the carrier frequency of the interfering signal. For simplicity, the interference will be regarded as an unmodulated carrier of amplitude A_i and the phase angle between the interference and the desired modulated wave will be ignored. Thus, the interference $r_i(t)$ at the receiver input is

$$r_i(t) = A_i \cos[2\pi(f_c + \Delta f)t] \tag{4.53}$$

Interference in Amplitude Modulation

In the case of AM, demodulation is performed by passing the received signal through an envelope detector. The received signal, which consists of the desired AM signal with carrier amplitude A_c, and the interference is given by

$$\begin{aligned} r(t) &= [A_c + m(t)] \cos(2\pi f_c t) + A_i \cos 2\pi(f_c + \Delta f)t \\ &= [A_c + m(t) + A_i \cos 2\pi \Delta f t] \cos(2\pi f_c t) - A_i \sin(2\pi \Delta f t) \sin(2\pi f_c t) \\ &= E(t) \cos[2\pi f_c t + \phi(t)] \end{aligned} \tag{4.54}$$

In the last equation, $E(t)$ is the envelope and $\phi(t)$ is the phase of the received signal. The phase is not relevant to the present discussion. The envelope is given by

$$E(t) = \left\{ [A_c + m(t) + A_i \cos(2\pi \Delta f t)]^2 + A_i^2 \sin^2(2\pi \Delta f t) \right\}^{\frac{1}{2}} \tag{4.55}$$

From the expression for the envelope, it is easy to see that if $A_i \ll A_c$, the envelope is approximated by

$$E(t) \approx A_c + m(t) + A_i \cos 2\pi \Delta f t \tag{4.56}$$

Note that A_c in the last equation can be removed by DC blocking to yield the demodulator output $y(t)$. Let M be the peak amplitude of the message signal. Then it is observed that M/A_i, the ratio of the peak message signal amplitude to the peak interference amplitude, is equal at the input and the output of the receiver. Thus, for the case of small interference, the interference is additive at the demodulator output, and demodulation does not increase or decrease the strength of the interference relative to the message signal. The case for small interference is illustrated in Figure. 4.14. Fig. 4.14(a) illustrates the spectrum of the modulated signal and the interference at the receiver input. Figure 4.14(b) illustrates the spectrum of the envelope detector output.

For the case in which the interference amplitude is much larger than the carrier amplitude for the desired channel, the envelope $E(t)$ can be approximated by

$$E(t) \approx [A_c + m(t)] \cos 2\pi \Delta f t + A_i \tag{4.57}$$

In this case, after the constant A_i is blocked with a capacitor, the demodulated output consists of the message signal multiplied by the interference, $\cos 2\pi \Delta f t$ and an additive component of the interference, $A_c \cos 2\pi \Delta f t$. The effect of interference is much worse in this case because the multiplicative interference results in serious distortion.

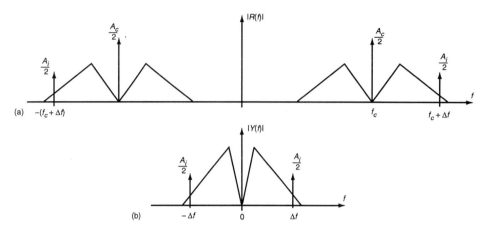

FIGURE 4.14 Illustration of interference in AM: (a) amplitude spectrum of AM wave plus interfrence, (b) amplitude spectrum of demodulated message signal plus interference.

Interference in DSB-SC AM

Consider the synchronous demodulation of a DSB-SC signal plus the interfering sinusoid previously described. The same receiver of Fig. 4.9 is employed, but in this case, the phase and frequency errors between the received and the local carriers are ignored. The received signal plus interference $r(t)$, the product modulator output $x(t)$, and the demodulated output of the low-pass filter $y(t)$, are given, respectively, by

$$r(t) = m(t)\cos 2\pi f_c t + A_i \cos[2\pi(f_c + \Delta f)t] \tag{4.58}$$

$$x(t) = \frac{1}{2}m(t)[1 + \cos 4\pi f_c t] + \frac{1}{2}A_i[\cos 2\pi \Delta f t + \cos(4\pi f_c t + 2\pi \Delta f t)] \tag{4.59}$$

$$y(t) = \frac{1}{2}m(t) + \frac{1}{2}A_i \cos 2\pi \Delta f t \tag{4.60}$$

The last equation shows that for DSB-SC, the effect of interference is additive to the message signal at the demodulator output. Moreover, the ratio of the peak message output to the interference output M/A_i is equal at both the input and the output of the receiver.

Interference in SSB-SC

The analysis and the results in this case are very similar to those for DSB-SC. The same receiver is employed for both cases. In this case, the received signal plus interference, the product modulator output, and the demodulated output of the low-pass filter are given, respectively, by

$$r(t) = m(t)\cos 2\pi f_c t \mp m_h(t)\sin 2\pi f_c t + A_i \cos 2\pi(f_c + \Delta f)t \tag{4.61}$$

$$x(t) = \frac{1}{2}m(t)[1 + \cos 4\pi f_c t] \mp \frac{1}{2}M_h(t)\sin 4\pi f_c t +$$
$$\frac{1}{2}A_i[\cos 2\pi \Delta f t + \cos(4\pi f_c t + 2\pi \Delta f t)] \tag{4.62}$$

$$y(t) = \frac{1}{2}m(t) + \frac{1}{2}A_i(t)\cos 2\pi \Delta f t \tag{4.63}$$

The demodulated output in the last equation shows that the effect of interference on SSB-SC is the same as formerly described for DSB-SC. Comparing these results with the AM case, it is easy to see that for small interference, the effect is about the same for AM as for DSB-SC and SSB-SC. However, AM performance is quite inferior to the other two in the case of large interference.

Interference in Angle Modulation

Consider for simplicity, only the unmodulated carrier of an angle modulated wave of carrier frequency f_c given by $A_c \cos 2\pi f_c t$. For the sake of simplicity, the interference is also considered to be the unmodulated sinusoid $A_i \cos 2\pi (f_c + \Delta f)t$. The signal at the input to the receiver is

$$\begin{aligned} r(t) &= A_c \cos 2\pi f_c t + A_i \cos 2\pi (f_c + \Delta f)t \\ &= [A_c + A_i \cos 2\pi \Delta f t] \cos 2\pi f_c t - A_i \sin 2\pi \Delta f t \sin 2\pi f_c t \\ &= E(t) \cos[2\pi f_c t + \phi(t)] \end{aligned} \quad (4.64)$$

In the last equation, $E(t)$ is the envelope of the received signal, whereas $\phi(t)$ is its phase. In angle modulated waves, the phase, but not the envelope, is relevant to the demodulator output. The phase is given by

$$\begin{aligned} \phi(t) &= \tan^{-1}\left(\frac{-A_i \sin 2\pi \Delta f t}{A_c + A_i \cos 2\pi \Delta f t}\right) \\ &\approx -\frac{A_i \sin 2\pi \Delta f t}{A_c} \quad \text{for } A_i \ll A_c \end{aligned} \quad (4.65)$$

For PM, the demodulated signal is given by

$$y(t) = \phi(t) = -\frac{A_i}{A_c} \sin 2\pi \Delta f t \quad (4.66)$$

For FM, the demodulated signal is given by

$$y(t) = \phi'(t) = -\frac{2\pi \Delta f A_i}{A_c} \cos 2\pi \Delta f t \quad (4.67)$$

The last two equations show that for FM, the amplitude of the interference in the demodulated output is proportional to Δf, the difference between the carrier frequencies of the desired channel and the interference. On the other hand, the interference amplitude is constant for all values of Δf in the case of PM. These differences are illustrated in Fig. 4.15.

The last two equations also show that for both FM and PM, the interference at the demodulator output is inversely proportional to A_c, the amplitude of the carrier of the desired channel. Thus, the stronger the desired channel and the weaker the interference, the smaller is the effect of the interference.

This is superior in performance to amplitude modulated systems in which, at the very best, the interference is independent of the carrier amplitude. In effect, angle modulated systems suppress weak interference. This statement gives a clue to the behavior of angle modulated systems in the presence of large interference. When the interference is large relative to the desired signal, the so-called **capture effect** sets in. The interference now assumes the role of the desired channel and effectively suppresses the weaker desired channel, in the same way that a weak interference would have been suppressed.

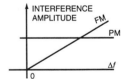

FIGURE 4.15 Effect of interference in angle modulation.

Defining Terms

Capture effect: In angle modulation, the demodulated output of the weaker of two interfering signals is inversely proportional to the amplitude of the stronger signal. Consequently, the stronger signal or channel suppresses the weaker channel. If the stronger signal is noise or an undesired channel, it is said to have captured the desired channel.

Crosstalk: Interference between adjacent communication channels in which an interfering adjacent channel is heard or received in the desired channel.

Dispersion: The variation of wave propagation velocity (or time delay) in a medium with the frequency of the wave, which results in the spreading of the output signal over a time duration greater than the duration of the input signal.

Ionosphere: The ionosphere consists of several layers of the upper atmosphere, ranging in altitude from 100 to 300 km, containing ionized particles. The ionization is due to the effect of solar radiation, particularly ultraviolet and cosmic radiations, on gas particles. The ionosphere serves as reflecting layers for radio (sky) waves.

Pilot carrier: A means of providing a carrier at the receiver, which matches the received carrier in phase and frequency. In this method, which is employed in suppressed carrier modulation systems, a carrier of very small amplitude is inserted into the modulated signal prior to transmission, extracted and amplified in the receiver, and then employed as a matching carrier for coherent detection.

Radio horizon: The maximum range, from transmitter to receiver on the Earth's surface, of direct (line-of-sight) radio waves. This is greater than the optical horizon because the radio waves follow a curved path as a result of the continuous refraction it undergoes in the atmosphere.

Troposphere: The region of the atmosphere within about 10 km of the Earth's surface. Within this region, the wireless radio channel is modified relative to free space conditions by weather conditions, pollution, dust particles, etc. These inhomogeneities act as point scatterers which deflect radio waves downward to reach the receiving antennas, thereby providing tropospheric scatter propagation.

References

Bultitude, R.J.C., Melancon, P., Zaghloul, H., Morrison, G., and Prokiki, M. 1993. The dependence of indoor radio channel multipath characteristics on transmit receive ranges. *IEEE J. Selec. Areas Commun.* 11(7):979–990.

Couch, L.W., II. 1990. *Digital and Analog Communication Systems*, 3rd ed. Macmillan, New York.

Fechtel, S.A. and Meyer, H. 1994. Optimal parametric feedforward estimation of frequency-selective fading radio channels. *IEEE Trans. Commun.* 42(2):1639–1650.

Gibson, J.D. 1993. *Principles of Digital and Analog Communications*, 2nd ed. Macmillan, New York.

Hashemi, H. 1993. Impulse response modeling of indoor radio propagation channels. *IEEE J. Selec. Areas Commun.* 11(7):967–978.

Haykin, S. 1994. *Communication Systems*, 3rd ed. J. Wiley, New York.

Lathi, B.P. 1989. *Modern Digital and Analog Communication Systems*, 2nd ed. Holt, Rinehart and Winston, Philadelphia, PA.

Porakis, J.G. and Salehi, M. 1994. *Communication Systems Engineering*. Prentice–Hall, Englewood Cliffs, NJ.

Roddy, D. and John Coolen, J. 1981. *Electronic Communications*, 2nd ed. Reston Pub., Reston, VA.

Serboyakov, I.Y. 1989. An estimate of the effectiveness of using an amplitude-frequency equalizer to compensate selective fading on digital radio relay links. *Telecommun. Radio Eng.* 44(6):112–115.

Stremler, F.G. 1992. *Introduction to Communication Systems*, 3rd ed. Addison–Wesley, Reading, MA.

Taub, H. and Schilling, D. 1986. *Principles of Communication Systems*, 2nd ed. McGraw–Hill, New York.

Voronkov, Y.V. 1992. The generation of an AM oscillation with low nonlinear envelope distortion and low attendant angle modulation. *Telecomm. Radio Eng.* 47(7):10–12.

Ziemer, R.E. and Tranter, W.H. 1990. *Principles of Communications Systems, Modulation and Noise*, 3rd ed. Houghton Mifflin, Boston, MA.

Further Information

Electronics and Communication in Japan

IEEE Transactions on Antenna and Propagation

IEEE Transactions on Communications

IEEE Journal of Selected Areas in Communications

Telecommunications and Radio Engineering, a technical journal published by Scripta Technica, Inc., Wiley Co.

5
Digital Test Equipment and Measurement Systems

Jerry C. Whitaker
Editor-in-Chief

5.1 Introduction 57
5.2 Logic Analyzer 57
5.3 Signature Analyzer 60
5.4 Manual Probe Diagnosis 60
5.5 Checking Integrated Circuits 60
5.6 Emulative Tester 61
5.7 Protocol Analyzer 62
5.8 Automated Test Instruments 62
     Computer-Instrument Interface • Software Considerations • Applications
5.9 Digital Oscilloscope 63
     Operating Principles • Digital Storage Oscilloscope (DSO) Features • Capturing Transient Waveforms • Triggering

5.1 Introduction

As the equipment used by consumers and industry becomes more complex, the requirements for highly skilled maintenance technicians also increases. Maintenance personnel today require advanced test equipment and must think in a systems mode to troubleshoot much of the hardware now in the field. New technologies and changing economic conditions have reshaped the way maintenance professionals view their jobs. As technology drives equipment design forward, maintenance difficulties will continue to increase. Such problems can be met only through improved test equipment and increased technician training.

Servicing computer-based professional equipment typically involves isolating the problem to the board level and then replacing the defective printed wiring board (PWB). Taken on a case-by-case basis, this approach seems efficient. The inefficiency in the approach, however (which is readily apparent), is the investment required to keep a stock of spare boards on hand. Further, because of the complex interrelation of circuits today, a PWB that appeared to be faulty may actually turn out to be perfect. The ideal solution is to troubleshoot down to the component level and replace the faulty device instead of swapping boards. In many cases, this approach requires sophisticated and expensive test equipment. In other cases, however, simple test instruments will do the job.

Although the cost of most professional equipment has been going up in recent years, maintenance technicians have seen a buyer's market in test instruments. The semiconductor revolution has done more than given consumers low-cost computers and disposable calculators. It has also helped to spawn a broad variety of inexpensive test instruments with impressive measurement capabilities.

5.2 Logic Analyzer

The logic analyzer is really two instruments in one: a *timing analyzer* and a *state analyzer*. The timing analyzer is analogous to an oscilloscope. It displays information in the same general form as a scope, with the horizontal axis representing time and the vertical axis representing voltage amplitude. The timing analyzer samples the input waveform to determine whether it is high or low. The instrument cares about only one voltage threshold. If the signal is above the threshold when it is sampled, it will be displayed as a 1 (or high); any signal below the threshold is displayed as a 0 (or low). From these sample points, a list of ones and zeros is generated to represent a picture of the input waveform. This data is stored in memory and used to reconstruct the input waveform, as shown in Fig. 5.1. A block diagram of a typical timing analyzer is shown in Fig. 5.2.

Sample points for the timing analyzer are developed by an internal clock. The period of the sample can be selected by the user. Because the analyzer samples asynchronously to the *unit under test* (UUT) under the direction

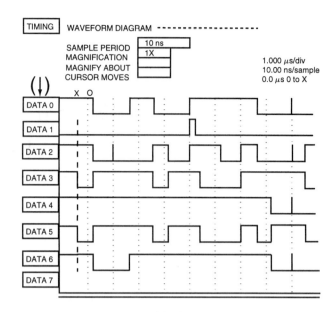

FIGURE 5.1 Typical display of a timing analyzer.

of the internal clock, a long sample period results in a more accurate picture of the data bus. A sample period should be selected that permits an accurate view of data activity, while not filling up the instrument's memory with unnecessary data.

Accurate sampling of data lines requires a trigger source to begin data acquisition. A fixed delay may be inserted between the trigger point and the *trace point* to allow for bus settling. This concept is illustrated in Fig. 5.3. A variety of trigger modes are commonly available on a timing analyzer, including:

- **Level triggering**. Data acquisition and/or display begins when a logic high or low is detected.
- **Edge triggering**. Data acquisition/display begins on the rising or falling edge of a selected signal. Although many logic devices are level dependent, clock and control signals are often edge sensitive.
- **Bus state triggering**. Data acquisition/display is triggered when a specific code is detected (specified in binary or hexadecimal).

The timing analyzer is constantly taking data from the monitored bus. Triggering, and subsequent generation of the trace point, controls the *data window* displayed to the technician. It is possible, therefore, configure the analyzer to display data that precedes the trace point. (See Fig. 5.4.) This feature can be a powerful troubleshooting and developmental tool.

The timing analyzer is probably the best method of detecting **glitches** in computer-based equipment. (A *glitch* being defined as any transition that crosses a logic threshold more than once between clock periods. See Fig. 5.5). The triggering input of the analyzer is set to the bus line that is experiencing random glitches. When the analyzer detects a glitch, it displays the bus state preceding, during, or after occurrence of the disturbance.

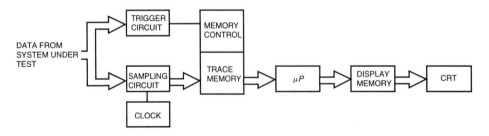

FIGURE 5.2 Block diagram of a timing analyzer.

The state analyzer is the second-half of a logic analyzer. It is used most often to trace the execution of instructions through a microprocessor system. Data, address, and status codes are captured and displayed as they occur on the microprocessor bus. A **state** is a sample of the bus when the data are valid. A state is usually displayed in a tabular format of hexadecimal, binary, octal, or assembly language. Because some microprocessors multiplex data and addresses on the same lines, the analyzer must be able to clock-in information at different clock rates. The analyzer, in essence, acts as a demultiplexer to capture an address at the proper time, and then to capture data present on the same bus at a different point in time. A state analyzer also gives the operator the ability to *qualify* the data stored. Operation of the instrument may be triggered by a specific logic pattern on the bus. State analyzers usually offer a *sequence term* feature that aids in triggering. A sequence term allows the operator to qualify data storage more accurately than would be possible with a single trigger point. A sequence term usually takes the following form:

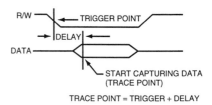

FIGURE 5.3 Use of a delay period between the trigger point and the trace point of a timing analyzer.

```
Find     xxxx
Then find yyyy
Start on  zzzz
```

A sequence term is useful for probing a subroutine from a specific point in a program. It also makes possible selective storage of data, as shown in Fig. 5.6.

To make the acquired data easier to understand, most state analyzers include software packages that interpret the information. Such *disassemblers* (also known as *inverse assemblers*) translate hex, binary, or octal codes into assembly code (or some other format) to make then easier to read.

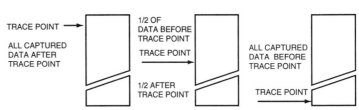

FIGURE 5.4 Data capturing options available from a timing analyzer.

The logic analyzer is used routinely by design engineers to gain an in-depth look at signals within digital circuits. A logic analyzer can operate at 100 MHz and beyond, making the instrument ideal for detecting glitches resulting from timing problems. Such faults are usually associated with design flaws, not with manufacturing defects or failures in the field.

Although the logic analyzer has benefits for the service technician, its use is limited. Designers require the ability to verify hardware and software implementations with test equipment; service technicians simply need to quickly isolate a fault. It is difficult and costly to automate test procedures for a given PWB using a logic analyzer. The technician must examine a long data stream and decide if the data are good or bad. Writing programs to validate state analysis data is possible, but—again—is time consuming.

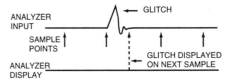

FIGURE 5.5 Use of a timing analyzer for detecting glitches on a monitored line.

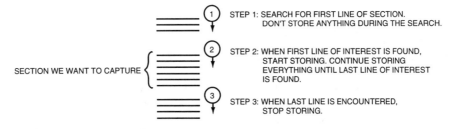

FIGURE 5.6 Illustration of the selective storage capability of state analyzer.

5.3 Signature Analyzer

Signature analysis is a common troubleshooting tool based on the old analog method of signal tracing. In analog troubleshooting, the technician followed an annotated schematic that depicted waveforms and voltages that should be observed with an oscilloscope and voltmeter. The signature analyzer captures a complex data stream from a test point on the PWB and converts it into a hexadecimal signature. These hexadecimal signatures are easy to annotate on schematics, permitting their use as references for comparison against signatures obtained from a UUT. The signature by itself means nothing; the value of the signature analyzer comes from comparing a known-good signature with the UUT.

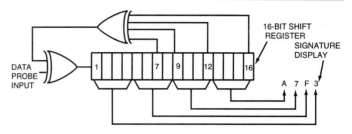

FIGURE 5.7 Simplified block diagram of a signature analyzer.

A block diagram of a simplified signature analyzer is shown in Fig. 5.7. As shown, a digital bit stream, accepted through a data probe during a specified measurement window, is marched through a 16-b linear feedback shift register. Whatever data is left over in the register after the specified measurement window has closed is converted into a 4-b hexadecimal readout. The feedback portion of the shift register allows just one faulty bit in a digital bit stream to create an entirely different signature than would be expected in a properly operating system. Hewlett-Packard, which developed this technique, terms it a *pseudorandom binary sequence* (PRBS) generator. This method allows the maintenance technician to identify a signal bit error in a digital bit stream with 99.998% certainty, even when picking up errors that are timing related.

To function properly, the signature analyzer requires start and stop signals, a clock input, and the data input. The start and stop inputs are derived from the circuit being tested and are used to bracket the beginning and end of the measurement window. During this gate period, data is input through the data probe. The clock input controls the sample rate of data entering the analyzer. The clock is most often taken from the clock input pin of the microprocessor. The start, stop, and clock inputs may be configured by the technician to trigger on the rising or falling edges of the input signals.

5.4 Manual Probe Diagnosis

Manual probe diagnosis integrates the logic probe, logic analyzer, and signature analyzer to troubleshoot computer-based hardware. The manual probe technique employs a database consisting of nodal-level signature measurements. Each node of a known-good unit under test is probed, and the signature information saved. This known-good data is then compared to the data from a faulty UUT. Efficient use of the manual probe technique requires a skilled operator to determine the proper order for probing circuit nodes, or a results-driven software package that directs the technician to the next logical step.

5.5 Checking Integrated Circuits

An integrated circuit tester identifies chip faults by generating test patterns that exercise all possible input-state combinations of the unit under test. For example, the NAND gate shown in Fig. 5.8 has a maximum of four input patterns that the instrument must generate to completely test the device. The input test patterns, and the resulting device outputs, are arranged in a truth table. The table represents the response that should be obtained with specified input stimulus. The operator inputs the device type number, which configures the instrument to the proper pattern and truth table. As illustrated in Fig. 5.9, devices that match the table are certified as good; those failing are rejected.

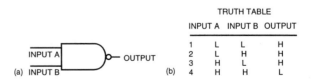

FIGURE 5.8 Truth table for a NAND gate.

Such conventional functional tests are well suited for out-of-circuit device qualification. Under out-of-circuit conditions, the instrument is free to toggle each input high or low, thereby executing all patterns of the truth table. For many in-circuit applications, however, inputs are wired in ways that prevent the instrument from executing the predetermined test pattern. Some instruments are able to circumvent the problem of nonstandard (in-circuit) configurations by including a *learning mode* in the system. Using this mode, the IC tester exercises the input of a gate that is known to be operating properly in a nonstandard configuration with the normal set of input signals. The resulting truth table is stored in memory. The instrument is then connected to an in-circuit device to be checked.

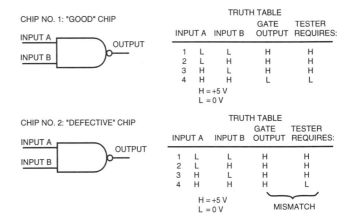

FIGURE 5.9 Component accept/reject process based on comparing observed states and an integrated circuit logic truth table.

5.6 Emulative Tester

Guided diagnostics can be used on certain types of hardware to facilitate semiautomated circuit testing. The test connector typically taps into one of the system board input/output ports and, armed with the proper software, the instrument goes through a step-by-step test routine that exercises the computer circuits. When the system encounters a fault condition, it outputs an appropriate message.

Emulative testers are a variation on the general guided diagnostics theme. Such an instrument plugs into the microprocessor socket of the computer and runs a series of diagnostic tests to identify the cause of a fault condition. An emulative tester emulates the board's microprocessor while verifying circuit operation. Testing the board inside-out allows easy access to all parts of the circuit; synchronization with the various data and address cycles of the PWB is automatic. Test procedures, such as read/write cycles from the microprocessor are generic, so that a high-quality functional test can be quickly created for any board. Signature analysis is used to verify that circuits are operating correctly. Even with long streams of data, there is no maximum memory depth.

Several different types of emulative testers are available. Some are designed to check only one brand of computer, or only computers based on one type of microprocessor. Other instruments can check a variety of microcomputer systems using so-called *personality* modules that adapt the instrument to the system being serviced.

Guided-fault isolation (GFI) is practical with an emulative tester because the instrument maintains control over the entire system. Automated tests can isolate faults to the node level. All board information and test procedures are resident within the emulative test instrument, including prompts to the operator on what to do next. Using the microprocessor test connection combined with movable probes or clips allows a closed-loop test of any part of a circuit. Input/output capabilities of emulative testers range from single point probes to well over a hundred I/O lines. These lines can provide stimulus as well as measurement capabilities.

The principal benefit of the guided probe over the manual probe is derived from the creation of a *topology database* for the UUT. The database describes devices on the UUT, their internal fault-propagating characteristics, and their interdevice connections. In this way, the tester can guide the operator down the proper logic path to isolate the fault. A guided probe accomplishes the same analysis as a conventional fault tree, but differs in that troubleshooting is based on a generic algorithm that uses the circuit model as its input. First, the system compares measurements taken from the failed UUT against the expected results. Next, it searches for a database representation of the UUT to determine the next logical place to take a measurement. The guided probe system automatically determines which nodes impact other nodes. The algorithm tracks its way through the logic system of the board to the source of the fault. Programming of a guided probe system requires the following steps:

- Development of a stimulus routine to exercise and verify the operating condition of the UUT (go/no-go status).
- Acquisition and storage of measurements for each node affected by the stimulus routine. The programmer divides the system or board into measurement sets (MSETs) of nodes having a common timebase. In this way, the user can take maximum advantage of the time domain feature of the algorithm.
- Programming a representation of the UUT that depicts the interconnection between devices, and the manner in which signals can propagate through the system. Common *connectivity* libraries are developed and reused from one application to another.
- Programming a measurement set cross reference database. This allows the guided probe instrument to cross MSET boundaries automatically without operator intervention.
- Implementation of a test program control routine that executes the stimulus routines. When a failure is detected, the guided probe database is invoked to determine the next step in the troubleshooting process.

5.7 Protocol Analyzer

Testing a local area network (LAN)—whether copper, fiber optic, or wireless—presents special challenges to the design engineer or maintenance technician. The protocol analyzer is commonly used to service LANs and other communications systems. The protocol analyzer performs data monitoring, terminal simulation, and bit error rate tests (BERTs). Sophisticated analyzers provide *high-level decide* capability. This ability refers to the open system interconnection (OSI) network seven layer model. Typically, a sophisticated analyzer based on a personal computer can decide to level 3. PC-based devices also may incorporate features that permit performance analysis of wide area networks (WANs). Network statistics, including response time and use, are measured and displayed graphically. The characteristics of a line can be easily viewed and measured. These functions enable the engineer to observe activity of a communications link and exercise the facility to verify proper operation. Simulation capability is usually available to emulate almost any data terminal or data communications equipment.

LAN test equipment functions include monitoring of protocols and emulation of a network node, bridge, or terminal. Statistical information is provided to isolate problems to particular devices, or to high periods of activity. Statistical data includes use, packet rates, error rates, and collision rates.

5.8 Automated Test Instruments

Computers can function as powerful troubleshooting tools. Advanced technology personal computers, coupled with add-on interface cards and applications software, provide a wide range of testing capabilities. Computers can also be connected to many different instruments to provide for automated testing. There are two basic types of stand-alone automated test instruments (ATEs): *functional* and *in-circuit*.

Functional testing exercises the unit under test to identify faults. Individual PWBs or subassemblies may be checked using this approach. Functional testing provides a fast go/no-go qualification check. Dynamic faults are best discovered through this approach. Functional test instrument are well suited to high volume testing of PWB subassemblies; however, programming is a major undertaking. An intimate knowledge of the subassembly is needed to generate the required test patterns. An in-circuit emulator is one type of functional tester.

In-circuit testing is primarily a diagnostic tool. It verifies the functionality of the individual components of a subassembly. Each device is checked and failing parts are identified. In-circuit testing, although valuable for detailed component checking, does not always operate at the clock rate of the subassembly. Propagation delays, race conditions, and other abnormalities may go undetected. Access to key points on the PWB may be accomplished in one of two ways:

- **Bed of nails**. The subassembly is placed on a dedicated test fixture and held in place by a vacuum or mechanical means. Probes access key electrical traces on the subassembly to check individual devices. Through a technique known as *backdriving*, the inputs of each device are isolated from the associated circuitry and the component is tested for functionality. This type of testing is expensive and is only practical for high-volume subassembly qualification testing and rework.

- PWB clips. Intended for lower volume applications than a bed of nails instrument, PWB clips replace the dedicated test fixture. The operator places the clips on the board as directed by the instrument. As access to all components simultaneously is not required, the tester is less hardware intensive and programming is simplified. Many systems include a library of software routines designed to test various classes of devices. Test instruments using PWB clips tend to be slow. Test times for an average PWB may range from 8 to 20 min vs 1 min or less for a bed of nails.

Computer-Instrument Interface

There are two primary nonproprietary types of computer interface systems used to connect test instruments and computers: IEEE-488 and RS-232. The IEEE-488 format is also called the *general purpose interface bus* (GPIB) or the *Hewlett Packard interface bus* (HPIB). RS-232 is the *standard serial interface* used on many computers. The two interface systems each have their own advantages. Both provide for connection of a computer to one or more measuring instruments. Both are bidirectional, which allows the computer to either send information or receive it from the outside world. Some systems provide both interfaces, but most have one or the other.

Test instruments utilizing the GPIB interface greatly outnumber those with RS-232. Several thousand test instruments are available with GPIB interfacing as an option. Some plotters and printers also accept a GPIB input.

By comparison, RS-232 is more common than GPIB in computer applications. Printers, plotters, scanners, and modems often use the standard serial interface. Test instruments incorporating RS-232 typically are those used for remote sensing, such as RF signal-strength meters or thermometers. Neither RS-232 nor GPIB is ideal for every application. Each protocol works well in some uses and marginally in others. The decision of which protocol to use for a particular application must be based on what the system needs to do. Because RS-232 is already built into personal computers, many users want to use it for automation. Yet GPIB is the preferred protocol for most test equipment applications.

Software Considerations

Most automated test instrument packages include software that permits the user to customize test instruments and procedures to meet the required test. The software, in effect, writes software. The user enters the codes needed by each automatic instrument, selects the measurements to be performed, and tells the computer where to store the test data. The program then puts the final software together after the user answers key configuration questions. The software looks up the operational codes for each instrument and compiles the software to perform the required tests. Automatic generation of programming greatly reduces the need to have experienced programmers on staff. Once installed in the computer, programming becomes as simple as fitting graphic symbols together on the screen.

Applications

The most common applications for computer-controlled testing are data gathering, product go/no-go qualification, and troubleshooting. All depend on software to control the instruments. Acquiring data can often be accomplished with a computer and a single instrument. The computer collects dozens or hundreds of readings until the occurrence of some event. The event may be a preset elapsed time or the occurrence of some condition at a test point, such as exceeding a preset voltage or dropping below a preset voltage. Readings from a variety of test points are then stored in the computer. Under computer direction, test instruments can also run checks that might be difficult or time consuming to perform manually. The computer may control several test instruments such as power supplies, signal generators, and frequency counters. The data is stored for later analysis.

5.9 Digital Oscilloscope

The digital storage oscilloscope (DSO) offers a number of significant advantages beyond the capabilities of analog instruments. A DSO can store in memory the signal being observed, permitting in-depth analysis impossible with previous technology. Because the waveform resides in memory, the data associated with the waveform can be transferred to a computer for real-time processing, or for processing at a later time.

FIGURE 5.10 Simplified block diagram of a digital storage oscilloscope.

Operating Principles

Figure 5.10 shows a block diagram of a DSO. Instead of being amplified and directly applied to the deflection plates of a cathode ray tube (CRT), the waveform is first converted into a digital form and stored in memory. To reproduce the waveform on the CRT, the data is sequentially read and converted back into an analog signal for display.

Although a DSO is specified by its maximum sampling rate, the actual rate used in acquiring a given waveform is usually dependent on the time-per-division setting of the oscilloscope. The record length (samples recorded over a given period of time) defines a finite number of sample points available for a given acquisition. The DSO must, therefore, adjust its sampling rate to fill a given

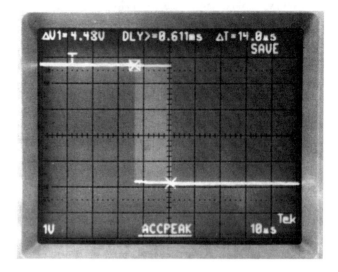

FIGURE 5.11 Use of the peak accumulation sampling mode to capture variations in pulse width. (*Source:* Tektronix, Beaverton, OR.)

record over the period set by the sweep control. To determine the sampling rate for a given sweep speed, the number of displayed points per division is divided into the sweep rate per division. Two additional features can modify the actual sampling rate:

- Use of an external clock for pacing the digitizing rate. With the internal digitizing clock disabled, the digitizer will be paced at a rate defined by the operator.
- Use of a peak detection (or glitch capture) mode. Peak detection allows the digitizer to sample at the full digitizing rate of the DSO, regardless of the time-base setting. The minimum and maximum values found between each normal sample interval are retained in memory. These minimum and maximum values are used to reconstruct the waveform display with the help of an algorithm that recreates a smooth trace along with any captured glitches. Peak detection allows the DSO to capture glitches even at its slowest sweep speed. For higher performance, a technique known as **peak-accumulation** (or *envelope mode*) may be used. With this approach, the instrument accumulates and displays the maximum and minimum excursions of a waveform for a given point in time. This builds an envelope of activity that can reveal infrequent noise spikes, long-term amplitude or time drift, and pulse jitter extremes. Figure 5.11 illustrates the advantage of peak accumulation when variations in data pulse width are monitored.

Achieving adequate samples for a high-frequency waveform places stringent requirements on the sampling rate. High-frequency waveforms require high sampling rates. **Equivalent-time sampling** is often used to provide high-bandwidth capture. This technique relies on sampling a repetitive waveform at a low rate to build up sample density. This concept is illustrated in Fig. 5.12. When the waveform to be acquired triggers the DSO, several samples are taken over the duration of the waveform. The next repetition of the waveform triggers the instrument again, and more samples are taken at different points on the waveform. Over many repetitions, the number of stored samples can be built up to the equivalent of a high sampling rate.

System operating speed also has an effect on the display update rate. Update performance is critical when measuring

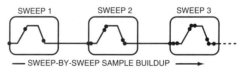

FIGURE 5.12 Increasing sample density through equivalent-time sampling.

waveform or voltage changes. If the display does not track the changes, adjustment of the circuit may be difficult.

There may be cases when, using the fastest sweep speed on a DSO, high-speed sampling still does not provide as quick a display update as necessary. In this event, a conventional analog scope may provide the best performance. Some DSO instruments offer a switchable analog display path for such applications.

Digital Storage Oscilloscope (DSO) Features

The digital oscilloscope has become a valuable tool in troubleshooting both analog and computer-based products. Advanced components and construction techniques have led to lower costs for digital scopes and higher performance. Digital scopes can capture and analyze transient signals, such as race conditions, clock jitter, glitches, dropouts, and intermittent faults. Automated features reduce testing and troubleshooting costs through the use of recallable instrument setups, direct parameter readout, and unattended monitoring. Digital oscilloscopes have inherent benefits not available on most analog oscilloscopes. These benefits include:

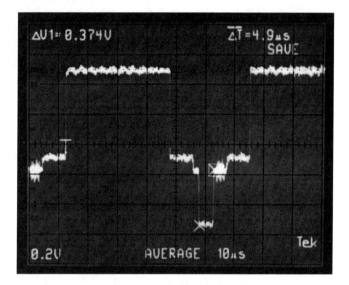

FIGURE 5.13 A digitized waveform "frozen" on the display to permit measurement and examination of detail. (*Source:* Tektronix, Beaverton, OR.)

- Increased resolution (determined by the quality of the analog-to-digital converter).
- Memory storage of digitized waveforms. Figure 5.13 shows a complex waveform captured by a DSO.
- Automatic setup for repetitive signal analysis. For complex multichannel configurations that are used often, front-panel storage/recall can save dozens of manual selections and adjustments. When multiple memory locations are available, multiple front-panel setups can be stored to save even more time.
- Autoranging. Many instruments will automatically adjust for optimum sweep, input sensitivity, and triggering. The instrument's micropro cessor automatically configures the front panel for optimum display. Such features permit the operator to concentrate on making measurements, not on adjusting the scope.
- Instant hardcopy output from printers and plotters.
- Remote programmability via GPIB for automated test applications.
- Trigger flexibility. Single-shot digitizing oscilloscopes capture transient signals and allow the user to view the waveform which preceded the trigger point. Figure 5.14 illustrates the use of pre/posttrigger for waveform analysis.
- Signal analysis. Intelligent scopes can make key measurements and comparisons. Display capabilities include voltage peak, mean voltage, RMS value, rise time, fall time, and frequency. Figure 5.15 shows the voltage measurement options available on one instrument.
- Cursor measurement. Advanced oscilloscopes permit the operator to take measurements or perform comparative operations on data appearing on the display. A measurement cursor consists of a pair of lines or dots that can be moved around the screen as needed. Figure 5.16 shows one such example. Cursors have been placed at different points on the displayed waveform. The instruments automatically determines the phase difference between the measurement points. The cursor readout follows the relative position of each cursor.
- Trace quality. Eye fatigue is reduced noticeably with a DSO when viewing low-repetition signals. For example, a 60-Hz waveform can be difficult to view for extended periods of time on a conventional scope because

the display tends to flicker. A DSO overcomes this problem by writing waveforms to the screen at the same rate regardless of the input signal. (See Fig. 5.17).

Digital memory storage offers a number of benefits, including:

- Reference memory. A previously acquired waveform can be stored in memory and compared with a sampled waveform. This feature is especially useful for repetitive testing or calibration of a device to a standard waveform pattern. Nonvolatile battery-backed memory permits reference waveforms to be transported to field sites.
- Simple data transfers to a host computer for analysis or archive.
- Local data analysis through the use of a built-in microprocessor.
- Cursors capable of providing a readout of delta and absolute voltage and time.
- No CRT blooming for display of fast transients.
- Full bandwidth capture of long duration waveforms, thus storing all of the signal details. The waveform can be expanded after capture to expose the details of a particular section.

Capturing Transient Waveforms

Single-shot digitizing makes it possible to capture and clearly display transient and intermittent signals. Waveforms such as signal glitches, dropouts, logic race conditions, intermittent faults, clock jitter, and power-up sequences can be examined with the help of a digital oscilloscope. With single-shot digitizing, the waveform is captured the first time it occurs, on the first trigger. It can then be displayed immediately or held in memory for analysis at a later date. In contrast, most analog oscilloscopes can only capture repetitive signals. Hundreds of cycles of the signal are needed to construct a representation of the waveshape. Figure 5.18 illustrates the benefits of digital storage in capturing transient waveforms. An analog scope will often fail to detect a transient pulse that a DSO can clearly display.

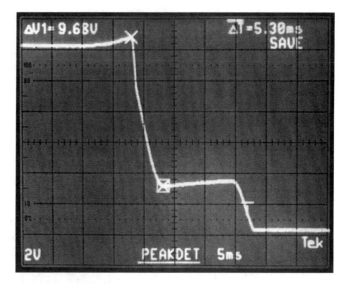

FIGURE 5.14 Use of the pre/post trigger function for waveform analysis. (*Source:* Tektronix, Beaverton, OR.)

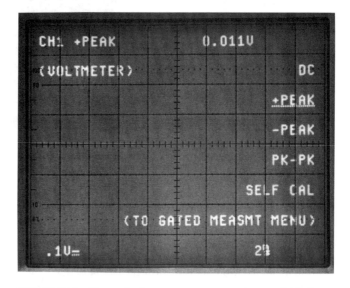

FIGURE 5.15 Menu of voltage measurement options available from a DSO. (*Source:* Tektronix, Beaverton, OR.)

Triggering

Basic triggering modes available on a digital oscilloscope permit the user to select the desired source, its coupling, level, and slope. More advanced digital scopes contain triggering circuitry similar to that found in a logic analyzer.

These powerful features let the user trigger on elusive conditions, such as pulse widths less than or greater than expected, intervals less than or greater than expected, and specified logic conditions. The logic triggering can include digital pattern, state qualified, and time/event qualified conditions. Many trigger modes are further enhanced by allowing the user to hold off the trigger by a selectable time or number of events. Hold off is especially useful when the input signal contains bursts of data or follows a repetitive pattern.

Pulse-width triggering lets the operator quickly check for pulses narrower than expected or wider than expected. The pulse-width trigger circuit checks the time from the trigger source transition of a given slope (typically the rising edge) to the next transition of opposite slope (typically the falling edge). The operator can interactively set the pulse-width threshold for the trigger. For example, a glitch can be considered any signal narrower than $\frac{1}{2}$ of a clock period. Conditions preceding the trigger can be displayed to show what events led up to the glitch.

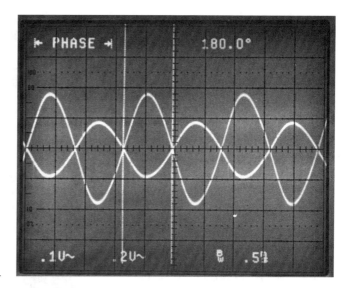

FIGURE 5.16 Use of cursors to measure the phase difference between two points on a waveform. (*Source:* Tektronix, Beaverton, OR.)

Interval triggering lets the operator quickly check for intervals narrower than expected or wider than expected. Typical applications include monitoring for transmission phase changes in the output of a modem or for signal dropouts, such as missing bits on a computer hard disk.

Pattern triggering lets the user trigger on the logic state (high, low, or either) of several inputs. The inputs can be external triggers or the input channels themselves. The trigger can be generated either upon entering or exiting the pattern. Applications include triggering on a particular address select or data bus condition. Once the pattern trigger is established, the operator can probe throughout the circuit, taking measurement synchronous with the trigger.

State qualified triggering enables the oscilloscope to trigger on one source, such as the input signal itself, only after the occurrence of a specified logic pattern. The pattern acts as an *enable* or *disable* for the source.

Advanced Features

Some digital oscilloscopes provide enhanced triggering modes that permit the user to select the level and slope for each input. This flexibility makes it easy to look for odd pulse shapes in the pulse-width trigger mode and for subtle dropouts in the interval trigger mode. It also simplifies testing different logic types (TTL, CMOS, and ECL, for example), and testing analog/digital combinational circuits with dual logic triggering. Additional flexibility is available on multiple channel scopes.

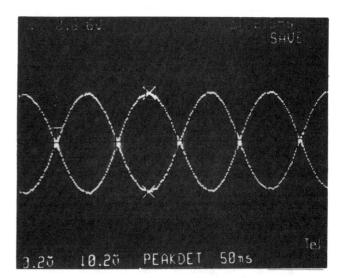

FIGURE 5.17 The DSO provides a stable, flicker-free display for viewing low-repetition signals. (*Source:* Tektronix, Beaverton, OR.)

Once a trigger has been sensed, multiple simultaneously sampled inputs permit the user to monitor conditions at several places in the unit under test, with each channel synchronized to the trigger. Additional useful trigger features for monitoring jitter or drift on a repetitive signal include:

- Enveloping
- Extremes
- Waveform delta
- Roof/floor

Each of these functions are related, and in some cases describe the same general operating mode. Various scope manufacturers use different nomenclature to describe proprietary triggering methods. Generally speaking, as the waveshape changes with respect to the trigger, the scope generates upper and lower traces. For every Nth sample point with respect to the trigger, the maximum and minimum values are saved. Thus, any jitter or drift is displayed in the envelope.

Advanced triggering features provide the greatest benefit in conjunction with single-shot sampling. Repetitive sampling scopes can only capture and display signals that repeat precisely from cycle to cycle. Several cycles of the waveform are required to create a digitally reconstructed representation of the input. If the signal varies from cycle to cycle, such scopes can be less effective than a standard analog oscilloscope for accurately viewing a waveform.

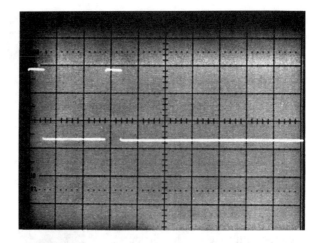
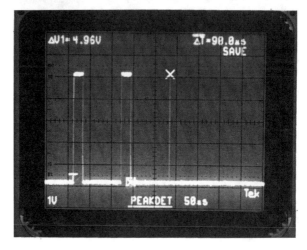

FIGURE 5.18 The benefits of a DSO in capturing transient signals: (a) Analog display of a pulsed signal with transient present, but not visible; (b) DSO display of the same signal clearly showing the transient. (*Source:* Tektronix, Beaverton, OR.)

Defining Terms

Bed of nails: A test fixture for automated circuit qualification in which a printed wiring board is placed in contact with a fixture that contacts the board at certain nodes required for exercising the assembly.
Bus state triggering: A data acquisition mode initiated when a specific digital code is detected.
Edge triggering: A data acquisition mode initiated by the rising or falling edge of a selected signal.
Equivalent-time sampling: An operating feature of a digital storage oscilloscope in which a repetitive waveform is sampled at a low rate to build up sample density.
Glitch: An undesired condition in a digital system in which a logic threshold transition occurs more than once between fixed or specified clock periods.
Level triggering: A data acquisition mode initiated when a logic high or low is detected.
Peak accumulation mode: An operating feature of digital storage oscilloscope in which the maximum and minimum excursions of the waveform are displayed for a given point in time.
State: In a digital system, a sample of the data bus when the data are valid.

References

Albright, J.R. 1989. Waveform analysis with professional-grade oscilloscopes. *Electronic Servicing and Technology*. Intertec, Overland Park, KS, April.

Breya, M. 1988. New scopes make faster measurements. *Mobile Radio Technology*. Intertec, Overland Park, KS, Nov.

Carey, G.D. 1989. Automated test instruments. *Broadcast Engineering*. Intertec, Overland Park, KS, Nov.

Harris, B. 1988. The digital storage oscilloscope: Providing the competitive edge. *Electronic Servicing and Technology*. Intertec, Overland Park, KS, June.

Harris, B. 1989. Understanding DSO accuracy and measurement performance. *Electronic Servicing and Technology*. Intertec, Overland Park, KS, April.

Hoyer, M. 1990. Bandwidth and rise time: Two keys to selecting the right oscilloscope. *Electronic Servicing and Technology*. Intertec, Overland Park, KS, April.

Montgomery, S. 1989. Advanced in digital oscilloscopes. *Broadcast Engineering*. Intertec, Overland Park, KS, Nov.

Persson, C. 1987. Oscilloscope: The eyes of the technician. *Electronic Servicing and Technology*. Interec, Overland Park, KS, April.

Persson, C. 1987. Test equipment for personal computers. *Electronic Servicing and Technology*. Intertec, Overland Park, KS, July.

Persson, C. 1989. The new breed of test instrument. *Broadcast Engineering*. Intertec, Overland Park, KS, Nov.

Persson, C. 1990. Oscilloscope special report. *Electronic Servicing and Technology*. Intertec, Overland Park, KS, April.

Siner, T.A. 1987. Guided probe diagnosis: affordable Automation. *Microservice Management*. Intertec, Overland Park, KS, March.

Sokol, F. 1989. Specialized test equipment. *Microservice Management*. Intertec, Overland Park, KS, Aug.

Toorens, H. 1990. Oscilloscopes: From looking glass to high-tech. *Electronic Servicing and Technology*. Intertec, Overland Park, KS, April.

Wickstead, M. 1985. Signature analyzers. *Microservice Management*. Intertec, Overland Park, KS, Oct.

Whitaker, J.C. 1989. *Maintaining Electronic Systems*. CRC Press, Boca Raton, FL.

Further Information

Most test equipment manufacturers provide detailed operational information on the use of their instruments. In addition, some companies offer applications booklets and engineering notes that explain test and measurement objectives, requirements, and procedures. Most of this information is available at little or no cost. The following book is also recommended:

Whitaker, J.C. 1989. *Maintaining Electronic Systems*. CRC Press, Boca Raton, FL, 1989.

6
Fourier Waveform Analysis

Jerry C. Hamann
University of Wyoming

John W. Pierre
University of Wyoming

6.1 Introduction 70
6.2 The Mathematical Preliminaries for Fourier Analysis 71
6.3 The Fourier Series for Continuous-Time Periodic Functions 74
6.4 The Fourier Transform for Continuous-Time Aperiodic Functions ... 75
6.5 The Fourier Series for Discrete-Time Periodic Functions 76
6.6 The Fourier Transform for Discrete-Time Aperiodic Functions 76
6.7 Example Applications of Fourier Waveform Techniques 77
Using the DFT/FFT in Fourier Analysis • Total Harmonic Distortion Measures

6.1 Introduction

Fourier waveform analysis originated in the early 1800s when Baron Jean Baptiste Joseph Fourier (1768–1830), a French mathematical physicist, developed these methods for investigating the conduction of heat in solid bodies. Fourier contended that rather complex **continuous-time waveforms** could be decomposed into a summation of simple trigonometric functions, sines and cosines of differing frequencies and magnitudes. When Fourier announced his results at a meeting of the French Academy of Scientists in 1807, his claims drew strong criticism from some of the most prominent mathematicians of the time, most notably Lagrange. Publication of his work in-full was delayed until 1822 when his seminal book, *Theorie Analytique de la Chaleur (The Analytical Theory of Heat),* was finally released. Technical doubts regarding Fourier's methods were largely dispelled by the late 19th century through the efforts of Dirichlet, Riemann, and others.

Following this somewhat rocky beginning, the concepts of Fourier analysis have found application in many fields of science, mathematics, and engineering. Indeed, the study of Fourier techniques is now a traditional undertaking within undergraduate electrical engineering programs. With the availability of electronic computers, Fourier techniques have matured dramatically in the analysis of **discrete-time waveforms**. The now ubiquitous **fast Fourier transform (FFT)** is a somewhat general title applied to efficient algorithms for the machine computation of *discrete Fourier transforms* (DFTs). The probability is high that a given electronics technician has one or more instruments at his disposal with the capability of carrying out Fourier analysis, for example, see Fig. 6.1.

FIGURE 6.1 Fourier analyzers, spectrum analyzers, dynamic signal analyzers, and distortion analyzers are now common examples of benchtop instrumentation. Fourier capability is also popular in the new generation of digitizing oscilloscopes. (*Source:* Photograph courtesy of Hewlett-Packard Inc., Palo Alto, CA.)

6.2 The Mathematical Preliminaries for Fourier Analysis

The analysis techniques described in the following four sections can be conveniently categorized based on the variety of waveform or function to which they apply. A summary of useful properties of the analysis methods is provided in Table 6.1. We proceed herein to examine the mathematical preliminaries for the Fourier methods.

Most if not all **periodic** and **aperiodic** continuous-time **waveforms** $x(t)$, of practical interest, have Fourier series or Fourier transform counterparts. However, to aid in the existence question, the following sufficient technical constraints on the function $x(t)$, known generally as the *Dirichlet conditions*, guarantee convergence of the Fourier technique:

1. The function must be **single-valued**.
2. The function must have a finite number of discontinuities in the periodic interval, or, if aperiodic, about the entire real line.
3. The function must remain finite and have a finite number of maxima and minima in the periodic interval, or, if aperiodic, about the entire real line.
4. The function must be absolutely integrable, that is, if periodic,

$$\int_{t_0}^{t_0+T} |x(t)| dt$$

must be finite or if aperiodic,

$$\int_{-\infty}^{+\infty} |x(t)| dt$$

must be finite.

The techniques of Fourier analysis are perhaps best appreciated initially by an example. Consider the ideal continuous-time square wave of period T s, having a maximum value of A and a minimum value of 0 as shown at the top of Fig. 6.2. Because of the periodic nature of the waveform, the technique of Fourier series (FS) analysis applies, and the resulting decomposition into an infinite sum of simple trigonometric functions is given by

$$x(t) = \frac{A}{2} + \sum_{k=1,3,5,\ldots}^{\infty} \frac{2A}{k\pi} (-1)^{(k-1)/2} \cos(k\Omega_0 t) \qquad (6.1)$$

In Eq. (6.1), the average value or DC term of the waveform is given by $A/2$, where as the **fundamental frequency** is described by $\Omega_0 = 2\pi/T$. Because the waveform is an **even function** only cosine terms are present. Because the waveform, less its average value, is characterized by **half-wave symmetry**, only the **odd harmonics** are present. To examine the reconstruction of the function $x(t)$ from the given decomposition, the middle plot of Fig. 6.2 displays the DC term, and the first and third harmonics. The sum of these terms is overlayed with the ideal square wave at the bottom of Fig. 6.2. This progressive reconstruction is continued in Fig. 6.3 with successive odd harmonics added as per Eq. (6.1). The ultimate goal of reconstructing the ideal square wave theoretically demands completion of the infinite sum, an unrealistic task of computation or analog signal summation. By truncating the summation at a large enough value of the harmonic number k, we are assured of having a best approximation of the ideal square wave in a least-square-error sense: that is, the truncated Fourier series is the best fit at any given truncation level $k = k_{\max} < \infty$. The ringing or overshoot, which forms near the discontinuity of the ideal square wave, is described by the **Gibbs phenomenon**, in tribute to an early investigator of the effects of truncating Fourier series summations.

In practice, we are seldom presented with ideal functions such as the square wave of Fig. 6.2. Indeed, an analytical description of the waveform of interest, as is implied by the use of $x(t)$ in formulas of the next two

TABLE 134.1 Summary of Mathematical Properties of Fourier Analysis Methods

Fourier Series (FS)	Fourier Transform (FT)	Discrete-Time Fourier Transform (DTFT)	Discrete-Time Fourier Series (DTFS)
time domain: continuous and periodic	time domain: continuous and aperiodic	time domain: discrete and aperiodic	time domain: discrete and periodic
frequency domain: discrete and aperiodic	frequency domain: continuous and aperiodic	frequency domain: continuous and periodic	frequency domain: discrete and periodic
linearity	linearity	linearity	linearity
$r_1 x_1(t) + r_2 x_2(t) \leftrightarrow r_1 C_{1k} + r_2 C_{2k}$	$r_1 x_1(t) + r_2 x_2(t) \leftrightarrow r_1 X_1(\Omega) + r_2 X_2(\Omega)$	$r_1 x_1(n) + r_2 x_2(n) \leftrightarrow r_1 X_1(\omega) + r_2 X_2(\omega)$	$r_1 x_1(n) + r_2 x_2(n) \leftrightarrow r_1 c_{1k} + r_2 c_{2k}$
shift properties	shift properties	shift properties	shift properties
$x(t-\tau) \leftrightarrow e^{-j\Omega_0 k\tau} C_k$	$x(t-\tau) \leftrightarrow e^{-j\Omega\tau} X(\Omega)$	$x(n-m) \leftrightarrow e^{-j\omega m} X(\omega)$	$x(n-m) \leftrightarrow e^{-j\omega_0 km} c_k$
$e^{j\Omega_0 mt} x(t) \leftrightarrow C_{k-m}$	$e^{j\Omega_0 t} x(t) \leftrightarrow X(\Omega - \Omega_0)$	$e^{j\omega_0 n} x(n) \leftrightarrow X(\omega - \omega_0)$	$e^{j\omega_0 mn} x(n) \leftrightarrow c_{k-m}$
convolution	convolution	convolution	convolution
$x_1(t) * x_2(t) \leftrightarrow C_{1k} C_{2k}$	$x_1(t) * x_2(t) \leftrightarrow X_1(\Omega) X_2(\Omega)$	$x_1(n) * x_2(n) \leftrightarrow X_1(\omega) X_2(\omega)$	$x_1(n) * x_2(n) \leftrightarrow c_{1k} c_{2k}$
$x_1(t) x_2(t) \leftrightarrow C_{1k} * C_{2k}$	$x_1(t) x_2(t) \leftrightarrow 1/2\pi X_1(\Omega) * X_2(\Omega)$	$x_1(n) x_2(n) \leftrightarrow 1/2\pi X_1(\omega) * X_2(\omega)$	$x_1(n) x_2(n) \leftrightarrow c_{1k} * c_{2k}$
modulation	modulation	modulation	modulation
$x(t) \cos(\Omega_0 mt)$	$x(t) \cos(\Omega_0 t)$	$x(n) \cos(\omega_0 n)$	$x(n) \cos(\omega_0 mn)$
$\leftrightarrow \frac{1}{2}[C_{k+m} + C_{k-m}]$	$\leftrightarrow \frac{1}{2}[X(\Omega + \Omega_0) + X(\Omega - \Omega_0)]$	$\leftrightarrow \frac{1}{2}[X(\omega + \omega_0) + X(\omega - \omega_0)]$	$\leftrightarrow \frac{1}{2}[c_{k+m} + c_{k-m}]$
Parseval's Theorem	Parseval's Theorem	Parseval's Theorem	Parseval's Theorem
$\frac{1}{T}\int_{t_0}^{t_0+T} x_1(t) x_2^*(t)\, dt$	$\int_{-\infty}^{\infty} x_1(t) x_2^*(t)\, dt$	$\sum_{n=-\infty}^{\infty} x_1(n) x_2^*(n)$	$\frac{1}{N} \sum_{n=0}^{N-1} x_1(n) x_2^*(n)$
$= \sum_{k=-\infty}^{\infty} C_{1k} C_{2k}^*$	$= \frac{1}{2\pi} \int_{-\infty}^{\infty} X_1(\Omega) X_2^*(\Omega)\, d\Omega$	$= \frac{1}{2\pi} \int_{-\pi}^{\pi} X_1(\omega) X_2^*(\omega)\, d\omega$	$= \sum_{k=0}^{N-1} c_{1k} c_{2k}^*$
real signals	real signals	real signals	real signals
$C_k = C_{-k}^*$	$X(\Omega) = X^*(-\Omega)$	$X(\omega) = X^*(-\omega)$	$c_k = c_{-k}^*$
$\mathrm{Re}\, C_k = \mathrm{Re}\, C_{-k}$	$\mathrm{Re}\, X(\Omega) = \mathrm{Re}\, X(-\Omega)$	$\mathrm{Re}\, X(\omega) = \mathrm{Re}\, X(-\omega)$	$\mathrm{Re}\, c_k = \mathrm{Re}\, c_{-k}$
$\mathrm{Im}\, C_k = -\mathrm{Im}\, C_{-k}$	$\mathrm{Im}\, X(\Omega) = -\mathrm{Im}\, X(-\Omega)$	$\mathrm{Im}\, X(\omega) = -\mathrm{Im}\, X(-\omega)$	$\mathrm{Im}\, c_k = -\mathrm{Im}\, c_{-k}$
$|C_k| = |C_{-k}|$	$|X(\Omega)| = |X(-\Omega)|$	$|X(\omega)| = |X(-\omega)|$	$|c_k| = |c_{-k}|$
$\arg C_k = -\arg C_{-k}$	$\arg X(\Omega) = -\arg X(-\Omega)$	$\arg X(\omega) = -\arg X(-\omega)$	$\arg c_k = -\arg c_{-k}$
$x(t)$ real and even C_k real and even	$x(t)$ real and even $X(\Omega)$ real and even	$x(n)$ real and even $X(\omega)$ real and even	$x(n)$ real and even c_k real and even
$x(t)$ real and odd C_k imag. and odd	$x(t)$ real and odd $X(\Omega)$ imag. and odd	$x(n)$ real and odd $X(\omega)$ imag. and odd	$x(n)$ real and odd c_k imag. and odd

Fourier Waveform Analysis

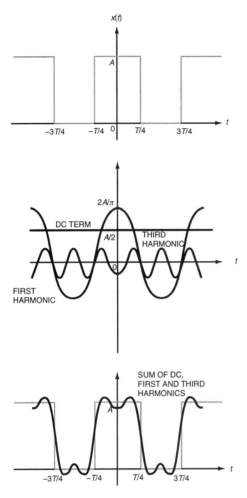

FIGURE 6.2 Example reconstruction of ideal square wave from Fourier series summation.

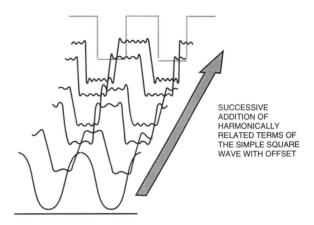

FIGURE 6.3 Continued reconstruction of ideal square wave from Fourier series summation.

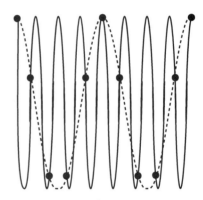

FIGURE 6.4 Example of aliasing wherein uniformly spaced samples from a high-frequency signal are indistinguishable from samples of a low-frequency signal.

sections, is rarely available. The discrete-time waveform counterparts of the Fourier series and Fourier transform provide viable alternatives for estimating the frequency content of signals if discrete measurements of the desired waveforms are available. This avenue of analysis, which is currently popular in benchtop instrumentation, is considered in further detail in Sec. 6.7.

One critical mathematical preliminary for the processing of discrete-time waveforms, which originate from sampling continous-time waveforms, is the issue of **sampling rate** and **aliasing**. The *Nyquist sampling theorem* can be summarized as follows:

If $x(t)$ is a **bandlimited** continuous-time waveform having no frequency content greater than or equal F_N Hz (the **Nyquist frequency**), then $x(t)$ is uniquely determined by a discrete-time waveform $x(n)$, which results from uniform sampling of $x(t)$ at a rate F_S, which is greater than or equal $2F_N$ (the **Nyquist rate**).

Failure to obey the Nyquist sampling theorem results in aliasing, that is, the effect of frequencies greater than the Nyquist frequency are reflected into the band of frequencies between DC and one-half the sampling rate. This situation is demonstrated in Fig. 6.4, where the samples of the high-frequency sinusoid are indistinguishable from a uniform sampling of the low-frequency sinusoid.

6.3 The Fourier Series for Continuous-Time Periodic Functions

Three forms of the FS are commonly encountered for continous-time waveforms $x(t)$ of period T. The first is the *trigonometric form*,

$$x(t) = \frac{a_0}{2} + \sum_{k=1}^{\infty} \left[a_k \cos\left(\frac{2\pi k}{T}t\right) + b_k \sin\left(\frac{2\pi k}{T}t\right) \right] \tag{6.2}$$

where the coefficients a_k and b_k are calculated via the integrals

$$a_k = \frac{2}{T} \int_{t_0}^{t_0+T} x(t) \cos\left(\frac{2\pi k}{T}t\right) dt \tag{6.3}$$

$$b_k = \frac{2}{T} \int_{t_0}^{t_0+T} x(t) \sin\left(\frac{2\pi k}{T}t\right) dt \tag{6.4}$$

with the representation for a_0 often appearing without the multiplier and divider of two [and therefore directly representing the average value of $x(t)$]. Symmetry properties of the particular function $x(t)$ can lead to simplifications of Eqs. (6.3) and (6.4).

A slight adaptation of this leads to the *alternative trigonometric form*,

$$x(t) = \frac{a_0}{2} + \sum_{k=1}^{\infty} A_k \cos\left(\frac{2\pi k}{T}t + \theta_k\right) \tag{6.5}$$

where the magnitude and phase terms are obtained via the respective identities

$$A_k = \sqrt{a_k^2 + b_k^2} \tag{6.6}$$

$$\theta_k = -\arctan\left(\frac{b_k}{a_k}\right) \tag{6.7}$$

Finally, the *complex exponential form* is given by

$$x(t) = \sum_{k=-\infty}^{\infty} C_k e^{j\frac{2\pi k}{T}t} \tag{6.8}$$

where the coefficients C_k are calculated via

$$C_k = \frac{1}{T} \int_{t_0}^{t_0+T} x(t) e^{-j\frac{2\pi k}{T}t} dt \tag{6.9}$$

It should be noted that the C_k are, in general, complex valued. They may be obtained from the trigonometric form data as follows:

$$|C_k| = |C_{-k}| = \frac{A_k}{2} \quad \text{for } k = 0, 1, 2, \ldots \tag{6.10}$$

$$\arg C_k = -\arg C_{-k} = \theta_k \quad \text{for } k = 0, 1, 2, \ldots \tag{6.11}$$

Common graphical representations of this data include plots of $|C_k|$ (or A_k) and θ_k, which are referred to as the *magnitude* and *phase spectra*, respectively. As is implied by the indexing of the spectral data by the integer k, the Fourier series of a continuous-time waveform is itself discrete in character, that is, it takes on nonzero values at only discrete, isolated points of the frequency axis, namely, the frequencies $\Omega_k = 2\pi k/T$.

The average power in a periodic waveform can be found in one of two ways, with the equality noted below commonly attributed to Parseval:

$$P = \frac{1}{T} \int_{t_0}^{t_0+T} |x(t)|^2 dt = \sum_{k=-\infty}^{\infty} |C_k|^2. \tag{6.12}$$

Fourier Waveform Analysis

Because of the right-hand side of Eq. (6.12), a plot of $|C_k|^2$ is commonly referred to as the *power spectral density*.

6.4 The Fourier Transform for Continuous-Time Aperiodic Functions

The Fourier transform (FT) for an aperiodic, continuous-time waveform $x(t)$ is given by

$$X(\Omega) = \mathcal{F}\{x(t)\} = \int_{-\infty}^{\infty} x(t)e^{-j\Omega t} dt \tag{6.13}$$

where the implication is that the frequency-domain counterpart $X(\Omega)$ of $x(t)$ is continuous in frequency and can be used to reconstruct $x(t)$ via

$$x(t) = \mathcal{F}^{-1}\{X(\Omega)\} = \frac{1}{2\pi} \int_{-\infty}^{\infty} X(\Omega)e^{j\Omega t} d\Omega \tag{6.14}$$

If the integration described in Eq. (6.14) is carried out in hertz rather than radians per second, the division by the factor 2π is removed.

The transformation given by Eq. (6.13) can also be applied to some familiar periodic functions as well as functions that are not absolutely integrable. Example Fourier transform pairs are shown in Table 6.2.

The total energy in an aperiodic waveform can be found in one of two ways, with the equality noted commonly

TABLE 134.2 Example Fourier Transform Pairs

$x(t)$	$x(t)$	$X(\Omega)$	$|X(\Omega)|$
	$\delta(t)$	1	
	1	$2\pi\delta(\Omega)$	
	$u(t)$	$\pi\delta(\Omega) + \dfrac{1}{j\Omega}$	
	$e^{-at}u(t), a > 0$	$\dfrac{1}{a + j\Omega}$	
	$te^{-at}u(t), a > 0$	$\dfrac{1}{(a + j\Omega)^2}$	
	$e^{-at}\cos(\Omega_0 t)u(t),\ a>0$	$\dfrac{a + j\Omega}{(a + j\Omega)^2 + \Omega_0^2}$	
	$\cos(\Omega_0 t)$	$\pi[\delta(\Omega - \Omega_0) + \delta(\Omega + \Omega_0)]$	
	$\begin{array}{l}1, |t| < \tau/2 \\ 0, |t| > \tau/2\end{array}$	$\dfrac{\sin(\Omega\tau/2)}{\Omega/2}$	
	$\begin{array}{l}1 - |t|/\tau, |t| < \tau \\ 0, |t| > \tau\end{array}$	$\dfrac{1}{\tau}\dfrac{\sin^2(\Omega\tau/2)}{(\Omega/2)^2}$	
	$\dfrac{\sin(\Omega_0 t/2)}{\pi t}$	$\begin{array}{l}1, |\Omega| < \Omega_0 \\ 0, |\Omega| > \Omega_0\end{array}$	

attributed to Parseval,

$$E = \int_{-\infty}^{\infty} |x(t)|^2 dt = \frac{1}{2\pi} \int_{-\infty}^{\infty} |X(\Omega)|^2 d\Omega \qquad (6.15)$$

Because of the right-hand side of Eq. (6.15), a plot of $|X(\Omega)|^2$ is commonly referred to as the *energy spectral density*. One advantage to the integration vs frequency is the ability to determine the amount of energy in a signal due to frequency components within a particular range of frequencies.

6.5 The Fourier Series for Discrete-Time Periodic Functions

The discrete-time Fourier series (DTFS) for discrete-time waveforms $x(n)$ of period N can also be given in three forms; however, the *complex exponential form* is by far the most common,

$$x(n) = \sum_{k=0}^{N-1} c_k e^{j\frac{2\pi k}{N}n} \qquad (6.16)$$

where the coefficients c_k are calculated via the finite summation

$$c_k = \frac{1}{N} \sum_{n=0}^{N-1} x(n) e^{-j\frac{2\pi k}{N}n} \qquad (6.17)$$

Because of the periodicity of the exponential in Eq. (6.17), the DTFS coefficients c_k form a discrete, periodic sequence of period N if the index k is extended beyond the fundamental range of 0 to $N-1$.

The average power in a periodic, discrete-time waveform can be found in one of two ways, with the equality noted commonly attributed to Parseval,

$$P = \frac{1}{N} \sum_{n=0}^{N-1} |x(n)|^2 = \sum_{k=0}^{N-1} |c_k|^2 \qquad (6.18)$$

Because of the right-hand side of Eq. (6.18), a plot of $|c_k|^2$ is commonly referred to as the power spectral density.

6.6 The Fourier Transform for Discrete-Time Aperiodic Functions

The discrete-time Fourier transform (DTFT) for aperiodic, finite energy discrete-time waveforms $x(n)$ is given by

$$X(\omega) = \sum_{n=-\infty}^{\infty} x(n) e^{-j\omega n} \qquad (6.19)$$

where the implication is that the frequency-domain counterpart $X(\omega)$ of $x(n)$ is continuous in the discrete frequency variable ω and can be used to reconstruct $x(n)$ via

$$x(n) = \frac{1}{2\pi} \int_{-\pi}^{\pi} X(\omega) e^{j\omega n} d\omega \qquad (6.20)$$

If the integration described in Eq. (6.20) is carried out in cycles per sample rather than radian per sample frequency, the division by the factor 2π is removed.

The transformation given by Eq. (6.19) can also be applied to some periodic functions as well as functions that are not absolutely summable. Examples of discrete-time Fourier transform pairs are shown in Table 6.3.

The total energy in an aperiodic, discrete-time waveform can be found in one of two ways, with the equality noted commonly attributed to Parseval,

$$E = \sum_{n=-\infty}^{\infty} |x(n)|^2 = \frac{1}{2\pi} \int_{\pi}^{-\pi} |X(\omega)|^2 d\omega \qquad (6.21)$$

Because of the right-hand side of Eq. (6.21), a plot of $|X(\omega)|^2$ is commonly referred to as the energy spectral density.

TABLE 134.3 Example Discrete-Time Fourier Transform Pairs

$x(n)$	$x(n)$	$X(\omega)$	$\|X(\omega)\|$
	$\delta(n)$	1	
	1	$\sum_{k=-\infty}^{\infty} 2\pi \delta(\omega + 2\pi k)$	
	$u(n)$	$\dfrac{1}{1 - e^{-j\omega}} + \sum_{k=-\infty}^{\infty} \pi \delta(\omega + 2\pi k)$	
	$e^{-an}u(n), a > 0$	$\dfrac{1}{1 - e^{-(a+j\omega)}}$	
	$ne^{-an}u(n), a > 0$	$\dfrac{e^{-(a+j\omega)}}{\left(1 - e^{-(a+j\omega)}\right)^2}$	
	$e^{-an}\cos(\omega_0 n)u(n), a > 0$	$\dfrac{1 - \cos(\omega_0)e^{-(a+j\omega)}}{1 - 2\cos(\omega_0)e^{-(a+jw)} + e^{-2(a+j\omega)}}$	
	$\cos(\omega_0 n)$	$\pi \sum_{k=-\infty}^{\infty} [\delta(\omega - \omega_0 + 2\pi k) + \delta(\omega + \omega_0 + 2\pi k)]$	
	$\begin{array}{l} 1, \|n\| < L/2 \\ 0, \|n\| > (L-1)/2 \\ L \text{ an odd integer} \end{array}$	$\dfrac{\sin(\omega L/2)}{\sin(\omega/2)}$	
	$\begin{array}{l} 1 - \|n\|/L, \|n\| < L \\ 0, \|n\| \geq L \\ L \text{ an odd integer} \end{array}$	$\dfrac{1}{L} \dfrac{\sin^2(\omega L/2)}{\sin^2(\omega/2)}$	
	$\dfrac{\sin(\omega_0 n)}{\pi n}$	$\begin{array}{l} 1, \|\omega\| < \omega_0 \\ 0, \omega_0 < \|\omega\| < \pi \end{array}$	

6.7 Example Applications of Fourier Waveform Techniques

Using the DFT/FFT in Fourier Analysis

The FFT is arguably the most significant advancement in signal analysis in recent decades. The FFT is actually the name given to many algorithms used to efficiently compute the DFT. The DFT of $x(n)$ is defined as

$$X(k) = DFT\{x(n)\} = \sum_{n=0}^{N-1} x(n) e^{-j\frac{2\pi k}{N} n} \quad \text{for } k = 0, 1, 2, \ldots, N - 1 \tag{6.22}$$

and the inverse transform is given by

$$x(n) = IDFT\{X(k)\} = \frac{1}{N} \sum_{k=0}^{N-1} X(k) e^{j\frac{2\pi n}{N} k} \quad \text{for } n = 0, 1, 2, \ldots, N - 1 \tag{6.23}$$

The DFT/FFT can be used to compute or approximate the four Fourier methods described in Secs. 6.3–6.6, that is the FS, FT, DTFS, and DTFT. Algorithms for the FFT are discussed extensively in the literature (see the **Further Information** entry for this topic).

The DTFS of a periodic, discrete-time waveform can be directly computed using the DFT. If $x(n)$ is one period of the desired signal, then the DTFS coefficients are given by

$$c_k = \frac{1}{N} DFT x(n) \quad \text{for } k = 0, 1, \ldots, N-1 \tag{6.24}$$

The DTFT of a waveform is a continuous function of the discrete or sample frequency ω. The DFT can be used to compute or approximate the DTFT of an aperiodic waveform at uniformly separated *sample frequencies* $k\omega_0 = k2\pi/N$, where k is any integer between 0 and $N - 1$. If N samples fully describe a finite duration $x(n)$, then

$$X(\omega)|_{\omega=\frac{2\pi}{N}k} = DFT x(n) \quad \text{for } k = 0, 1, \ldots, N-1 \tag{6.25}$$

If the $x(n)$ in Eq. (6.25) is not of finite duration, then the equality should be changed to an approximately equal. In many practical cases of interest, $x(n)$ is not of finite duration. The literature contains many approaches to truncating and **windowing** such $x(n)$ for Fourier analysis purposes.

The continuous-time FS coefficients can also be approximated using the DFT. If a continuous-time waveform $x(t)$ of period T is sampled at a rate of T_S samples per second to obtain $x(n)$, in accordance with the Nyquist sampling theorem, then the FS coefficients are approximated by

$$C_k \approx \frac{T_s}{T} DFT\{x(n)\} \quad \text{for } k = 0, 1, \ldots, N-1 \tag{6.26}$$

where N samples of $x(n)$ should represent precisely one period of the original $x(t)$. Using the DFT in this manner is equivalent to computing the integral in Eq. (6.9) with a simple rectangular or Euler approximate.

In a similar fashion, the continuous-time FT can be approximated using the DFT. If an aperiodic, continuous-time waveform $x(t)$ is sampled at a rate of T_S samples per second to obtain $x(n)$, in accordance with the Nyquist sampling theorem, then uniformally separated samples of the FT are approximated by

$$X(\Omega)|_{\Omega=\frac{2\pi}{NT_s}k} \approx T_s(DFT\{x(n)\}) \quad \text{for } k = 0, 1, \cdots, N-1 \tag{6.27}$$

For $x(t)$ that are not of finite duration, truncating and windowing can be applied to improve the approximation.

Total Harmonic Distortion Measures

The applications of Fourier analysis are extensive in the electronics and instrumentation industry. One typical application, the computation of total harmonic distortion (THD), is described herein. This application provides a measure of the nonlinear distortion, which is introduced to a pure sinusoidal signal when it passes through a system of interest, perhaps an amplifier. The root-mean-square (rms) total harmonic distortion (THD$_{rms}$) is defined as the ratio of the rms value of the sum of the harmonics, not including the fundamental, to the rms value of the fundamental,

$$\text{THD}_{rms} = \frac{\sqrt{\sum_{k=2}^{\infty} A_k^2}}{A_1} \tag{6.28}$$

As an example, consider the clipped sinusoidal waveform shown in Fig. 6.5. The Fourier series representation of this waveform is given by.

$$v(t) = A\left(\frac{1}{2} + \frac{1}{\pi}\right) \cos\left(\frac{2\pi}{T}t\right) + \frac{2\sqrt{2}A}{\pi} \sum_{k=3,5,7,\ldots}^{\infty} \frac{\sin\left(\frac{\pi}{4}k\right) - k \cos\left(\frac{\pi}{4}k\right)}{k(1-k^2)} \cos\left(\frac{2\pi}{T}kt\right) \tag{6.29}$$

The magnitude spectrum for this waveform is shown in Fig. 6.6. Because of symmetry, only the odd harmonics are present. Clearly the fundamental is the dominant component, but due to the clipping, many additional harmonics are present. The THD$_{rms}$ is readily computed from the coefficients in Eq. (6.29), yielding

Fourier Waveform Analysis

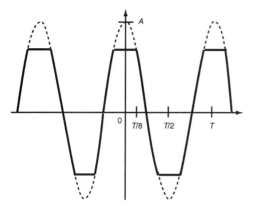

FIGURE 6.5 A clipped sinusoidal waveform for total harmonic distortion example.

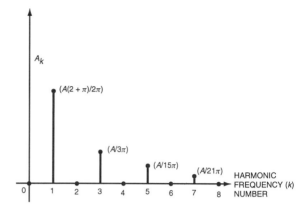

FIGURE 6.6 The magnitude spectrum, through eighth harmonic, for the clipped sinusoid total harmonic distortion example.

13.42% for this example. In practice, the coefficients of the Fourier series are typically approximated via the DFT/FFT, as described at the beginning of this section.

Defining Terms

Aliasing: Refers to an often detrimental phenomenon associated with the sampling of continous-time waveforms at a rate below the Nyquist rate. Frequencies greater than one-half the sampling rate become indistinguishable from frequencies in the fundamental bandwidth, that is, between DC and one-half the sampling rate.

Aperiodic waveform: This phrase is used to describe a waveform that does not repeat itself in a uniform, periodic manner. Compare with periodic waveform.

Bandlimited: A waveform is described as bandlimited if the frequency content of the signal is constrained to lie within a finite band of frequencies. This band is often described by an upper limit, the Nyquist frequency, assuming frequencies from DC up to this upper limit may be present. This concept can be extended to frequency bands that do not include DC.

Continuous-time waveform: A waveform, herein represented by $x(t)$, that takes on values over the continuum of time t, the assumed independent variable. The Fourier series and Fourier transform apply to continuous-time waveforms. Compare with discrete-time waveform.

Discrete-time waveform: A waveform, herein represented by $x(n)$, that takes on values at a countable, discrete set of sample times or sample numbers n, the assumed independent variable. The discrete Fourier transform, the discrete-time Fourier series, and discrete-time Fourier transform apply to discrete-time waveforms. Compare with continuous-time waveform.

Even function: If a function $x(t)$ can be characterized as being a mirror image of itself horizontally about the origin, it is described as an even function. Mathematically, this demands that $x(t) = x(-t)$ for all t. The name arises from the standard example of an even function as a polynomial containing only even powers of the independent variable. A pure cosine is also an even function.

Fast Fourier transform (FFT): Title which is now somewhat loosely used to describe any efficient algorithm for the machine computation of the discrete Fourier transform. Perhaps the best known example of these algorithms is the seminal work by Cooley and Tukey in the early 1960s.

Fourier waveform analysis: Refers to the concept of decomposing complex waveforms into the sum of simple trigonometric or complex exponential functions.

Fundamental frequency: For a periodic waveform, the frequency corresponding to the smallest or fundamental period of repetition of the waveform is described as the fundamental frequency.

Gibbs phenomenon: Refers to an oscillatory behavior in the convergence of the Fourier transform or series in the vicinity of a discontinuity, typically observed in the reconstruction of a discontinuous $x(t)$. Formally stated, the Fourier transform does converge uniformly at a discontinuity, but rather, converges to the average value of the waveform in the neighborhood of the discontinuity.

Half-wave symmetry: If a periodic function satisfies $x(t) = -x(t - T/2)$, it is said to have half-wave symmetry.

Harmonic frequency: Refers to any frequency that is a positive integer multiple of the fundamental frequency of a periodic waveform.

Nyquist frequency: For a bandlimited waveform, the width of the band of frequencies contained within the waveform is described by the upper limit known as the Nyquist frequency.

Nyquist rate: To obey the Nyquist sampling theorem, a bandlimited waveform should be sampled at a rate which is at least twice the Nyquist frequency. This minimum sampling rate is known as the Nyquist rate. Failure to follow this restriction results in aliasing.

Odd function: If a function $x(t)$ can be characterized as being a reverse mirror image of itself horizontally about the origin, it is described as an odd function. Mathematically, this demands that $x(t) = -x(-t)$ for all t. The name arises from the standard example of an odd function as a polynomial containing only odd powers of the independent variable. A pure sine is also an odd function.

Periodic waveform: This phrase is used to describe a waveform that repeats itself in a uniform, periodic manner. Mathematically, for the case of a continuous-time waveform, this characteristic is often expressed as $x(t) = x(t \pm kT)$, which implies that the waveform described by the function $x(t)$ takes on the same value for any increment of time kT, where k is any integer and the characteristic value T, a real number greater than zero, describes the fundamental period of $x(t)$. For the case of a discrete-time waveform, we write $x(n) = x(n \pm kN)$, which implies that the waveform $x(n)$ takes on the same value for any increment of sample number kN, where k is any integer and the characteristic value N, an integer greater than zero, describes the fundamental period of $x(n)$.

Quarter-wave symmetry: If a periodic function displays half-wave symmetry and is *even* symmetric about the $\frac{1}{4}$ and $\frac{3}{4}$ period points (the negative and positive lobes of the function), then it is said to have quarter-wave symmetry.

Sampling rate: Refers to the frequency at which a continuous-time waveform is sampled to obtain a corresponding discrete-time waveform. Values are typically given in hertz.

Single-valued: For a function of a single variable, such as $x(t)$, single-valued refers to the quality of having one and only one value $y_0 = x(t_0)$ for any t_0. The square root is an example of a function which is not single-valued.

Windowing: A term used to describe various techniques for preconditioning a discrete-time waveform before processing by algorithms such as the discrete Fourier transform. Typical applications include extracting a finite duration approximation of an infinite duration waveform.

References

Bracewell, R.N. 1989. The Fourier transform. *Sci. Amer.* (June):86–95.
Brigham, E.O. 1974. *The Fast Fourier Transform.* Prentice–Hall, Englewood Cliffs, NJ.
Nilsson, J.W. 1993. *Electric Circuits,* 4th ed. Addison–Wesley, Reading, MA.
Oppenheim, A.V. and Schafer, R. 1989. *Discrete-Time Signal Processing.* Prentice–Hall, Englewood Cliffs, NJ.
Proakis, J.G. and Manolakis, D.G. 1992. *Digital Signal Processing: Principles, Algorithms, and Applications,* 2nd ed. Macmillan, New York.
Ramirez, R.W. 1985. *The FFT, Fundamentals and Concepts.* Prentice–Hall, Englewood Cliffs, NJ.

Further Information

For in-depth descriptions of practical instrumentation incorporating Fourier analysis capability, consult the following sources:

Witte, Robert A. *Spectrum and Network Measurements.* Prentice–Hall, Englewood Cliffs, NJ, 1993.
Witte, Robert A. *Electronic Test Instruments: Theory and Applications.* Prentice–Hall, Englewood Cliffs, NJ, 1993.

For further investigation of the history of Fourier and his transform, the following source should prove interesting:

Herviel, J. 1975. *Joseph Fourier: The Man and the Physicist.* Clarendon Press.

A thoroughly engaging introduction to Fourier concepts, accessible to the reader with only a fundamental background in trigonometry, can be found in the following publication:

Who is Fourier? A Mathematical Adventure. Transnational College of LEX, English translation by Alan Gleason, Language Research Foundation, Boston, MA, 1995.

Perspectives on the FFT algorithms and their application can be found in the following sources:

Cooley, J.W. and Tukey, J.W. 1965. An algorithm for the machine calculation of complex Fourier series. *Mathematics of Computation* 19(April):297–301.

Cooley, J.W. 1992. How the FFT gained acceptance. *IEEE Signal Processing Magazine* (Jan.):10–13, 1992.

Heideman, M.T., Johnson, D.H., and Burrus, C.S. 1984. Gauss and the history of the fast fourier transform. *IEEE ASSP Magazine* 1:14–21.

Kraniauskas, P. 1994. A plain man's guide to the FFT. *IEEE Signal Processing Magazine* (April):24–35.

Press, W.H., Flannery, B.P., Teukolsky, S.A., and Vetterling, W.T. 1988. *Numerical Recipes in C: The Art of Scientific Computing.* Cambridge Univ. Press.

7
Computer-Based Signal Analysis

Rodger E. Ziemer
*University of Colorado,
Colorado Springs*

7.1 Introduction 82
7.2 Signal Generation and Analysis 82
    Signal Generation • Curve Fitting • Statistical Data Analysis • Signal Processing
7.3 Symbolic Mathematics 87

7.1 Introduction

In recent years several mathematical packages have appeared for signal and system analysis on computers. Among these are **Mathcad**® [Math Soft 1995], **MATLAB**® [1995], and **Mathematica**® [Wolfram 1991], to name three general purpose packages. More specialized computer simulation programs include **SPW**® [Alta Group 1993] and **System View**® [Elanix 1995]. More specialized to electronic circuit analysis and design is **PSpice** [Rashid 1990] and **Electronics Workbench**® [Interactive 1993]. The purpose of this section is to discuss some of these tools and their utility for signal and system analysis by means of computer. Attention will be focused on the more general tool MATLAB. Several text books are now available that make extensive use of MATLAB in student exercises [Etter 1993, Gottling 1995, Frederick and Chow 1995].

7.2 Signal Generation and Analysis

Signal Generation

MATLAB is a vector- or array-based program. For example, if one wishes to generate and plot a sinusoid by means of MATLAB, the statements involved would be:

t = 0:.01:10;
x = sin(2*pi*t);
plot(t,x,'−w'), xlabel('t'), ylabel('x(t)'),grid

The first line generates a vector of values for the independent variable starting at 0, ending at 10, and spaced by 0.01. The second statement generates a vector of values for the dependent variable, $x = \sin(2\pi t)$, and the third statement plots the vector **x** vs the vector **t**. The resulting plot is shown in Fig. 7.1.

In Matlab, one has the options of running a program stored in an **m-file**, invoking the statements from the **command window**, or writing a **function** to perform the steps in producing the sinewave or other operation. For example, the command window option would be invoked as follows:

≫ t = 0:.01:10;
≫ x = sin(2*pi*t);
≫ plot(t,x,'−w'), xlabel('t'), ylabel('x(t)'), grid

The command window prompt is ≫ and each line is executed as it is typed and entered.

An example of a function implementation is provided by the generation of a unit step:

Computer-Based Signal Analysis

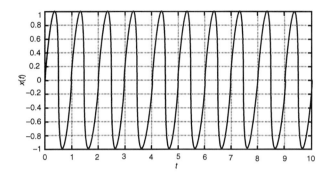

FIGURE 7.1 Plot of a sinusoid generated by MATLAB.

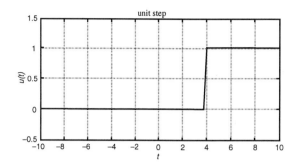

FIGURE 7.2 Unit step starting at **t** = 2 generated by the given step generation function.

```
%       Function for generation of a unit step
function u = stepfn(t)
L = length(t);
u = zeros(size(t));
for i = 1:L
        if t(i) >= 0
                u(i) = 1;
        end
end
```

The command window statements for generation of a unit step starting at **t** = 2 are given next and a plot is provided in Fig. 7.2,

```
≫ t = −10:0.1:10;
≫ u = stepfn(t−2);
≫ plot(t,u, '−w'), xlabel('t'), ylabel('u(t)'), grid, title('unit step'), axis([−10 10 −0.5 1.5])
```

Curve Fitting

MATLAB has several functions for fitting polynomials to data and a function for evaluation of a **polynomial fit** to the data. These functions include *table1, table2, spline,* and *polyfit*. The first one makes a linear fit to a set of data pairs, the second does a planar fit to data triples, and the third does a cubic fit to data pairs. The polyfit function does a **least-mean square-error** fit to data pairs. The uses of these are illustratrated by the following program:

```
x = [0 1 2 3 4 5 6 7 8 9];
y = [0 20 60 68 77 110 113 120 140 135];
newx = 0:0.1:9;
newy = spline(x, y, newx);
for n = 1:5
        X = polyfit(x,y,n);
        f(:,n) = polyval(X,newx)';
end
subplot(321), plot(x,y,'w', newx,newy,'w',x,y,'ow'),axis([0 10 0 150]),grid
subplot(322), plot(newx,f(:,1),'w',x,y,'ow'),axis([0 10 0 150]),grid
subplot(323), plot(newx,f(:,2),'w',x,y,'ow'),axis([0 10 0 150]),grid
subplot(324), plot(newx,f(:,3),'w',x,y,'ow'),axis([0 10 0 150]),grid
subplot(325), plot(newx,f(:,4),'w',x,y,'ow'),axis([0 10 0 150]),grid
subplot(326), plot(newx,f(:,5),'w',x,y,'ow'),axis([0 10 0 150]),grid
```

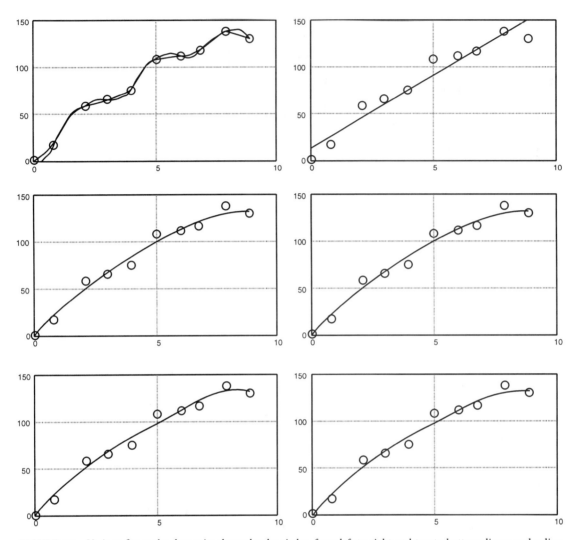

FIGURE 7.3 Various fits to the data pairs shown by the circles, from left to right and top to bottom: linear and spline fits, linear least-squares fit, quadratic least-squares fit, cubic least-squares fit, quartic least-squares fit, fifth-order least-squares fit.

Plots of the various fits are shown in Fig. 7.3. In the program, the linear interpolation is provided by the plotting routine itself, although the numerical value for the linear interpolation of a data point is provided by the statement $table1(x, y, x_0)$ where x_0 is the x value that an interpolated y value is desired. The polyfit statement returns the coefficients of the least-squares fit polynomial of specified degree n to the data pairs. For example, the coefficients of the fifth-order polynomial returned by polyfit are −0.0150 0.3024 −1.9988 3.3400 25.0124 −1.1105 from highest to lowest degree. The **polyval** statement provides an evaluation of the polynomial at the element values of the vector *newx*.

Statistical Data Analysis

MATLAB has several **statistical data analysis functions**. Among these are random number generation, sample mean and standard deviation computation, histogram plotting, and correlation coefficient computation for pairs

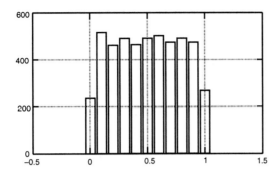 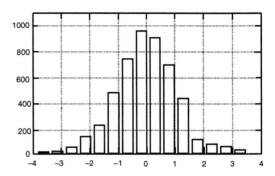

FIGURE 7.4 Histograms for 5000 pseudorandom numbers uniform in [0, 1] and 5000 Gaussian random numbers with mean zero and variance one.

of random data. The next program illustrates several of these functions,

```
X = rand(1, 5000);
Y = randn(1, 5000);
mean_X = mean(X)
std_dev_X = std(X)
mean_Y = mean(Y)
std_dev_Y = std(Y)
rho = corrcoef(X, Y)
subplot(211), hist(X, [0 .1 .2 .3 .4 .5 .6 .7 .8 .9 1]), grid
subplot(212), hist(Y, 15), grid
```

The computed values returned by the program (note the semicolons left off) are mean_X = 0.5000; std_dev_X = 0.2883; mean_Y = −0.0194; std_dev_Y = 0.9958;

rho = 1.0000 0.0216
       0.0216 1.0000

The theoretical values are 0.5, 0.2887, 0, and 1, respectively, for the first four. The correlation coefficient matrix should have 1s on the main diagonal and 0s off the main diagonal. Histograms for the two cases of **uniform** and **Gaussian variates** are shown in Fig. 7.4. In the first plot statement, a vector giving the centers of the desired **histogram** bins is given. The two end values at 0 and 1 will have, on the average, half the values in the other bins since the random numbers generated are uniform in [0, 1]. In the second histogram plot statement, the number of bins is specified at 15.

Signal Processing

MATLAB has several **toolboxes** available for implementing special computations involving such areas as filter design and image processing. In this section we discuss a few of the special functions available in the **signal processing toolbox**.

A linear system, or filter, can be specified by the numerator and denominator polynomials of its **transfer function**. This is the case for both **continuous-time** and **discrete-time linear systems**. For example, the **amplitude, phase,** and **impulse responses** of the continuous-time linear system with transfer function

$$H(s) = \frac{s^3 + 2.5s^3 + 10s + 1}{s^3 + 10s^3 + 10s + 1} \tag{7.1}$$

can be found with the following MATLAB program:

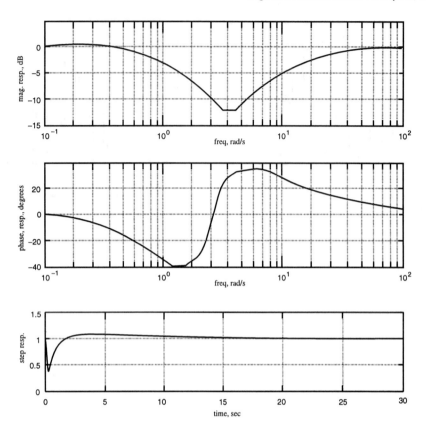

FIGURE 7.5 Magnitude, phase, and step responses for a continuous-time linear system.

```
%       Matlab example for creating Bode plots and step response for
%       a continuous-time linear system
%
num = [1 2.5 10 1];
den = [1 10 10 1];
[MAG,PHASE,W] = bode(num,den);
[Y,X,T] = step(num, den);
subplot(311),semilogx(W,20*log10(MAG), '-w'), xlabel('freq, rad/s'), ...
        ylabel('mag. resp.,   dB'), grid, axis([0.1 100 −15 5])

subplot(312),semilogx(W,PHASE, '-w'), xlabel('freq, rad/s'), ...
        ylabel('phase resp.,   degrees'), grid, axis([0.1 100 −40 40])
subplot(313),plot(T, Y, '-w'), grid, xlabel('time, sec'), ylabel('step resp.'), ...
        axis([0 30 0 1.5])
```

Plots for these three response functions are shown in Fig. 7.5.

In addition, MATLAB has several other programs that can be used for filter design and signal analysis. The ones discussed here are meant to just give a taste of the possiblities available.

7.3 Symbolic Mathematics

MATLAB can manipulate variables symbolically. This includes algebra with scalar expressions, matrix algebra, linear algebra, calculus, differential equations, and transform calculus. For example, to enter a symbolic expression in the command window in MATLAB, one can do one of the following:

```
≫ A = 'cos(x)'
A =
cos(x)
≫ B = sym('sin(x)')
B =
sin(x)
```

Once defined, it is a simple matter to perform symbolic operations on A and B. For example:

```
≫ diff(A)
ans =
−sin(x)
≫ int(B)
ans =
−cos(x)
```

To illustrate the Laplace transform capabilities of MATLAB, consider

```
≫ F = laplace('t*exp(−3*t)')
F =
1/(s+3)^2
≫ G = laplace('t^2*Heaviside(t)')
G =
2/s^2
≫ h = invlaplace(symmul(F, G))
h =
2/9*t*exp(−3*t)+4/27*exp(−3*t)+2/9*t−4/27
```

Alternatively, we could have carried out the symbolic multiply as a separate step:

```
≫ H = symmul(F, G)
H =
2/(s+3)^2/s^2
≫ h = invlaplace(H)
h =
2/9*t*exp(−3*t)+4/27*exp(−3*t)+2/9*t−4/27
```

Defining Terms

Amplitude (magnitude) response: The magnitude of the steady-state response of a fixed, linear system to a unit-amplitude input sinusoid.

Command window: The window in MATLAB in which the computations are done, whether with a direct command line or through an m-file.

Continuous-time fixed linear systems: A system that responds to continuous-time signals for which superposition holds and a time shift in the input results in the same output, but shifted by the amount of the time shift of the input.

Discrete-time fixed linear systems: A system that responds to discrete-time signals for which superposition holds and a time shift in the input results in the same output, but shifted by the amount of the time shift of the input.

Electronics workbench: An analysis/simulation computer program for electronic circuits, similar to Pspice.

Function: In MATLAB, a subprogram that implements a small set of statements that appear commonly enough to warrant their implementation.

Gaussian variates: Pseudorandom numbers in MATLAB, or other computational language, that obey a Gaussian, or bell-shaped, probability density function.

Histogram: A function in MATLAB, or other computational language, that provides a frequency analysis of random variates into contiguous bins.

Least-square-error fit: An algorithm or set of equations resulting from fitting a polynomial or other type curve, such as logarithmic, to data pairs such that the sum of the squared errors between data points and the curve is minimized.

Mathcad: A computer package like MATLAB that includes numerical analysis, programming, and symbolic manipulation capabilities.

Mathematica: A computer package like MATLAB that includes numerical analysis, programming, and symbolic manipulation capabilities.

MATLAB: A computer package that includes numerical analysis, programming, and symbolic manipulation capabilities.

M-file: A file in MATLAB that is the method for storing programs.

Phase response: The phase shift of the steady-state response of a fixed, linear system to a unit-amplitude input sinusoid relative to the input.

Polynomial fit: A function in MATLAB for fitting a polynomial curve to a set of data pairs. See least-squared-error fit. Another function fits a cubic spline to a set of data pairs.

Polyval: A function in MATLAB for evaluating a polynomial fit to a set of data pairs to a vector of abscissa values.

PSpice: An analysis/simulation computer program for electronic circuits, which originated at the University of California, Berkeley, as a batch processing program and was later adapted to personal computers using Windows.

Signal processing toolbox: A toolbox, or set of functions, in MATLAB for implementing signal processing and filter analysis and design.

SPW: A block-diagram oriented computer simulation package that is specifically for the analysis and design of signal processing and communications systems.

Statistical data analysis functions: Functions in MATLAB, or any other computer analysis package, that are specifically meant for analysis of random data. Functions include those for generation of pseudorandom variates, plotting histograms, computing sample mean and standard deviation, etc.

Step response: The response of a fixed, linear system to a unit step applied at time zero.

SystemView: A block-diagram oriented computer simulation package that is specifically for the analysis and design of signal processing and communications systems, but not as extensive as SPW.

Toolboxes: Sets of functions in MATLAB designed to facilitate the computer analysis and design of certain types of systems, such as communications, control, or image processing systems.

Transfer function: A ratio of polynomials in s that describes the input–output response characteristics of a fixed, linear system.

Uniform variates: Pseudorandom variates generated by computer that are equally likely to be anyplace within a fixed interval, usually [0, 1].

References

Alta Group. SPW.
Elanix. 1995. *SystemView® by Elanix: The Student Edition*. PWS Publishing, Boston, MA.
Etter, D.M. 1993. *Engineering Problem Solving Using MATLAB*. Prentice–Hall, Englewood Cliffs, NJ.
Frederick, D.K. and Chow, J.H. 1995. *Feedback Control Problems Using MATLAB and the Control System Toolbox*. PWS Publishing, Boston, MA.
Gottling, J.G. 1995. *Matrix Analysis of Circuits Using MATLAB*. Prentice–Hall, Englewood Cliffs, NJ.
Interactive. 1993. *Electronics Workbench, User's Guide*. Interactive Image Technologies, Ltd. Toronto, Ontario, Canada.
Mathsoft. 1995. *User's Guide—Mathcad®*. MathSoft, Inc. Cambridge, MA.
MATLAB. 1995. *The Student Edition of MATLAB®*. Prentice–Hall, Englewood Cliffs, NJ.

Rashid, M.H. 1990. *Spice for Circuits and Electronics Using PSpice.* Prentice–Hall, Englewood Cliffs, NJ.

Wolfram, S. 1991. *Mathematica: A System for Doing Mathematics by Computer,* 2nd ed. Addison–Wesley, New York.

Further Information

There are many books that can be referenced in regard to computer-aided analysis of signals and systems. Rather than add to the reference list, two mathematics books are suggested that give backgound pertinent to the development of many of the functions implemented in such computer analysis tools.

Kreyszig, E. 1988. *Advanced Engineering Mathematics,* 6th ed. Wiley, New York.

Smith, J.W. 1987. *Mathematical Modeling and Digital Simulation for Engineers and Scientists,* 2nd ed. Wiley, New York.

II

Conversion Factors, Standards and Constants

8	**Properties of Materials** *James F. Shackelford*..	92
	Introduction • Structure • Composition • Physical Properties • Mechanical Properties • Thermal Properties • Chemical Properties • Electrical and Optical Properties • Additional Data	
9	**Frequency Bands and Assignments** *Robert D. Greenberg*................................	118
	U.S. Table of Frequency Allocations	
10	**International Standards and Constants**	214
	International System of Units (SI) • Physical Constants	
11	**Conversion Factors** *Jerry C. Whitaker*..	222
	Introduction • Conversion Constants and Multipliers	
12	**General Mathematical Tables** *W.F. Ames and George Cain*............................	241
	Introduction to Mathematics Chapter • Elementary Algebra and Geometry • Trigonometry • Series • Differential Calculus • Integral Calculus • Special Functions • Basic Definitions: Linear Algebra Matrices • Basic Definitions: Vector Algebra and Calculus • The Fourier Transforms: Overview	
13	**Communications Terms: Abbreviations**	288

Jerry C. Whitaker
Editor-in-Chief

ONE OF THE GOALS OF THIS BOOK is to provide easy reference to a wide variety od data that engineere need on the job. Section II represents the culmination of that effort. Extensive tables and other data are given on a wide range of topics, as listed above. Particular attention should be given to Chapter 9, **Frequency Bands and Assignments,** where a detailed listing of frequency allocation and communications services is given. Other chapters in this section provide important reference data on the properties of materials, international standards, conversion factors, general mathematical tables, and abbrevations for common terms in the communication industry.

8
Properties of Materials

8.1	Introduction	92
8.2	Structure	92
8.3	Composition	92
8.4	Physical Properties	93
8.5	Mechanical Properties	93
8.6	Thermal Properties	95
8.7	Chemical Properties	95
8.8	Electrical and Optical Properties	95
8.9	Additional Data	96
	Electrical Resistivity • Properties of Semiconductors	

James F. Shackelford
University of California, Davis

8.1 Introduction

The term *materials science and engineering* refers to that branch of engineering dealing with the processing, selection, and evaluation of solid-state materials [Shackelford 1996]. As such, this is a highly interdisciplinary field. This chapter reflects the fact that engineers outside of the specialized field of materials science and engineering largely need guidance in the selection of materials for their specific applications. A comprehensive source of property data for engineering materials is available in *The CRC Materials Science and Engineering Handbook* [Shackelford, Alexander, and Park 1994]. This brief chapter will be devoted to defining key terms associated with the properties of engineering materials and providing representative tables of such properties. Because the underlying principle of the fundamental understanding of solid-state materials is the fact that atomic- or microscopic-scale structure is responsible for the nature of materials properties, we shall begin with a discussion of structure. This will be followed by a discussion of the importance of specifying the chemical composition of commercial materials. These discussions will be followed by the definition of the main categories of materials properties.

8.2 Structure

A central tenet of materials science is that the behavior of materials (represented by their **properties**) is determined by their structure on the atomic and microscopic scales [Shackelford 1996]. Perhaps the most fundamental aspect of the structure–property relationship is to appreciate the basic skeletal arrangement of atoms in **crystalline** solids. Table 8.1 illustrates the fundamental possibilities, known as the 14 **Bravais lattices**. All crystalline structures of real materials can be produced by decorating the unit cell patterns of Table 8.1 with one or more atoms and repetitively stacking the unit cell structure through three-dimensional space.

8.3 Composition

The properties of commercially available materials are determined by chemical composition as well as structure [Shackelford 1996]. As a result, extensive numbering systems have been developed to label materials, especially metal **alloys**. Table 8.2 gives an example for gray cast irons.

Properties of Materials

TABLE 8.1 The 14 Bravais Lattices

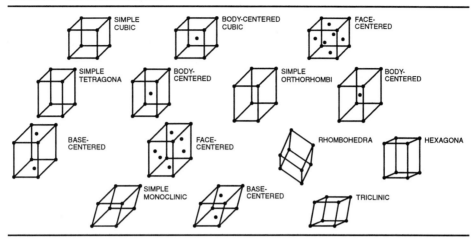

Source: Shackelford, J.F. 1996. *Introduction to Materials Science for Engineers*, 4th ed., p. 58. Prentice–Hall, Upper Saddle River, NJ.

8.4 Physical Properties

Among the most basic and practical characteristics of engineering materials are their physical properties. Table 8.3 gives the **density** of a wide range of materials in units of mega gram per cubic meter (= gram per cubic centimeter), whereas Table 8.4 gives the **melting points** for several common metals and ceramics.

8.5 Mechanical Properties

Central to the selection of materials for structural applications is their behavior in response to mechanical loads. A wide variety of mechanical properties are available to help guide materials selection [Shackelford, Alexander, and Park 1994]. The most basic of the mechanical properties are defined in terms of the **engineering stress** and the **engineering strain**. The engineering stress σ is defined as

$$\sigma = \frac{P}{A_0} \qquad (8.1)$$

where P is the load on the sample with an original (zero stress) cross-sectional area A_0. The engineering strain ε is defined as

$$\varepsilon = \frac{[l - l_0]}{l_0} = \frac{\triangle l}{l_0} \qquad (8.2)$$

where l is the sample length at a given load and l_0 is the original (zero stress) length. The maximum engineering stress that can be withstood by the material during its load history is termed the *ultimate tensile strength*, or simply **tensile strength** (TS). An example of the tensile strength for selected wrought (meaning worked, as opposed to cast) aluminum alloys is given in Table 8.5. The stiffness of a material is indicated by the linear relationship between engineering stress and engineering strain for relatively small levels of load application. The **modulus of elasticity** E, also known as **Young's modulus**, is

TABLE 8.2 Composition Limits of Selected Gray Cast Irons (%)

UNS	SAE Grade	C	Mn	Si	P	S
F10004	G1800	3.40 to 3.70	0.50 to 0.80	2.80 to 2.30	0.15	0.15
F10005	G2500	3.20 to 3.50	0.60 to 0.90	2.40 to 2.00	0.12	0.15
F10009	G2500	3.40 min	0.60 to 0.90	1.60 to 2.10	0.12	0.12
F10006	G3000	3.10 to 3.40	0.60 to 0.90	2.30 to 1.90	0.10	0.16
F10007	G3500	3.00 to 3.30	0.60 to 0.90	2.20 to 1.80	0.08	0.16
F10010	G3500	3.40 min	0.60 to 0.90	1.30 to 1.80	0.08	0.12
F10011	G3500	3.50 min	0.60 to 0.90	1.30 to 1.80	0.08	0.12
F10008	G4000	3.00 to 3.30	0.70 to 1.00	2.10 to 1.80	0.07	0.16
F10012	G4000	3.10 to 3.60	0.60 to 0.90	1.95 to 2.40	0.07	0.12

Source: Data from ASM. 1984. *ASM Metals Reference Book,* 2nd ed., p. 166. American Society for Metals, Metals Park, OH.

given by the ratio

$$E = \frac{\sigma}{\varepsilon} \qquad (8.3)$$

Table 8.6 gives values of Young's modulus for selected compositions of glass materials. The ductility of a material is indicated by the percent elongation at failure (= $100 \times \varepsilon_{\text{failure}}$), representing the general ability of the material to be plastically (i.e., permanently) deformed. The percent elongation at failure for selected polymers is given in Table 8.7.

TABLE 8.3 Density of Selected Materials, Mg/m³

Metal		Ceramic		Glass		Polymer	
Ag	10.50	Al$_2$O$_3$	3.97–3.986	SiO$_2$	2.20	ABS	1.05–1.07
Al	2.7	BN (cub)	3.49	SiO$_2$ 10 wt% Na$_2$O	2.291	Acrylic	1.17–1.19
Au	19.28	BeO	3.01–3.03	SiO$_2$ 19.55 wt% Na$_2$O	2.383	Epoxy	1.80–2.00
Co	8.8	MgO	3.581	SiO$_2$ 29.20 wt% Na$_2$O	2.459	HDPE	0.96
Cr	7.19	SiC(hex)	3.217	SiO$_2$ 39.66 wt% Na$_2$O	2.521	Nylon, type 6	1.12–1.14
Cu	8.93	Si$_3$N$_4$ (α)	3.184	SiO$_2$ 39.0 wt% CaO	2.746	Nylon 6/6	1.13–1.15
Fe	7.87	Si$_3$N$_4$ (β)	3.187			Phenolic	1.32–1.46
Ni	8.91	TiO$_2$ (rutile)	4.25			Polyacetal	1.425
Pb	11.34	UO$_2$	10.949–10.97			Polycarbonate	1.2
Pt	21.44	ZrO$_2$ (CaO)	5.5			Polyester	1.31
Ti	4.51	Al$_2$O$_3$ MgO	3.580			Polystyrene	1.04
W	19.25	3Al$_2$O$_3$ 2SiO$_2$	2.6–3.26			PTFE	2.1–2.3

TABLE 8.4 Melting Point of Selected Metals and Ceramics

Metal	M.P. (°C)	Ceramic	M.P. (°C)
Ag	962	Al$_2$O$_3$	2049
Al	660	BN	2727
Au	1064	B$_2$O$_3$	450
Co	1495	BeO	2452
Cr	1857	NiO	1984
Cu	1083	PbO	886
Fe	1535	SiC	2697
Ni	1453	Si$_3$N$_4$	2442
Pb	328	SiO$_2$	1723
Pt	1772	WC	2627
Ti	1660	ZnO	1975
W	3410	ZrO$_2$	2850

TABLE 8.5 Tensile Strength of Selected Wrought Aluminum Alloys

Alloy	Temper	TS (MPa)
1050	0	76
1050	H16	130
2024	0	185
2024	T361	495
3003	0	110
3003	H16	180
5050	0	145
5050	H34	195
6061	0	125
6061	T6, T651	310
7075	0	230
7075	T6, T651	570

TABLE 8.6 Young's Modulus of Selected Glasses, GPa

Type	E
SiO$_2$	72.76–74.15
SiO$_2$ 20 mol % Na$_2$O	62.0
SiO$_2$ 30 mol % Na$_2$O	60.5
SiO$_2$ 35 mol % Na$_2$O	60.2
SiO$_2$ 24.6 mol % PbO	47.1
SiO$_2$ 50.0 mol % PbO	44.1
SiO$_2$ 65.0 mol % PbO	41.2
SiO$_2$ 60 mol % B$_2$O$_3$	23.3
SiO$_2$ 90 mol % B$_2$O$_3$	20.9
B$_2$O$_3$	17.2–17.7
B$_2$O$_3$ 10 mol % Na$_2$O	31.4
B$_2$O$_3$ 20 mol % Na$_2$O	43.2

TABLE 8.7 Total Elongation at Failure of Selected Polymers

Polymer	Elongation
ABS	5–20
Acrylic	2–7
Epoxy	4.4
HDPE	700–1000
Nylon, type 6	30–100
Nylon 6/6	15–300
Phenolic	0.4–0.8
Polyacetal	25
Polycarbonate	110
Polyester	300
Polypropylene	100–600
PTFE	250–350

[a]% in 50 mm section.

Source: Selected data for Tables 8.3–8.7 from Shackelford, J.F., Alexander, W., and Park, J.S., eds. 1994. *CRC Materials Science and Engineering Handbook*, 2nd ed. CRC Press, Boca Raton, FL.

8.6 Thermal Properties

Many applications of engineering materials depend on their response to a thermal environment. The **thermal conductivity** k is defined by **Fourier's law**

$$k = \frac{-\left[\dfrac{dQ}{dt}\right]}{\left[A\left(\dfrac{dT}{dx}\right)\right]} \qquad (8.4)$$

where dQ/dt is the rate of heat transfer across an area A due to a temperature gradient dT/dx. It is also important to note that the dimensions of a material will, in general, increase with temperature. Increases in temperature lead to greater thermal vibration of the atoms and an increase in average separation distance of adjacent atoms. The **linear coefficient of thermal expansion** α is given by

TABLE 8.8 Thermal Conductivity and Thermal Expansion of Alloy Cast Irons

Alloy	Thermal Conductivity, W/(m K)	Thermal Expansion Coefficient, 10^{-6} mm/(mm °C)
Low-C white iron	22[a]	12[b]
Martensitic nickel-chromium iron	30[a]	8–9[b]
High-nickel gray iron	38–40	8.1–19.3
High-nickel ductile iron	13.4	12.6–18.7
Medium-silicon iron	37	10.8
High-chromium iron	20	9.3–9.9
High-nickel iron	37–40	8.1–19.3
Nickel-chromium-silicon iron	30	12.6–16.2
High-nickel (20%) ductile iron	13	18.7

[a] Estimated.
[b] 10–260°C. (*Source:* Data from ASM. 1984. *ASM Metals Reference Book*, 2nd ed., p. 172. American Society for Metals, Metals Park, OH.)

$$\alpha = \frac{dl}{l\,dT} \qquad (8.5)$$

with α having units of millimeter per millimeter degree Celsius. Examples of thermal conductivity and thermal expansion coefficient for alloy cast irons are given in Table 8.8.

8.7 Chemical Properties

A wide variety of data is available to characterize the nature of the reaction between engineering materials and their chemical environments [Shackelford, Alexander, and Park 1994]. Perhaps no such data are more fundamental and practical than the **electromotive force series** of metals shown in Table 8.9. The voltage associated with various half-cell reactions in standard aqueous environments are arranged in order, with the materials associated with more anodic reactions tending to be corroded in the presence of a metal associated with a more cathodic reaction.

TABLE 8.9 Electromotive Force Series of Metals

Metal	Potential, V	Metal	Potential, V	Metal	Potential, V
Anodic or Corroded End					
Li	−3.04	Al	−1.70	Pb	−0.13
Rb	−2.93	Mn	−1.04	H	0.00
K	−2.92	Zn	−0.76	Cu	0.52
Ba	−2.90	Cr	−0.60	Ag	0.80
Sr	−2.89	Cd	−0.40	Hg	0.85
Ca	−2.80	Ti	−0.33	Pd	1.0
Na	−2.71	Co	−0.28	Pt	1.2
Mg	−2.37	Ni	−0.23	Au	1.5
Be	−1.70	Sn	−0.14		
				Cathodic or Noble Metal End	

Source: Data compiled by J.S. Park from Bolz, R.E. and Tuve, G.L., eds. 1973. *CRC Handbook of Tables for Applied Engineering Science*. CRC Press, Boca Raton, FL.

8.8 Electrical and Optical Properties

To this point, we have concentrated on various properties dealing largely with the structural applications of engineering materials. In many cases the electromagnetic nature of the materials may determine their engineering

TABLE 8.10 Electrical Resistivity of Selected Materials

Metal (Alloy Cast Iron)	$\rho, \Omega \cdot m$	Ceramic	$\rho, \Omega \cdot m$	Polymer	$\rho, \Omega \cdot m$
Low-C white cast iron	$0.53 \cdot 10^{-6}$	Al_2O_3	$>10^{13}$	ABS	$2-4 \cdot 10^{13}$
Martensitic Ni-Cr iron	$0.80 \cdot 10^{-6}$	B_4C	$0.3-0.8 \cdot 10^{-2}$	Acrylic	$>10^{13}$
High-Si iron	$0.50 \cdot 10^{-6}$	BN	$1.7 \cdot 10^{11}$	HDPE	$>10^{15}$
High-Ni iron	$1.4-1.7 \cdot 10^{-6}$	BeO	$>10^{15}$	Nylon 6/6	$10^{12}-10^{13}$
Ni-Cr-Si iron	$1.5-1.7 \cdot 10^{-6}$	MgO	$1.3 \cdot 10^{13}$	Phenolic	$10^{7}-10^{11}$
High-Al iron	$2.4 \cdot 10^{-6}$	SiC	$1-1 \cdot 10^{10}$	Polyacetal	10^{13}
Medium-Si ductile iron	$0.58\text{-}0.87 \cdot 10^{-6}$	Si_3N_4	10^{11}	Polypropylene	$>10^{15}$
High-Ni (20%) ductile iron	$1.02 \cdot 10^{-6}$	SiO_2	10^{16}	PTFE	$>10^{16}$

Source: Selected data from Shackelford, J.F., Alexander, W., and Park, J.S., eds. 1994. *CRC Materials Science and Engineering Handbook,* 2nd ed. CRC Press, Boca Raton, FL.

TABLE 8.11 Refractive Index of Selected Polymers

Polymer	n
Acrylic	1.485–1.500
Cellulose Acetate	1.46–1.50
Epoxy	1.61
HDPE	1.54
Polycarbonate	1.586
PTFE	1.35
Polyester	1.50–1.58
Polystyrene	1.6
SAN	1.565–1.569
Vinylidene Chloride	1.60–1.63

Source: Data compiled by J.S. Park from Lynch, C.T., ed. 1975. *CRC Handbook of Materials Science, Vol. 3.* CRC Press, Inc., Boca Raton, FL; and ASM. 1988. *Engineered Materials Handbook, Vol. 2, Engineering Plastics.* ASM International, Metals Park, OH.

applications. Perhaps the most fundamental relationship in this regard is **Ohm's law,** which states that the magnitude of current flow I through a circuit with a given resistance R and voltage V is related by

$$V = IR \tag{8.6}$$

where V is in units of volts, I is in amperes, and R is in ohms. The resistance value depends on the specific sample geometry. In general, R increases with sample length l and decreases with sample area A. As a result, the property more characteristic of a given material and independent of its geometry is **resistivity** ρ defined as

$$\rho = \frac{[RA]}{l} \tag{8.7}$$

The units for resistivity are ohm meter ($\Omega \cdot m$). Table 8.10 gives the values of electrical resistivity for various materials, indicating that metals typically have low resistivities (and correspondingly high electrical conductivities) and ceramics and polymers typically have high resistivities (and correspondingly low conductivities).

An important aspect of the electromagnetic nature of materials is their optical properties. Among the most fundamental optical characteristics of a light transmitting material is the **index of refraction** n defined as

$$n = \frac{v_{\text{vac}}}{v} \tag{8.8}$$

where v_{vac} is the speed of light in vacuum (essentially equal to that in air) and v is the speed of light in a transparent material. The index of refraction for a variety of polymers is given in Table 8.11.

8.9 Additional Data

The following, more detailed tabular data are reprinted from: Dorf, R.C. 1993. *The Electrical Engineering Handbook.* CRC Press, Boca Raton, FL, pp. 2527–2544.

Electrical Resistivity

Electrical Resistivity of Pure Metals

The first part of this table gives the electrical resistivity, in units of $10^{-8} \, \Omega \cdot m$, for 28 common metallic elements as a function of temperature. The data refer to polycrystalline samples. The number of significant figures indicates the accuracy of the values. However, at low temperatures (especially below 50 K) the electrical resistivity is extremely sensitive to sample purity. Thus the low-temperature values refer to samples of specified purity and treatment.

The second part of the table gives resistivity values in the neighborhood of room temperature for other metallic elements that have not been studied over an extended temperature range.

Properties of Materials

Electrical Resistivity in $10^{-8}\ \Omega \cdot m$

T/K	Aluminum	Barium	Beryllium	Calcium	Cesium	Chromium	Copper
1	0.000100	0.081	0.0332	0.045	0.0026		0.00200
10	0.000193	0.189	0.0332	0.047	0.243		0.00202
20	0.000755	0.94	0.0336	0.060	0.86		0.00280
40	0.0181	2.91	0.0367	0.175	1.99		0.0239
60	0.0959	4.86	0.067	0.40	3.07		0.0971
80	0.245	6.83	0.075	0.65	4.16		0.215
100	0.442	8.85	0.133	0.91	5.28	1.6	0.348
150	1.006	14.3	0.510	1.56	8.43	4.5	0.699
200	1.587	20.2	1.29	2.19	12.2	7.7	1.046
273	2.417	30.2	3.02	3.11	18.7	11.8	1.543
293	2.650	33.2	3.56	3.36	20.5	12.5	1.678
298	2.709	34.0	3.70	3.42	20.8	12.6	1.712
300	2.733	34.3	3.76	3.45	21.0	12.7	1.725
400	3.87	51.4	6.76	4.7		15.8	2.402
500	4.99	72.4	9.9	6.0		20.1	3.090
600	6.13	98.2	13.2	7.3		24.7	3.792
700	7.35	130	16.5	8.7		29.5	4.514
800	8.70	168	20.0	10.0		34.6	5.262
900	10.18	216	23.7	11.4		39.9	6.041

T/K	Gold	Hafnium	Iron	Lead	Lithium	Magnesium	Manganese
1	0.0220	1.00	0.0225		0.007	0.0062	7.02
10	0.0226	1.00	0.0238		0.008	0.0069	18.9
20	0.035	1.11	0.0287		0.012	0.0123	54
40	0.141	2.52	0.0758		0.074	0.074	116
60	0.308	4.53	0.271		0.345	0.261	131
80	0.481	6.75	0.693	4.9	1.00	0.557	132
100	0.650	9.12	1.28	6.4	1.73	0.91	132
150	1.061	15.0	3.15	9.9	3.72	1.84	136
200	1.462	21.0	5.20	13.6	5.71	2.75	139
273	2.051	30.4	8.57	19.2	8.53	4.05	143
293	2.214	33.1	9.61	20.8	9.28	4.39	144
298	2.255	33.7	9.87	21.1	9.47	4.48	144
300	2.271	34.0	9.98	21.3	9.55	4.51	144
400	3.107	48.1	16.1	29.6	13.4	6.19	147
500	3.97	63.1	23.7	38.3		7.86	149
600	4.87	78.5	32.9			9.52	151
700	5.82		44.0			11.2	152
800	6.81		57.1			12.8	
900	7.86					14.4	

T/K	Molybdenum	Nickel	Palladium	Platinum	Potassium	Rubidium	Silver
1	0.00070	0.0032	0.0200	0.002	0.0008	0.0131	0.00100
10	0.00089	0.0057	0.0242	0.0154	0.0160	0.109	0.00115
20	0.00261	0.0140	0.0563	0.0484	0.117	0.444	0.0042
40	0.0457	0.068	0.334	0.409	0.480	1.21	0.0539
60	0.206	0.242	0.938	1.107	0.90	1.94	0.162
80	0.482	0.545	1.75	1.922	1.34	2.65	0.289
100	0.858	0.96	2.62	2.755	1.79	3.36	0.418
150	1.99	2.21	4.80	4.76	2.99	5.27	0.726
200	3.13	3.67	6.88	6.77	4.26	7.49	1.029
273	4.85	6.16	9.78	9.6	6.49	11.5	1.467
293	5.34	6.93	10.54	10.5	7.20	12.8	1.587
298	5.47	7.12	10.73	10.7	7.39	13.1	1.617
300	5.52	7.20	10.80	10.8	7.47	13.3	1.629
400	8.02	11.8	14.48	14.6			2.241

Electrical Resistivity in $10^{-8}\ \Omega \cdot m$ (continued)

T/K	Molybdenum	Nickel	Palladium	Platinum	Potassium	Rubidium	Silver
500	10.6	17.7	17.94	18.3			2.87
600	13.1	25.5	21.2	21.9			3.53
700	15.8	32.1	24.2	25.4			4.21
800	18.4	35.5	27.1	28.7			4.91
900	21.2	38.6	29.4	32.0			5.64

T/K	Sodium	Strontium	Tantalum	Tungsten	Vanadium	Zinc	Zirconium
1	0.0009	0.80	0.10	0.000016		0.0100	0.250
10	0.0015	0.80	0.102	0.000137	0.0145	0.0112	0.253
20	0.016	0.92	0.146	0.00196	0.039	0.0387	0.357
40	0.172	1.70	0.751	0.0544	0.304	0.306	1.44
60	0.447	2.68	1.65	0.266	1.11	0.715	3.75
80	0.80	3.64	2.62	0.606	2.41	1.15	6.64
100	1.16	4.58	3.64	1.02	4.01	1.60	9.79
150	2.03	6.84	6.19	2.09	8.2	2.71	17.8
200	2.89	9.04	8.66	3.18	12.4	3.83	26.3
273	4.33	12.3	12.2	4.82	18.1	5.46	38.8
293	4.77	13.2	13.1	5.28	19.7	5.90	42.1
298	4.88	13.4	13.4	5.39	20.1	6.01	42.9
300	4.93	13.5	13.5	5.44	20.2	6.06	43.3
400		17.8	18.2	7.83	28.0	8.37	60.3
500		22.2	22.9	10.3	34.8	10.82	76.5
600		26.7	27.4	13.0	41.1	13.49	91.5
700		31.2	31.8	15.7	47.2		104.2
800		35.6	35.9	18.6	53.1		114.9
900			40.1	21.5	58.7		123.1

Electrical Resistivity of Other Metallic Elements in the Neighborhood of Room Temperature

Element	T/K	Electrical Resistivity $10^{-8}\ \Omega \cdot m$	Element	T/K	Electrical Resistivity $10^{-8}\ \Omega \cdot m$
Antimony	273	39	Polonium	273	40
Bismuth	273	107	Praseodymium	290–300	70.0
Cadmium	273	6.8	Promethium	290–300	75
Cerium	290–300	82.8	Protactinium	273	17.7
Cobalt	273	5.6	Rhenium	273	17.2
Dysprosium	290–300	92.6	Rhodium	273	4.3
Erbium	290–300	86.0	Ruthenium	273	7.1
Europium	290–300	90.0	Samarium	290–300	94.0
Gadolinium	290–300	131	Scandium	290–300	56.2
Gallium	273	13.6	Terbium	290–300	115
Holmium	290–300	81.4	Thallium	273	15
Indium	273	8.0	Thorium	273	14.7
Iridium	273	4.7	Thulium	290–300	67.6
Lanthanum	290–300	61.5	Tin	273	11.5
Lutetium	290–300	58.2	Titanium	273	39
Mercury	273	94.1	Uranium	273	28
Neodymium	290–300	64.3	Ytterbium	290–300	25.0
Niobium	273	15.2	Yttrium	290–300	59.6
Osmium	273	8.1			

Electrical Resistivity of Selected Alloys

Values of the resistivity are given in units of 10^{-8} Ω·m. General comments in the preceding table for pure metals also apply here.

Alloy—Aluminum-Copper

Wt % Al	273 K	293 K	300 K	350 K	400 K
99[a]	2.51	2.74	2.82	3.38	3.95
95[a]	2.88	3.10	3.18	3.75	4.33
90[b]	3.36	3.59	3.67	4.25	4.86
85[b]	3.87	4.10	4.19	4.79	5.42
80[b]	4.33	4.58	4.67	5.31	5.99
70[b]	5.03	5.31	5.41	6.16	6.94
60[b]	5.56	5.88	5.99	6.77	7.63
50[b]	6.22	6.55	6.67	7.55	8.52
40[c]	7.57	7.96	8.10	9.12	10.2
30[c]	11.2	11.8	12.0	13.5	15.2
25[f]	16.3[aa]	17.2	17.6	19.8	22.2
15[h]	—	12.3	—	—	—
19[g]	1.8[aa]	11.0	11.1	11.7	12.3
5[e]	9.43	9.61	9.68	10.2	10.7
1[b]	4.46	4.60	4.65	5.00	5.37

Alloy—Aluminum-Magnesium

	273 K	293 K	300 K	350 K	400 K
99[c]	2.96	3.18	3.26	3.82	4.39
95[c]	5.05	5.28	5.36	5.93	6.51
90[c]	7.52	7.76	7.85	8.43	9.02
85	—	—	—	—	—
80	—	—	—	—	—
70	—	—	—	—	—
60	—	—	—	—	—
50	—	—	—	—	—
40	—	—	—	—	—
30	—	—	—	—	—
25	—	—	—	—	—
15	—	—	—	—	—
10[b]	17.1	17.4	17.6	18.4	19.2
5[b]	13.1	13.4	13.5	14.3	15.2
1[a]	5.92	6.25	6.37	7.20	8.03

Alloy—Copper-Gold

Wt % Cu	273 K	293 K	300 K	350 K	400 K
99[c]	1.73	1.86[aa]	1.91[aa]	2.24[aa]	2.58[aa]
95[c]	2.41	2.54[aa]	2.59[aa]	2.92[aa]	3.26[aa]
90[c]	3.29	4.42[aa]	3.46[aa]	3.79[aa]	4.12[aa]
85[c]	4.20	4.33	4.38[aa]	4.71[aa]	5.05[aa]
80[c]	5.15	5.28	5.32	5.65	5.99
70[c]	7.12	7.25	7.30	7.64	7.99
60[c]	9.18	9.13	9.36	9.70	10.05
50[c]	11.07	11.20	11.25	11.60	11.94
40[c]	12.70	12.85	12.90[aa]	13.27[aa]	13.65[aa]
30[c]	13.77	13.93	13.99[aa]	14.38[aa]	14.78[aa]
25[c]	13.93	14.09	14.14	14.54	14.94
15[c]	12.75	12.91	12.96[aa]	13.36[aa]	13.77
10[c]	10.70	10.86	10.91	11.31	11.72
5[c]	7.25	7.41[aa]	7.46	7.87	8.28
1[c]	3.40	3.57	3.62	4.03	4.45

Alloy—Copper-Nickel

Wt % Cu	273 K	293 K	300 K	350 K	400 K
99[c]	2.71	2.85	2.91	3.27	3.62
95[c]	7.60	7.71	7.82	8.22	8.62
90[c]	13.69	13.89	13.96	14.40	14.81
85[c]	19.63	19.83	19.90	2032	20.70
80[c]	25.46	25.66	25.72	26.12[aa]	26.44[aa]
70[i]	36.67	36.72	36.76	36.85	36.89
60[i]	45.43	45.38	45.35	45.20	45.01
50[i]	50.19	50.05	50.01	49.73	49.50
40[c]	47.42	47.73	47.82	48.28	48.49
30[i]	40.19	41.79	42.34	44.51	45.40
25[c]	33.46	35.11	35.69	39.67[aa]	42.81[aa]
15[c]	22.00	23.35	23.85	27.60	31.38
10[c]	16.65	17.82	18.26	21.51	25.19
5[c]	11.49	12.50	12.90	15.69	18.78
1[c]	7.23	8.08	8.37	10.63[aa]	13.18[aa]

Alloy—Copper-Palladium

Wt % Cu	273 K	293 K	300 K	350 K	400 K
99[c]	2.10	2.23	2.27	2.59	2.92
95[c]	4.21	4.35	4.40	4.74	5.08
90[c]	6.89	7.03	7.08	7.41	7.74
85[c]	9.48	9.61	9.66	10.01	10.36
80[c]	11.99	12.12	12.16	12.51[aa]	12.87
70[c]	16.87	17.01	17.06	17.41	17.78
60[c]	21.73	21.87	21.92	22.30	22.69
50[c]	27.62	27.79	27.86	28.25	28.64
40[c]	35.31	35.51	35.57	36.03	36.47
30[c]	46.50	46.66	46.71	47.11	47.47
25[c]	46.25	46.45	46.52	46.99[aa]	47.43[aa]
15[c]	36.52	36.99	37.16	38.28	39.35
10[c]	28.90	29.51	29.73	31.19[aa]	32.56[aa]
5[c]	20.00	20.75	21.02	22.84[aa]	24.54[aa]
1[c]	11.90	12.67	12.93[aa]	14.82[aa]	16.68[aa]

Alloy—Copper-Zinc

Wt % Cu	273 K	293 K	300 K	350 K	400 K
99[b]	1.84	1.97	2.02	2.36	2.71
95[b]	2.78	2.92	2.97	3.33	3.69
90[b]	3.66	3.81	3.86	4.25	4.63
85[b]	4.37	4.54	4.60	5.02	5.44
80[b]	5.01	5.19	5.26	5.71	6.17
70[b]	5.87	6.08	6.15	6.67	7.19
60	—	—	—	—	—
50	—	—	—	—	—
40	—	—	—	—	—
30	—	—	—	—	—
25	—	—	—	—	—
15	—	—	—	—	—
10	—	—	—	—	—
5	—	—	—	—	—
1	—	—	—	—	—

Electrical Resistivity of Selected Alloys (*continued*)

	273 K	293 K	300 K	350 K	400 K
Alloy—Gold-Palladium					
Wt % Au					
99[c]	2.69	2.86	2.91	3.32	3.73
95[c]	5.21	5.35	5.41	5.79	6.17
90[i]	8.01	8.17	8.22	8.56	8.93
85[b]	10.50[aa]	10.66	10.72[aa]	11.100[aa]	11.48[aa]
80[b]	12.75	12.93	12.99	13.45	13.93
70[c]	18.23	18.46	18.54	19.10	19.67
60[b]	26.70	26.94	27.02	27.63[aa]	28.23[aa]
50[a]	27.23	27.63	27.76	28.64[aa]	29.42[aa]
40[a]	24.65	25.23	25.42	26.74	27.95
30[b]	20.82	21.49	21.72	23.35	24.92
25[b]	18.86	19.53	19.77	21.51	23.19
15[a]	15.08	15.77	16.01	17.80	19.61
10[a]	13.25	13.95	14.20[aa]	16.00[aa]	17.81[aa]
5[a]	11.49[aa]	12.21	12.46[aa]	14.26[aa]	16.07[aa]
1[a]	10.07	10.85[aa]	11.12[aa]	12.99[aa]	14.80[aa]
Alloy—Gold-Silver					
Wt % Au					
99[b]	2.58	2.75	2.80[aa]	3.22[aa]	3.63[aa]
95[a]	4.58	4.74	4.79	5.19	5.59
90[i]	6.57	6.73	6.78	7.19	7.58
85[j]	8.14	8.30	8.36[aa]	8.75	9.15
80[j]	9.34	9.50	9.55	9.94	10.33
70[j]	10.70	10.86	10.91	11.29	11.68[aa]
60[j]	10.92	11.07	11.12	11.50	11.87
50[j]	10.23	10.37	10.42	10.78	11.14
40[j]	8.92	9.06	9.11	9.46[aa]	9.81
30[a]	7.34	7.47	7.52	7.85	8.19
25[a]	6.46	6.59	6.63	6.96	7.30[aa]
15[a]	4.55	4.67	4.72	5.03	5.34
10[a]	3.54	3.66	3.71	4.00	4.31
5[i]	2.52	2.64[aa]	2.68[aa]	2.96[aa]	3.25[aa]
1[b]	1.69	1.80	1.84[aa]	2.12[aa]	2.42[aa]
Alloy—Iron-Nickel					
Wt % Fe					
99[a]	10.9	12.0	12.4	—	18.7
95[c]	18.7	19.9	20.2	—	26.8
90[c]	24.2	25.5	25.9	—	33.2
85[c]	27.8	29.2	29.7	—	37.3
80[c]	30.1	31.6	32.2	—	40.0
70[b]	32.3	33.9	34.4	—	42.4
60[c]	53.8	57.1	58.2	—	73.9
50[d]	28.4	30.6	31.4	—	43.7
40[d]	19.6	21.6	22.5	—	34.0
30[c]	15.3	17.1	17.7	—	27.4
25[b]	14.3	15.9	16.4	—	25.1
15[c]	12.6	13.8	14.2	—	21.1
10[c]	11.4	12.5	12.9	—	18.9
5[c]	9.66	10.6	10.9	—	16.1[aa]
1[b]	7.17	7.94	8.12	—	12.8
Alloy—Silver-Palladium					
Wt % Ag					
99[b]	1.891	2.007	2.049	2.35	2.66
95[b]	3.58	3.70	3.74	4.04	4.34
90[b]	5.82	5.94	5.98	6.28	6.59

	273 K	293 K	300 K	350 K	400 K
Alloy—Silver-Palladium					
85[k]	7.92[aa]	8.04[aa]	8.08	8.38[aa]	8.68[aa]
80[k]	10.01	10.13	10.17	10.47	10.78
70[k]	14.53	14.65	14.69	14.99	15.30
60[i]	20.9	21.1	21.2	21.6	22.0
50[k]	31.2	31.4	31.5	32.0	32.4
40[m]	42.2	42.2	42.2	42.3	42.3
30[b]	40.4	40.6	40.7	41.3	41.7
25[k]	36.67[aa]	37.06	37.19	38.1[aa]	38.8[aa]
15[i]	27.08[aa]	26.68[aa]	27.89[aa]	29.3[aa]	30.6[aa]
10[i]	21.69	22.39	22.63	24.3	25.9
5[b]	15.98	16.72	16.98	18.8[aa]	20.5[aa]
1[a]	11.06	11.82	12.08[aa]	13.92[aa]	15.70[aa]

[a] Uncertainty in resistivity is ±2%.
[b] Uncertainty in resistivity is ±3%.
[c] Uncertainty in resistivity is ±5%.
[d] Uncertainty in resistivity is ±7% below 300 K and ±5% at 300 and 400 K.
[e] Uncertainty in resistivity is ±7%.
[f] Uncertainty in resistivity is ±8%.
[g] Uncertainty in resistivity is ±10%.
[h] Uncertainty in resistivity is ±12%.
[i] Uncertainty in resistivity is ±4%.
[j] Uncertainty in resistivity is ±1%.
[k] Uncertainty in resistivity is ±3% up to 300 K and ±4% above 300 K.
[m] Uncertainty in resistivity is ±2% up to 300 K and ±4% above 300 K.
[a] Crystal usually a mixture of α-hcp and fcc lattice.
[aa] In temperature range where no experimental data are available.

Electrical Resistivity of Selected Alloy Cast Irons

Description	Electrical resistivity, $\mu\Omega \cdot m$
Abrasion-resistant white irons	
Low-C white iron	0.53
Martensitic nickel–chromium iron	0.80
Corrosion-resistant irons	
High-silicon iron	0.50
High-chromium iron	
High-nickel gray iron	1.0[a]
High-nickel ductile iron	1.0[a]
Heat-resistant gray irons	
Medium-silicon iron	
High-chromium iron	
High-nickel iron	1.4–1.7
Nickel-chromium–silicon iron	1.5–1.7
High-aluminum iron	2.4
Heat-resistant ductile irons	
Medium-silicon ductile iron	0.58–0.87
High-nickel ductile (20 Ni)	1.02
High-nickel ductile (23 Ni)	1.0[a]

[a] Estimated. (*Source:* Data from 1984. *ASM Metals Reference Book*, 2nd ed. American Society for Metals, Metals Park, OH.)

Resistivity of Selected Ceramics (Listed by Ceramic)

Ceramic	Resistivity, $\Omega \cdot$ cm
Borides	
Chromium diboride (CrB_2)	21×10^{-6}
Hafnium diboride (HfB_2)	$10\text{--}12 \times 10^{-6}$ at room temp.
Tantalum diboride (TaB_2)	68×10^{-6}
Titanium diboride (TiB_2) (polycrystalline)	
85% dense	$26.5\text{--}28.4 \times 10^{-6}$ at room temp.
85% dense	9.0×10^{-6} at room temp.
100% dense, extrapolated values	$8.7\text{--}14.1 \times 10^{-6}$ at room temp.
	3.7×10^{-6} at liquid air temp.
Titanium diboride (TiB_2) (monocrystalline)	
Crystal length 5 cm, 39 deg. and 59 deg. orientation with respect to growth axis	$6.6 \pm 0.2 \times 10^{-6}$ at room temp.
Crystal length 1.5 cm, 16.5 deg. and 90 deg. orientation with respect to growth axis	$6.7 \pm 0.2 \times 10^{-6}$ at room temp.
Zirconium diboride (ZrB_2)	9.2×10^{-6} at 20°C
	1.8×10^{-6} at liquid air temp.
Carbides: boron carbide (B_4C)	0.3–0.8

Dielectric Constants

Dielectric Constants of Solids: Temperatures in the Range 17–22°C.

Material	Freq., Hz	Dielectric constant	Material	Freq., Hz	Dielectric Constant
Acetamide	4×10^8	4.0	Phenanthrene	4×10^8	2.80
Acetanilide	–	2.9	Phenol (10°C)	4×10^8	4.3
Acetic acid (2°C)	4×10^8	4.1	Phosphorus, red	10^8	4.1
Aluminum oleate	4×10^8	2.40	Phosphorus, yellow	10^8	3.6
Ammonium bromide	10^8	7.1	Potassium aluminum		
Ammonium chloride	10^8	7.0	sulfate	10^6	3.8
Antimony trichloride	10^8	5.34	Potassium carbonate		
Apatite $\perp$ optic axis	3×10^8	9.50	(15°C)	10^8	5.6
Apatite $\parallel$ optic axis	3×10^8	7.41	Potassium chlorate	6×10^7	5.1
Asphalt	$<3 \times 10^6$	2.68	Potassium chloride	10^4	5.03
Barium chloride (anhyd.)	6×10^7	11.4	Potassium chromate	6×10^7	7.3
Barium chloride ($2H_2O$)	6×10^7	9.4	Potassium iodide	6×10^7	5.6
Barium nitrate	6×10^7	5.9	Potassium nitrate	6×10^7	5.0
Barium sulfate (15°C)	10^8	11.4	Potassium sulfate	6×10^7	5.9
Beryl $\perp$ optic axis	10^4	7.02	Quartz $\perp$ optic axis	3×10^7	4.34
Beryl $\parallel$ optic axis	10^4	6.08	Quartz $\parallel$ optic axis	3×10^7	4.27
Calcite $\perp$ optic axis	10^4	8.5	Resorcinol	4×10^8	3.2
Calcite $\parallel$ optic axis	10^4	8.0	Ruby $\perp$ optic axis	10^4	13.27
Calcium carbonate	10^6	6.14	Ruby $\parallel$ optic axis	10^4	11.28
Calcium fluoride	10^4	7.36	Rutile $\perp$ optic axis	10^8	86
Calcium sulfate ($2H_2O$)	10^4	5.66	Rutile $\parallel$ optic axis	10^8	170
Cassiterite $\perp$ optic axis	10^{12}	23.4	Selenium	10^8	6.6
Cassiterite $\parallel$ optic axis	10^{12}	24	Silver bromide	10^6	12.2
d-Cocaine	5×10^8	3.10	Silver chloride	10^6	11.2
Cupric oleate	4×10^8	2.80	Silver cyanide	10^6	5.6
Cupric oxide (15°C)	10^8	18.1	Smithsonite $\perp$ optic axis	10^{12}	9.3
Cupric sulfate (anhyd.)	6×10^7	10.3			
Cupric sulfate ($5H_2O$)	6×10^7	7.8	Smithsonite $\parallel$ optic axis	10^{10}	9.4
Diamond	10^8	5.5			
Diphenylymethane	4×10^8	2.7	Sodium carbonate (anhyd.)	6×10^7	8.4

Dielectric Constants of Solids: Temperatures in the Range 17–22°C (continued)

Material	Freq., Hz	Dielectric constant	Material	Freq., Hz	Dielectric Constant
Dolomite ⊥ optic axis	10^8	8.0			
Dolomite ∥ optic axis	10^8	6.8	Sodium carbonate (10H$_2$O)	6×10^7	5.3
Ferrous oxide (15°C)	10^8	14.2			
Iodine	10^8	4	Sodium chloride	10^4	6.12
Lead acetate	10^6	2.6	Sodium nitrate	–	5.2
Lead carbonate (15°C)	10^8	18.6	Sodium oleate	4×10^8	2.75
Lead chloride	10^6	4.2	Sodium perchlorate	6×10^7	5.4
Lead monoxide (15°C)	10^8	25.9	Sucrose (mean)	3×10^8	3.32
Lead nitrate	6×10^7	37.7	Sulfur (mean)	–	4.0
Lead oleate	4×10^8	3.27	Thallium chloride	10^6	46.9
Lead sulfate	10^6	14.3	p-Toluidine	4×10^8	3.0
Lead sulfide (15°)	10^6	17.9	Tourmaline ⊥ optic axis	10^4	7.10
Malachite (mean)	10^{12}	7.2			
Mercuric chloride	10^6	3.2	Tourmaline ∥ optic axis	10^4	6.3
Mercurous chloride	10^6	9.4			
Naphthalene	4×10^8	2.52	Urea	4×10^8	3.5
			Zircon ⊥, ∥	10^8	12

Dielectric Constants of Ceramics

Material	Dielectric constant, 10^6 Hz	Dielectric strength V/mil	Volume resistivity $\Omega \cdot$ cm (23°C)	Loss factor[a]
Alumina	4.5–8.4	40–160	10^{11}–10^{14}	0.0002–0.01
Corderite	4.5–5.4	40–250	10^{12}–10^{14}	0.004–0.012
Forsterite	6.2	240	10^{14}	0.0004
Porcelain (dry process)	6.0–8.0	40–240	10^{12}–10^{14}	0.0003–0.02
Porcelain (wet process)	6.0–7.0	90–400	10^{12}–10^{14}	0.006–0.01
Porcelain, zircon	7.1–10.5	250–400	10^{13}–10^{15}	0.0002–0.008
Steatite	5.5–7.5	200–400	10^{13}–10^{15}	0.0002–0.004
Titanates (Ba, Sr, Ca, Mg, and Pb)	15–12.000	50–300	10^8–10^{15}	0.0001–0.02
Titanium dioxide	14–110	100–210	10^{13}–10^{18}	0.0002–0.005

[a] Power factor × dielectric constant equals loss factor.

Dielectric Constants of Glasses

Type	Dielectric constant at 100 MHz (20°C)	Volume resistivity (350°C M $\Omega \cdot$ cm)	Loss factor[a]
Corning 0010	6.32	10	0.015
Corning 0080	6.75	0.13	0.058
Corning 0120	6.65	100	0.012
Pyrex 1710	6.00	2,500	0.025
Pyrex 3320	4.71	–	0.019
Pyrex 7040	4.65	80	0.013
Pyrex 7050	4.77	16	0.017
Pyrex 7052	5.07	25	0.019
Pyrex 7060	4.70	13	0.018
Pyrex 7070	4.00	1,300	0.0048
Vycor 7230	3.83	–	0.0061
Pyrex 7720	4.50	16	0.014
Pyrex 7740	5.00	4	0.040
Pyrex 7750	4.28	50	0.011
Pyrex 7760	4.50	50	0.0081

Properties of Materials

Dielectric Constants of Glasses (*continued*)

Type	Dielectric constant at 100 MHz (20°C)	Volume resistivity (350°C M $\Omega \cdot$ cm)	Loss factor[a]
Vycor 7900	3.9	130	0.0023
Vycor 7910	3.8	1,600	0.00091
Vycor 7911	3.8	4,000	0.00072
Corning 8870	9.5	5,000	0.0085
G.E. Clear (silica glass)	3.81	4,000–30,000	0.00038
Quartz (fused)	3.75–4.1 (1 MHz)	–	0.0002 (1 MHz)

[a] Power factor × dielectric constant equals loss factor.

Properties of Semiconductors

Semiconducting Properties of Selected Materials

Substance	Minimum energy gap, eV R.T.	0 K	$\frac{dE_g}{dT} \times 10^4$, eV/°C	$\frac{dE_g}{dP} \times 10^6$, eV·cm²/kg	Density of states electron effective mass m_{d_n} (m_o)	Electron mobility μ_n, cm²/V·s	$-x$	Density of states hole effective mass m_{d_p} (m_o)	Hole mobility μ_p, cm²/V·s	$-x$
Si	1.107	1.153	−2.3	−2.0	1.1	1,900	2.6	0.56	500	2.3
Ge	0.67	0.744	−3.7	±7.3	0.55	3,800	1.66	0.3	1,820	2.33
αSn	0.08	0.094	−0.5		0.02	2,500	1.65	0.3	2,400	2.0
Te	0.33				0.68	1,100		0.19	560	
III–V Compounds										
AlAs	2.2	2.3				1,200			420	
AlSb	1.6	1.7	−3.5	−1.6	0.09	200	1.5	0.4	500	1.8
GaP	2.24	2.40	−5.4	−1.7	0.35	300	1.5	0.5	150	1.5
GaAs	1.35	1.53	−5.0	+9.4	0.068	9,000	1.0	0.5	500	2.1
GaSb	0.67	0.78	−3.5	+12	0.050	5,000	2.0	0.23	1,400	0.9
InP	1.27	1.41	−4.6	+4.6	0.067	5,000	2.0		200	2.4
InAs	0.36	0.43	−2.8	+8	0.022	33,000	1.2	0.41	460	2.3
InSb	0.165	0.23	−2.8	+15	0.014	78,000	1.6	0.4	750	2.1
II–VI Compounds										
ZnO	3.2		−9.5	+0.6	0.38	180	1.5			
ZnS	3.54		−5.3	+5.7		180			5(400°C)	
ZnSe	2.58	2.80	−7.2	+6		540			28	
ZnTe	2.26			+6		340			100	
CdO	2.5 ± 0.1		−6		0.1	120				
CdS	2.42		−5	+3.3	0.165	400		0.8		
CdSe	1.74	1.85	−4.6		0.13	650	1.0	0.6		
CdTe	1.44	1.56	−4.1	+8	0.14	1,200		0.35	50	
HgSe	0.30				0.030	20,000	2.0			
HgTe	0.15		−1		0.017	25,000		0.5	350	
Halite Structure Compounds										
PbS	0.37	0.28	+4		0.16	800		0.1	1,000	2.2
PbSe	0.26	0.16	+4		0.3	1,500		0.34	1,500	2.2
PbTe	0.25	0.19	+4	−7	0.21	1,600		0.14	750	2.2
Others										
ZnSb	0.50	0.56			0.15	10				1.5
CdSb	0.45	0.57	−5.4		0.15	300			2,000	1.5
Bi$_2$S$_3$	1.3					200			1,100	
Bi$_2$Se$_3$	0.27					600			675	
Bi$_2$Te$_3$	0.13		−0.95		0.58	1,200	1.68	1.07	510	1.95

Semiconducting Properties of Selected Materials (*continued*)

Substance	Minimum energy gap, eV R.T.	0 K	$\frac{dE_g}{dT}$ ×10^4, eV/°C	$\frac{dE_g}{dP}$ ×10^6, eV·cm^2/kg	Density of states electron effective mass m_{d_n} (m_o)	Electron mobility and temperature dependence μ_n, cm^2/V·s	$-x$	Density of states hole effective mass m_{d_p}, (m_o)	Hole mobility and temperature dependence μ_p, cm^2/V·s	$-x$
Mg$_2$Si		0.77	−6.4		0.46	400	2.5		70	
Mg$_2$Ge		0.74	−9			280	2		110	
Mg$_2$Sn	0.21	0.33	−3.5		0.37	320			260	
Mg$_3$Sb$_2$		0.32				20			82	
Zn$_3$As$_2$	0.93					10	1.1		10	
Cd$_3$As$_2$	0.55				0.046	100,000	0.88			
GaSe	2.05		3.8						20	
GaTe	1.66	1.80	−3.6			14	−5			
InSe	1.8					9000				
TlSe	0.57		−3.9		0.3	30		0.6	20	1.5
CdSnAs$_2$	0.23				0.05	25,000	1.7			
Ga$_2$Te$_3$	1.1	1.55	−4.8							
α-In$_2$Te$_3$	1.1	1.2			0.7				50	1.1
β-In$_2$Te$_3$	1.0								5	
Hg$_5$In$_2$Te$_8$	0.5								11,000	
SnO$_2$									78	

Band Properties of Semiconductors

Part A. Data on Valence Bands of Semiconductors (Room Temperature)

Substance	Band curvature effective mass (expressed as fraction of free electron mass)			Energy separation of split-off band, eV	Measured (light) hole mobility cm^2/V·s
	Heavy holes	Light holes	Split-off band holes		
Semiconductors with Valence Band Maximum at the Center of the Brillouin Zone ("Γ")					
Si	0.52	0.16	0.25	0.044	500
Ge	0.34	0.043	0.08	0.3	1,820
Sn	0.3				2,400
AlAs					
AlSb	0.4			0.7	550
GaP				0.13	100
GaAs	0.8	0.12	0.20	0.34	400
GaSb	0.23	0.06		0.7	1,400
InP				0.21	150
InAs	0.41	0.025	0.083	0.43	460
InSb	0.4	0.015		0.85	750
CbTe	0.35				50
HgTe	0.5				350

| | Semiconductors with Multiple Valence Band Maxima | | | | |
| | | Band curvature effective masses | | | Measured (light) |
Substance	Number of equivalent valleys and direction	Longitudinal m_L	Transverse m_T	Anisotropy, $K = m_L/m_T$	hole mobility, cm^2/V·s
PbSe	4 "L" [111]	0.095	0.047	2.0	1,500
PbTe	4 "L" [111]	0.27	0.02	10	750
Bi$_2$Te$_3$	6	0.207	∼0.045	4.5	515

Part B. Data on conduction Bands of Semiconductors (Room Temperature Data)

Single Valley Semiconductors

Substance	Energy gap, eV	Effective mass (m_o)	Mobility cm²/V · s	Comments
GaAs	1.35	0.067	8,500	3 (or 6?) equivalent [100] valleys 0.36 eV above this maximum with a mobility of ~50
InP	1.27	0.067	5,000	3 (or 6?) equivalent [100] valleys 0.4 eV above this minimum
InAs	0.36	0.022	33,000	Equivalent valleys ~1.0 eV above this minimum
InSb	0.165	0.014	78,000	
CdTe	1.44	0.11	1,000	4 (or 8?) equivalent [111] valleys 0.51 eV above this minimum

Multivalley Semiconductors

Substance	Energy gap	Number of equivalent valleys and direction	Band curvature effective mass Longitudinal m_L	Transverse m_T	Anisotropy $K = m_L/m_T$
Si	1.107	6 in [100] Δ	0.90	0.192	4.7
Ge	0.67	4 in [111] at L	1.588	0.0815	19.5
GaSb	0.67	as Ge (?)	~1.0	~0.2	~5
PbSe	0.26	4 in [111] at L	0.085	0.05	1.7
PbTe	0.25	4 in [111] at L	0.21	0.029	5.5
Bi₂Te₃	0.13	6			~0.05

Resistivity of Semiconducting Minerals

Mineral	$\rho, \Omega \cdot m$	Mineral	$\rho, \Omega \cdot m$
Diamond (C)	2.7	Gersdorffite, NiAsS	1 to 160 ×10⁻⁶
Sulfides		Glaucodote, (Co, Fe)AsS	5 to 100 ×10⁻⁶
Argentite, Ag₂S	1.5 to 2.0 ×10⁻³	Antimonide	
Bismuthinite, Bi₂S₃	3 to 570	Dyscrasite, Ag₃Sb	0.12 to 1.2 ×10⁻⁶
Bornite, Fe₂S₃ · nCu₂S	1.6 to 6000 ×10⁻⁶	Arsenides	
Chalcocite, Cu₂S	80 to 100 ×10⁻⁶	Allemonite, SbAs₂	70 to 60,000
Chalcopyrite, Fe₂S₃ · Cu₂S	150 to 9000 ×10⁻⁶	Lollingite, FeAs₂	2 to 270 ×10⁻⁶
Covellite, CuS	0.30 to 83 ×10⁻⁶	Nicollite, NiAs	0.1 to 2 ×10⁻⁶
Galena, PbS	6.8 ×10⁻⁶ to 9.0 ×10⁻²	Skutterudite, CoAs₃	1 to 400 ×10⁻⁶
Haverite, MnS₂	10 to 20	Smaltite, CoAs₂	1 to 12 ×10⁻⁶
Marcasite, FeS₂	1 to 150 ×10⁻³	Tellurides	
Metacinnabarite, 4HgS	2 ×10⁻⁶ to 1 ×10⁻³	Altaite, PbTe	20 to 200 ×10⁻⁶
Millerite, NiS	2 to 4 ×10⁻⁷	Calavarite, AuTe₂	6 to 12 ×10⁻⁶
Molybdenite, MoS₂	0.12 to 7.5	Coloradoite, HgTe	4 to 100 ×10⁻⁶
Pentlandite, (Fe, Ni)₉S₈	1 to 11 ×10⁻⁶	Hessite, Ag₂Te	4 to 100 ×10⁻⁶
Pyrrhotite, Fe₇S₈	2 to 160 ×10⁻⁶	Nagyagite, Pb₆Au(S, Te)₁₄	20 to 80 ×10⁻⁶
Pyrite, FeS₂	1.2 to 600 ×10⁻³	Sylvanite, AgAuTe₄	4 to 20 ×10⁻⁶
Sphalerite, ZnS	2.7 ×10⁻³ to 1.2 ×10⁴	Oxides	
Antimony-sulfur compounds		Braunite, Mn₂O₃	0.16 to 1.0
Berthierite, FeSb₂S₄	0.0083 to 2.0	Cassiterite, SnO₂	4.5 ×10⁻⁴ to 10,000
Boulangerite, Pb₅Sb₄S₁₁	2 ×10³ to 4 ×10⁴	Cuprite, Cu₂O	10 to 50
Cylindrite, Pb₃Sn₄Sb₂S₁₄	2.5 to 60	Hollandite, (Ba, Na, K)Mn₈O₁₆	2 to 100 ×10⁻³
Franckeite, Pb₅Sn₃Sb₂S₁₄	1.2 to 4	Ilmenite, FeTiO₃	0.001 to 4
Hauchecornite, Ni₉(Bi, Sb)₂S₈	1 to 83 ×10⁻⁶	Magnetite, Fe₃O₄	52 ×10⁻⁶
Jamesonite, Pb₄FeSb₆S₁₄	0.020 to 0.15	Manganite, MnO · OH	0.018 to 0.5
Tetrahedrite, Cu₃SbS₃	0.30 to 30,000	Melaconite, CuO	6000

Resistivity of Semiconducting Minerals (*continued*)

Mineral	$\rho, \Omega \cdot m$	Mineral	$\rho, \Omega \cdot m$
Arsenic-sulfur compounds		Psilomelane, $KMnO \cdot MnO_2 \cdot nH_2O$	0.04 to 6000
Arsenopyrite, FeAsS	20 to 300 $\times 10^{-6}$	Pyrolusite, MnO_2	0.007 to 30
Cobaltite, CoAsS	6.5 to 130 $\times 10^{-3}$	Rutile, TiO_2	29 to 910
Enargite, Cu_3AsS_4	0.2 to 40 $\times 10^{-3}$	Uraninite, UO	1.5 to 200

Source: Carmichael, R.S., ed. 1982. *Handbook of Physical Properties of Rocks*, Vol. I. CRC Press, Boca Raton, FL.

Properties of Magnetic Alloys

Name	Composition,[a] weight percent					Remanence, B_r, G	Coercive force, H_e, O	Maximum energy product, $(BH)_{max}$, G–O $\times 10^{-6}$
	Al	Ni	Co	Cu	Other			
U.S.								
Alnico I	12	20–22	5			7,100	440	1.4
Alnico II	10	17	12.5	6		7,200	540	1.6
Alnico III	12	24–26		3		6,900	470	1.35
Alnico IV	12	27–28	5			5,500	700	1.3
Alnico V[b]	8	14	24	3		12,500	600	5.0
Alnico V DG[b]	8	14	24	3		13,100	640	6.0
Alnico VI[b]	8	15	24	3	1.25 Ti	10,500	750	3.75
Alnico VII[b]	8.5	18	24	3	5 Ti	7,200	1,050	2.75
Alnico XII	6	18	35		8 Ti	5,800	950	1.6
Chromium steel					1 Mn 0.9 C 3.5 Cr	10,000 9,700	50 65	0.2 0.3
Cobalt steel			17		0.9 C 0.3 Mn 2.5 Cr 8 W 0.75 C	9,500	150	0.65
Cunico		21	29	50		3,400	660	0.80
Cunife		20		60		5,400	550	1.5
Ferroxdur 1		$BaFe_{12}O_{19}$				2,200	1,800	1.0
Ferroxdur 2		$BaF_{12}O_{19}$ (oriented)				3,840	2,000	3.5
Platinum-cobalt			23		77 Pt	6,000	4,300	7.5
Remalloy			12		17 Mo	10,500	250	1.1
Silmanol	4.4				86.6 Ag 8.8 Mn	550	6,000	0.075
Tungsten steel					5 W 0.3 Mn 0.7 C	10,000	70	0.32
Vicalloy I			52		10 V	8,800	300	1.0
Vicalloy II (wire)			52		14 V	10,000	510	3.5
Germany								
Alni 90	12	21				8,000	350	1.2
Alni 120	13	27				6,000	570	1.2
Alnico 130	12	23	5			6,300	620	1.4
Alnico 160	11	24	12	4		6,200	700	1.6
Alnico 190	12	21	15	4		7,000	700	1.8
Alnico 250	8	19	23	4	6 Ti	6,500	1,000	2.2
Alnico 400[b]	9	15	23	4		12,000	650	4.8
Alnico 580[b] (semicolumnar)	9	15	23	4		13,000	700	6.0

Properties of Magnetic Alloys (continued)

Name	Composition,[a] weight percent					Remanence, B_r, G	Coercive force, H_e, O	Maximum energy product, $(BH)_{max}$, G–O $\times 10^{-6}$
	Al	Ni	Co	Cu	Other			
Oerstit 800	9	18	19	4	4 Ti	6,600	750	1.95
Great Britain								
Alcomax I	7.5	11	25	3	1.5 Ti	12,000	475	3.5
Alcomax II	8	11.5	24	4.5		12,400	575	4.7
Alcomax IISC (semicolumnar)	8	11	22	4.5		12,800	600	5.15
Alcomax III	8	13.5	24	3	0.8 Nb	12,500	670	5.10
Alcomax IIISC (semicolumnar)	8	13.5	24	3	0.8 Nb	13,000	700	5.80
Alcomax IV	8	13.5	24	3	2.5 Nb	11,200	750	4.30
Alcomax IVSC (semicolumnar)	8	13.5	24	3	2.5 Nb	11,700	780	5.10
Alni, high B_r	13	24		3.5		6,200	480	1.25
Alni, normal						5,600	580	1.25
Alni, high H_e	12	32			0–0.5 Ti	5,000	680	1.25
Alnico, high B_r	10	17	12	6		8,000	500	1.70
Alnico, normal						7,250	560	1.70
Alnico, high H_e	10	20	13.5	6	0.25 Ti	6,600	620	1.70
Columax (columnar)	similar to Alcomax III or IV					13,000–14,000	700–800	7.0–8.5
Hycomax	9	21	20	1.6		9,500	830	3.3

[a]Remainder of unlisted composition is either iron or iron plus trace impurities.
[b]Cast anisotropic. Unmarked are cast isotropic.

Properties of Magnetic Alloys: High-Permeability Magnetic Alloys

Name	Composition[a], weight percent	Sp. gr., g/cm^3	Tensile Strength		Remark	Use
			kg/mm^{2b}	Form		
Silicon iron AISI M 15	Si 4	7.68–7.64	44.3	—	Annealed 4 h 802–1093°C	Low core losses
Silicon iron AISI M 8	Si 3	7.68–7.64	44.2	Grain oriented	Annealed 4 h 802–1204°C	
45 Permalloy	Ni 45; Mn 0.3	8.17	—	—	—	Audio transformer, coils, relays
Monimax	Ni 47; Mo 3	8.27	—	—	—	High-frequency coils
4–79 Permalloy	Ni 79; Mo 4; Mn 0.3	8.74	55.4	—	H$_2$ annealed 1121°C	Audio coils, transformers, magnetic shields
Sinimax	Ni 43; Si 3	7.70	—	—	—	High-frequency coils
Nu-metal	Ni 75; Cr 2; Cu 5	8.58	44.8	—	H$_2$ annealed 1221°C	Audio coils, magnetic shields, transformers
Supermalloy	Ni 79; Mo 5; Mn 0.3	8.77	—	—	—	Pulse transformers, magnetic amplifiers, coils
2-V Permendur	Co 40; V 2	8.15	46.3	—	—	DC electromagnets, pole tips

[a]Iron is additional alloying metal.
[b]kg/mm^2 $\times$ 1422.33 = lbs/in^2.

Properties of Magnetic Alloys: Cast Permanent Magnetic Alloys

Alloy name (country of manufacture[a])	Composition[b] weight percent	Sp.gr., g/cm^3	Thermal expansion Cm×10^{-6}/cm×°C	Thermal expansion Between °C	Tensile strength kg/mm^{2c}	Form	Remark[d]	Use
Alnico I (US)	Al 12; Ni 20–22; Co 5	6.9	12.6	20–300	2.9	Cast	i	Permanent magnets
Alnico II (US)	Al 10; Ni 17; Cu 6; Co 12.5	7.1	12.4	20–300	2.1 / 45.7	Cast / Sintered	i	Temperature controls magnetic toys and novelties
Alnico III (US)	Al 12; Ni 24–26; Cu 3	6.9	12	20–300	8.5	Cast	i	Tractor magnetos
Alnico IV (US)	Al 12; Ni 27–28; Co 5	7.0	13.1	20–300	6.3 / 42.1	Cast / Sintered	i	Application requiring high coercive force
Alnico V (US)	Al 8; Ni 14; Co 24; Cu 3	7.3	11.3	—	3.8 / 35	Cast / Sintered	a	Application requiring high energy
Alnico V DG (US)	Al 8; Ni 14; Co 24; Cu 3	7.3	11.3	—	—	—	a, c	
Alnico VI (US)	Al 8; Ni 15; Co 24; Cu 3; Ti 1.25	7.3	11.4	—	16.1	Cast	a	Application requiring high energy
Alnico VII (US)	Al 8.5; Ni 18; Cu 3; Co 24; Ti 5	7.17	11.4	—	—	—	a	
Alnico XII (US)	Al 6; Ni 18; Co 35; Ti 8	7.2	11	20–300	—	—	—	Permanent magnets
Comol (US)	Co 12; Mo 17	8.16	9.3	20–300	88.6	—	—	Permanent magnets
Cunife (US)	Cu 60; Ni 20	8.52	—	—	70.3	—	—	Permanent magnets
Cunico (US)	Cu 50; Ni 21	8.31	—	—	70.3	—	—	Permanent magnets
Barium ferrite Feroxdur (US)	BaFe$_{12}$O$_{19}$	4.7	10	—	70.3	—	—	Ceramics
Alcomax I (GB)	Al 7.5; Ni 11; Co 25; Cu 3; Ti 1.5	—	—	—	—	—	a	Permanent magnets
Alcomax II (GB)	Al 8; Ni 11.5; Co 24; Cu 4.5	—	—	—	—	—	a	Permanent magnets
Alcomax II SC (GB)	Al 8; Ni 11; Co 22; Cu 4.5	7.3	—	—	—	—	a, sc	
Alcomax III (GB)	Al 8; Ni 13.5; Co 24; Nb 0.8	7.3	—	—	—	—	a	Magnets for motors, loudspeakers
Alcomax IV (GB)	Al 8; Ni 13.5; Cu 3; Co 24; Nb 2.5	—	—	—	—	—	—	Magnets for cycle-dynamos
Columax (GB)	Similar to Alcomax III or IV	—	—	—	—	—	a, sc	Permanent magnets, heat treatable
Hycomax (GB)	Al 9; Ni 21; Co 20; Cu 1.6	—	—	—	—	—	a	Permanent magnets
Alnico (high H_c) (GB)	Al 10; Ni 20; Co 13.5; Cu 6; Ti 0.25	7.3	—	—	—	—	i	
Alnico (high B_r) (GB)	Al 10; Ni 17; Co 12; Cu 6	7.3	—	—	—	—	i	
Alni (high H_c) (GB)	Al 12; Ni 32; Ti 0–0.5	6.9	—	—	—	—	i	
Alni (high B_r) (GB)	Al 13; Ni 24; Cu 3.5	—	—	—	—	—	i	
Alnico 580 (Ger)	Al 9; Ni 15; Co 23; Cu 4	—	—	—	—	—	i	
Alnico 400 (Ger)	Al 9; Ni 15; Co 23; Cu 4	—	—	—	—	—	a	

Properties of Magnetic Alloys: Cast Permanent Magnetic Alloys (*continued*)

Alloy name (country of manufacture[a]	Composition[b] weight percent	Sp.gr., g/cm³	Thermal expansion Cm×10⁻⁶ / cm×°C	Between °C	Tensile strength kg/mm²[c]	Form	Remark[d]	Use
Oerstit 800 (Ger)	Al 9; Ni 18; Co 19; Cu 4; Ti 4	—	—	—	—	—	i	Permanent magnets
Alnico 250 (Ger)	Al 8; Ni 19; Co 23; Cu 4; Ti 6	—	—	—	—	—	i	
Alnico 190 (Ger)	Al 12; Ni 21; Cu 4; Co 15	—	—	—	—	—	i	
Alnico 160 (Austria)	Al 11; Ni 24; Co 12; Cu 4	—	—	—	—	—	i	Permanent magnets, sintered
Alnico 130 (Ger)	Al 12; Ni 23; Co 5	—	—	—	—	—	i	
Alni 120 (Ger)	Al 13; Ni 27	—	—	—	—	—	i	
Alni 90 (Ger)	Al 12; Ni 21	—	—	—	—	—	i	

[a] US, United States; GB, Great Britain; Ger, Germany.
[b] Iron is the additional alloying metal for each of the magnets listed.
[c] kg/mm² × 1422.33 = lb/in².
[d] i. = isotropic; a. = anisotropic; c. = columnar; sc. = semicolumnar.

Properties of Antiferromagnetic Compounds

Compound	Crystal Symmetry	θ_N[a] K	θ_P[b] K	$(P_A)_{eff}$[c] μ_B	P_A[d] μ_B
$CoCl_2$	Rhombohedral	25	−38.1	5.18	3.1 ± 0.6
CoF_2	Tetragonal	38	50	5.15	3.0
CoO	Tetragonal	291	330	5.1	3.8
Cr	Cubic	475			
Cr_2O_3	Rhombohedral	307	485	3.73	3.0
CrSb	Hexagonal	723	550	4.92	2.7
$CuBr_2$	Monoclinic	189	246	1.9	
$CuCl_2 \cdot 2H_2O$	Orthorhombic	4.3	4–5	1.9	
$CuCl_2$	Monoclinic	~70	109	2.08	
$FeCl_2$	Hexagonal	24	−48	5.38	4.4 ± 0.7
FeF_2	Tetragonal	79–90	117	5.56	4.64
FeO	Rhombohedral	198	507	7.06	3.32
α-Fe_2O_3	Rhombohedral	953	2940	6.4	5.0
α-Mn	Cubic	95			
$MnBr_2 \cdot 4H_2O$	Monoclinic	2.1	$\{2.5, 1.3\}$	5.93	
$MnCl_2 \cdot 4H_2O$	Monoclinic	1.66	1.8	5.94	
MnF_2	Tetragonal	72–75	113.2	5.71	5
MnO	Rhombohedral	122	610	5.95	5.0
β-MnS	Cubic	160	982	5.82	5.0
MnSe	Cubic	~173	361	5.67	
MnTe	Hexagonal	310–323	690	6.07	5.0
$NiCl_2$	Hexagonal	50	−68	3.32	
NiF_2	Tetragonal	78.5–83	115.6	3.5	2.0

Properties of Antiferromagnetic Compounds (*continued*)

Compound	Crystal Symmetry	θ_N[a] K	θ_P[b] K	$(P_A)_{\text{eff}}$[c] μ_B	P_A[d] μ_B
NiO	Rhombohedral	533–650	~2000	4.6	2.0
TiCl$_3$		100			
V$_2$O$_3$		170			

[a] θ_N = Néel temperature, determined from susceptibility maxima or from the disappearance of magnetic scattering.
[b] θ_P = a constant in the Curie-Weiss law written in the form $\chi_A = C_A/(T + \theta_P)$, which is valid for antiferromagnetic material for $T > \theta_N$.
[c] $(P_A)_{\text{eff}}$ = effective moment per atom, derived from the atomic Curie constant $C_A = (P_A)^2_{\text{eff}}(N^2/3R)$ and expressed in units of the Bohr magneton, $\mu_B = 0.9273 \times 10^{-20}$ erg G^{-1}.
[d] P_A = magnetic moment per atom, obtained from neutron diffraction measurements in the ordered state.

Properties of Magnetic Alloys: Saturation Constants and Curie Points of Ferromagnetic Elements

Element	σ_s[a] (20°C)	M_s[b] (20°C)	σ_s (0 K)	n_B[c]	Curie point, °C
Fe	218.0	1,714	221.9	2.219	770
Co	161	1,422	162.5	1.715	1,131
Ni	54.39	484.1	57.50	0.604	358
Gd	0	0	253.5	7.12	16

[a] σ_s = saturation magnetic moment/gram.
[b] M_s = saturation magnetic moment/cm^3, in cgs units.
[c] n_B = magnetic moment per atom in Bohr magnetons.
(*Source:* 1963. *American Institute of Physics Handbook*. McGraw–Hill, New York.)

Magnetic Properties of Transformer Steels

Ordinary Transformer Steel

B(Gauss)	H(Oersted)	Permeability = B/H
2,000	0.60	3,340
4,000	0.87	4,600
6,000	1.10	5,450
8,000	1.48	5,400
10,000	2.28	4,380
12,000	3.85	3,120
14,000	10.9	1,280
16,000	43.0	372
18,000	149	121

High Silicon Transformer Steels

B	H	Permeability
2,000	0.50	4,000
4,000	0.70	5,720
6,000	0.90	6,670
8,000	1.28	6,250
10,000	1.99	5,020
12,000	3.60	3,340
14,000	9.80	1,430
16,000	47.4	338
18,000	165	109

Initial Permeability of High Purity Iron for Various Temperatures

L. Alberts and B.J. Shepstone

Temperature °C	Permeability (Gauss/oersted)
0	920
200	1,040
400	1,440
600	2,550
700	3,900
770	12,580

Saturation Constants for Magnetic Substances

Substance	Field Intensity (For Saturation)	Induced Magnetization	Substance	Field Intensity (For Saturation)	Induced Magnetization
Cobalt	9,000	1,300	Nickel, hard	8,000	400
Iron, wrought	2,000	1,700	annealed	7,000	515
cast	4,000	1,200	Vicker's steel	15,000	1,600
Manganese steel	7,000	200			

Magnetic Materials: High-Permeability Materials

| Material | Form | Approximate % Composition | | | | | Typical Heat Treatment, °C | Permeability at $B = 20$, G | Maximum Permeability | Saturation flux density B, G | Hysteresis[a] loss, W_h, ergs/cm² | Coercive[a] force H_a, O | Resistivity, $\mu \cdot \Omega$cm | Density, g/cm³ |
		Fe	Ni	Co	Mo	Other								
Cold rolled steel	Sheet	98.5	—	—	—	—	950 Anneal	180	2,000	21,000	—	1.8	10	7.88
Iron	Sheet	99.91	—	—	—	—	950 Anneal	200	5,000	21,500	5,000	1.0	10	7.88
Purified iron	Sheet	99.95	—	—	—	—	1480 H₂ + 880	5,000	180,000	21,500	300	0.05	10	7.88
4% Silicon-iron	Sheet	96	—	—	—	4 Si	800 Anneal	500	7,000	19,700	3,500	0.5	60	7.65
Grain oriented[b]	Sheet	97	—	—	—	3 Si	800 Anneal	1,500	30,000	20,000	—	0.15	47	7.67
45 Permalloy	Sheet	54.7	45	—	—	0.3 Mn	1050 Anneal	2,500	25,000	16,000	1,200	0.3	45	8.17
45 permalloy[c]	Sheet	54.7	45	—	—	0.3 Mn	1200 H₂ Anneal	4,000	50,000	16,000	—	0.07	45	8.17
Hipernik	Sheet	50	50	—	—	—	1200 H₂ Anneal	4,500	70,000	16,000	220	0.05	50	8.25
Monimax	Sheet	—	—	—	—	—	1125 H₂ Anneal	2,000	35,000	15,000	—	0.1	80	8.27
Sinimax	Sheet	—	—	—	—	—	1125 H₂ Anneal	3,000	35,000	11,000	—	—	90	—
78 Permalloy	Sheet	21.2	78.5	—	—	0.3 Mn	1050 + 600 Q[d]	8,000	100,000	10,700	200	0.05	16	8.60
4–79 Permalloy	Sheet	16.7	79	—	4	0.3 Mn	1100 + Q	20,000	100,000	8,700	200	0.05	55	8.72
Mu metal	Sheet	18	75	—	—	2 Cr, 5 Cu	1175 H₂	20,000	100,000	6,500	—	0.05	62	8.58
Supermalloy	Sheet	15.7	79	—	5	0.3 Mn	1300 H₂ + Q	100,000	800,000	8,000	12,000	0.002	60	8.77
Permendur	Sheet	49.7	—	50	—	0.3 Mn	800 Anneal	800	5,000	24,500	6,000	2.0	7	8.3
2 V Permendur	Sheet	49	—	49	—	2 V	800 Anneal	800	4,500	24,000	—	2.0	26	8.2
Hiperco	Sheet	64	—	34	—	Cr	850 Anneal	650	10,000	24,200	—	1.0	25	8.0
2–81 Permalloy	Insulated powder	17	81	—	2	—	650 Anneal	125	130	8,000	—	<1.0	10⁶	7.8
Carbonyl iron	Insulated powder	99.9	—	—	—	—	—	55	132	—	—	—	—	7.86
Ferroxcube III	Sintered powder	\multicolumn{5}{l}{MnFe₂O₄ + ZnFe₂O₄}		—	1,000	1,500	2,500	—	0.1	10⁸	5.0

[a] At saturation.
[b] Properties in direction of rolling.
[c] Similar properties for Nicaloi, 4750 alloy, Carpenter 49, Armco 48.
[d] Q, quench or controlled cooling.

Magnetic Materials: Permanent Magnet Alloys

Material	% composition (remainder Fe)	Heat treatment[a] (temperature, °C)	Magnetizing force H_{max}, O	Coercive force H_c, O	Residual induction B_r, G	Energy product $BH_{max}.\times10^{-6}$	Method of fabrication[b]	Mechanical properties[c]	Weight lb/In.^{3}
Carbon steel	1 Mn, 0.9 C	Q 800	300	50	10,000	0.20	HR, M, P	H, S	0.280
Tungsten steel	5 W, 0.3 Mn, 0.7 C	Q 850	300	70	10,300	0.32	HR, M, P	H, S	0.292
Chromium steel	3.5 Cr, 0.9 C, 0.3 Mn	Q 830	300	65	9,700	0.30	HR, M, P	H, S	0.280
17% Cobalt steel	17 Co, 0.75 C, 2.5 Cr, 8 W	—	1,000	150	9,500	0.65	HR, M, P	H, S	—
36% Cobalt steel	36 Co, 0.7 C, 4 Cr, 5 W	Q 950	1,000	240	9,500	0.97	HR, M, P	H, S	0.296
Remalloy or Comol	17 Mo, 12 Co	—	1,000	250	10,500	1.1	HR, M, P	H	0.295
Alnico I	12 Al, 20 Ni, 5 Co	Q 1200, B 700	1,000	440	7,200	1.4	C, G	H, B	0.249
Alnico II	10 Al, 17 Ni, 2.5 Co, 6 Cu	A 1200, B 700	2,000	550	7,200	1.6	C, G	H, B	0.256
Alnico II (sintered)	10 Al, 17 Ni, 2.5 Co, 6 Cu	A 1200, B 600	2,000	520	6,900	1.4	Sn, G	H	0.249
Alnico IV	12 Al, 28 Ni, 5 Co	A 1300	3,000	700	5,500	1.3	Sn, C, G	H	0.253
Alnico V	8 Al, 14 Ni, 24 Co, 3 Cu	Q 1200, B 650	2,000	550	12,500	4.5	C, G	H, B	0.264
Alnico VI	8 Al, 15 Ni, 24 Co, 3 Cu, 1 Ti	AF 1300, B 600	3,000	750	10,000	3.5	C, G	H, B	0.268
Alnico XII	6 Al, 18 Ni, 35 Co, 8 Ti	—	3,000	950	5,800	1.5	C, G	H, B	0.26
Vicalloy I	52 Co, 10 V	B 600	1,000	300	8,800	1.0	C, CR, M, P	D	0.295
Vicalloy II (wire)	52 Co, 14 V	CW + B 600	2,000	510	10,000	3.5	C, CR, M, P	D	0.292
Cunife (wire)	60 Cu, 20 Ni	CW + B 600	2,400	550	5,400	1.5	C, CR, M, P	D, M	0.311
Cunico	50 Cu, 21 Ni, 29 Co	—	3,200	660	3,400	0.80	C, CR, M, P	D, M	0.300
Vectolite	30Fe$_2$O$_3$, 44Fe$_3$O$_4$, 26C$_2$O$_3$	—	3,000	1,000	1,600	0.60	Sn, G	W	0.113
Silmanal	86.8Ag, 8.8Mn, 4.4Al	—	20,000	6,000[d]	550	0.075	C, CR, M, P	D, M	0.325
Platinum-cobalt	77 Pt, 23 Co	Q 1200, B 650	15,000	3,600	5,900	6.5	C, CR, M	D	—
Hyflux	Fine powder	—	2,000	390	6,600	0.97	—	—	0.176

[a]Q, quenched in oil or water; A, air colled; B, baked; F, cooled in magnetic field; CW, cold worked.
[b]HR, hot rolled or froged; CR, cold rolled or drawn; M, machined; G, must be ground; P, punched; C, cast; Sn, sintered.
[c]H, hard; B, brittle; S, strong; D, ductile; M, malleable; W, weak.
[d]Value given is intrinsic H_c.

Properties of Materials

Resistance of Wires

The following table gives the approximate resistance of various metallic conductors. The values have been computed from the resistivities at 20°C, except as otherwise stated, and for the dimensions of wire indicated. Owing to differences in purity in the case of elements and of composition in alloys, the values can be considered only as approximations.

The following dimensions have been adopted in the computation.

B. & S. gauge	Diameter mm	mils 1 mil = 0.001 in	B. & S. gauge	Diameter mm	mils 1 mil = 0.001 in
10	2.588	101.9	26	0.4049	15.94
12	2.053	80.81	27	0.3606	14.20
14	1.628	64.08	28	0.3211	12.64
16	1.291	50.82	30	0.2546	10.03
18	1.024	40.30	32	0.2019	7.950
20	0.8118	31.96	34	0.1601	6.305
22	0.6438	25.35	36	0.1270	5.000
24	0.5106	20.10	40	0.07987	3.145

B. & S. No.	Ω/cm	Ω/ft	B. & S. No.	Ω/cm	Ω/ft	B. & S. No.	Ω/cm	Ω/ft
Advance[a] (0°C) $\varrho = 48. \times 10^{-6}$ Ω·cm			Aluminum $\varrho = 2.828 \times 10^{-6}$ Ω·cm			Brass $\varrho = 7.00 \times 10^{-6}$ Ω·cm		
10	0.000912	0.0278	10	0.0000538	0.00164	10	0.000133	0.00406
12	0.00145	0.0442	12	0.0000855	0.00260	12	0.000212	0.00645
14	0.00231	0.0703	14	0.000136	0.00414	14	0.000336	0.0103
16	0.00367	0.112	16	0.000216	0.00658	16	0.000535	0.0163
18	0.00583	0.178	18	0.000344	0.0105	18	0.000850	0.0259
20	0.00927	0.283	20	0.000546	0.0167	20	0.00135	0.0412
22	0.0147	0.449	22	0.000869	0.0265	22	0.00215	0.655
24	0.0234	0.715	24	0.00138	0.0421	24	0.00342	0.104
26	0.0373	1.14	26	0.00220	0.0669	26	0.00543	0.166
27	0.0470	1.43	27	0.00277	0.0844	27	0.00686	0.209
28	0.0593	1.81	28	0.00349	0.106	28	0.00864	0.263
30	0.0942	2.87	30	0.00555	0.169	30	0.0137	0.419
32	0.150	4.57	32	0.00883	0.269	32	0.0219	0.666
34	0.238	7.26	34	0.0140	0.428	34	0.0348	1.06
36	0.379	11.5	36	0.0223	0.680	36	0.0552	1.68
40	0.958	29.2	40	0.0564	1.72	40	0.140	4.26
Climax $\varrho = 87 \times 10^{-6}$ Ω·cm			Constantan (0°C) $\varrho = 44.1 \times 10^{-6}$ Ω·cm			Copper, annealed $\varrho = 1.724 \times 10^{-6}$ Ω·cm		
10	0.00165	0.0504	10	0.000838	0.0255	10	0.0000328	0.000999
12	0.00263	0.0801	12	0.00133	0.0406	12	0.0000521	0.00159
14	0.00418	0.127	14	0.00212	0.0646	14	0.0000828	0.00253
16	0.00665	0.203	16	0.00337	0.103	16	0.000132	0.00401
18	0.0106	0.322	18	0.00536	0.163	18	0.000209	0.00638
20	0.0168	0.512	20	0.00852	0.260	20	0.000333	0.0102
22	0.0267	0.815	22	0.0135	0.413	22	0.000530	0.0161
24	0.0425	1.30	24	0.0215	0.657	24	0.000842	0.0257
26	0.0675	2.06	26	0.0342	1.04	26	0.00134	0.0408
27	0.0852	2.60	27	0.0432	1.32	27	0.00169	0.0515
28	0.107	3.27	28	0.0545	1.66	28	0.00213	0.0649
30	0.171	5.21	30	0.0866	2.64	30	0.00339	0.103
32	0.272	8.28	32	0.138	4.20	32	0.00538	0.164
34	0.432	13.2	34	0.219	6.67	34	0.00856	0.261
36	0.687	20.9	36	0.348	10.6	36	0.0136	0.415
40	1.74	52.9	40	0.880	26.8	40	0.0344	1.05

[a]Trademark.

Resistance of Wires (*continued*)

B. & S. No.	Ω/cm	Ω/ft	B. & S. No.	Ω/cm	Ω/ft	B. & S. No.	Ω/cm	Ω/ft
Eureka[a] (0°C) $\varrho = 47. \times 10^{-6}$ Ω·cm			Excello $\varrho = 92. \times 10^{-6}$ Ω·cm			German silver $\varrho = 33 \times 10^{-6}$ Ω·cm		
10	0.000893	0.0272	10	0.00175	0.0533	10	0.000627	0.0191
12	0.00142	0.0433	12	0.00278	0.0847	12	0.000997	0.304
14	0.00226	0.0688	14	0.00442	0.135	14	0.00159	0.0483
16	0.00359	0.109	16	0.00703	0.214	16	0.00252	0.0768
18	0.00571	0.174	18	0.0112	0.341	18	0.00401	0.122
20	0.00908	0.277	20	0.0178	0.542	20	0.00638	0.194
22	0.0144	0.440	22	0.0283	0.861	22	0.0101	0.309
24	0.0230	0.700	24	0.0449	1.37	24	0.0161	0.491
26	0.0365	1.11	26	0.0714	2.18	26	0.0256	0.781
27	0.0460	1.40	27	0.0901	2.75	27	0.0323	0.985
28	0.0580	1.77	28	0.114	3.46	28	0.0408	1.24
30	0.0923	2.81	30	0.181	5.51	30	0.0648	1.97
32	0.147	4.47	32	0.287	8.75	32	0.103	3.14
34	0.233	7.11	34	0.457	13.9	34	0.164	4.99
36	0.371	11.3	36	0.726	22.1	36	0.260	0.794
40	0.938	28.6	40	1.84	56.0	40	0.659	20.1
Gold $\varrho = 2.44 \times 10^{-6}$ Ω·cm			Iron $\varrho = 10 \times 10^{-6}$ Ω·cm			Lead $\varrho = 22 \times 10^{-6}$ Ω·cm		
10	0.0000464	0.00141	10	0.000190	0.00579	10	0.000418	0.0127
12	0.0000737	0.00225	12	0.000302	0.00921	12	0.000665	0.0203
14	0.000117	0.00357	14	0.000481	0.0146	14	0.00106	0.0322
16	0.000186	0.00568	16	0.000764	0.0233	16	0.00168	0.0512
18	0.000296	0.00904	18	0.00121	0.0370	18	0.00267	0.0815
20	0.000471	0.0144	20	0.00193	0.0589	20	0.00425	0.130
22	0.000750	0.0228	22	0.00307	0.0936	22	0.00676	0.206
24	0.00119	0.0363	24	0.00489	0.149	24	0.0107	0.328
26	0.00189	0.0577	26	0.00776	0.237	26	0.0171	0.521
27	0.00239	0.728	27	0.00979	0.299	27	0.0215	0.657
28	0.00301	0.0918	28	0.0123	0.376	28	0.0272	0.828
30	0.00479	0.146	30	0.0196	0.598	30	0.0432	1.32
32	0.00762	0.232	32	0.0312	0.952	32	0.0687	2.09
34	0.0121	0.369	34	0.0497	1.51	34	0.109	3.33
36	0.0193	0.587	36	0.789	2.41	36	0.174	5.29
40	0.0487	1.48	40	0.200	6.08	40	0.439	13.4
Magnesium $\varrho = 4.6 \times 10^{-6}$ Ω·cm			Manganin $\varrho = 44 \times 10^{-6}$ Ω·cm			Molybdenum $\varrho = 5.7 \times 10^{-6}$ Ω·cm		
10	0.0000874	0.00267	10	0.000836	0.0255	10	0.000108	0.00330
12	0.000139	0.00424	12	0.00133	0.0405	12	0.000172	0.00525
14	0.000221	0.00674	14	0.00211	0.0644	14	0.000274	0.00835
16	0.000351	0.0107	16	0.00336	0.102	16	0.000435	0.0133
18	0.000559	0.0170	18	0.00535	0.163	18	0.000693	0.0211
20	0.000889	0.0271	20	0.00850	0.259	20	0.00110	0.0336
22	0.00141	0.0431	22	0.0135	0.412	22	0.00175	0.0534
24	0.00225	0.0685	24	0.0215	0.655	24	0.00278	0.0849
26	0.00357	0.109	26	0.0342	1.04	26	0.00443	0.135
27	0.00451	0.137	27	0.0431	1.31	27	0.00558	0.170
28	0.00568	0.173	28	0.0543	1.66	28	0.00704	0.215
30	0.00903	0.275	30	0.0864	2.63	30	0.0112	0.341
32	0.0144	0.438	32	0.137	4.19	32	0.0178	0.542
34	0.0228	0.696	34	0.218	6.66	34	0.0283	0.863
36	0.0363	1.11	36	0.347	10.6	36	0.0450	1.37
40	0.0918	2.80	40	0.878	26.8	40	0.114	3.47

[a]Trademark.

Properties of Materials

Resistance of Wires (*continued*)

B. & S. No.	Ω/cm	Ω/ft
Monel Metal $\rho = 42 \times 10^{-6}$ Ω·cm		
10	0.000798	0.0243
12	0.00127	0.0387
14	0.00202	0.0615
16	0.00321	0.0978
18	0.00510	0.156
20	0.00811	0.247
22	0.0129	0.393
24	0.0205	0.625
26	0.0326	0.994
27	0.0411	1.25
28	0.0519	1.58
30	0.0825	2.51
32	0.131	4.00
34	0.209	6.36
36	0.331	10.1
40	0.838	25.6

B. & S. No.	Ω/cm	Ω/ft
[a]Nichrome $\rho = 150 \times 10^{-6}$ Ω·cm		
10	0.0021281	0.06488
12	0.0033751	0.1029
14	0.0054054	0.1648
16	0.0085116	0.2595
18	0.0138383	0.4219
20	0.0216218	0.6592
22	0.0346040	1.055
24	0.0548088	1.671
26	0.0875760	2.670
28	0.1394328	4.251
30	0.2214000	6.750
32	0.346040	10.55
34	0.557600	17.00
36	0.885600	27.00
38	1.383832	42.19
40	2.303872	70.24

B. & S. No.	Ω/cm	Ω/ft
Nickel $\rho = 7.8 \times 10^{-6}$ Ω·cm		
10	0.000148	0.00452
12	0.000236	0.00718
14	0.000375	0.0114
16	0.000596	0.0182
18	0.000948	0.0289
20	0.00151	0.0459
22	0.00240	0.0730
24	0.00381	0.116
26	0.00606	0.185
27	0.00764	0.233
28	0.00963	0.294
30	0.0153	0.467
32	0.0244	0.742
34	0.0387	1.18
36	0.0616	1.88
40	0.156	4.75

Platinum $\rho = 10 \times 10^{-6}$ Ω·cm		
10	0.000190	0.00579
12	0.000302	0.00921
14	0.000481	0.0146
16	0.000764	0.0233
18	0.00121	0.0370
20	0.00193	0.0589
22	0.00307	0.0936
24	0.00489	0.149
26	0.00776	0.237
27	0.00979	0.299
28	0.0123	0.376
30	0.0196	0.598
32	0.0312	0.952
34	0.0497	1.51
36	0.0789	2.41
40	0.200	6.08

Silver (18°C) $\rho = 1.629 \times 10^{-6}$ Ω·cm		
10	0.0000310	0.000944
12	0.0000492	0.00150
14	0.0000783	0.00239
16	0.000124	0.00379
18	0.000198	0.00603
20	0.000315	0.00959
22	0.000500	0.0153
24	0.000796	0.0243
26	0.00126	0.0386
27	0.00160	0.0486
28	0.00201	0.0613
30	0.00320	0.0975
32	0.00509	0.155
34	0.00809	0.247
36	0.0129	0.392
40	0.325	0.991

Steel, piano wire (0°C) $\rho = 11.8 \times 10^{-6}$ Ω·cm		
10	0.000224	0.00684
12	0.000357	0.0109
14	0.000567	0.0173
16	0.000901	0.0275
18	0.00143	0.0437
20	0.00228	0.0695
22	0.00363	0.110
24	0.00576	0.176
26	0.00916	0.279
27	0.0116	0.352
28	0.0146	0.444
30	0.0232	0.706
32	0.0368	1.12
34	0.0586	1.79
36	0.0931	2.84
40	0.236	7.18

Steel, Invar (35% Ni) $\rho = 81 \times 10^{-6}$ Ω·cm		
10	0.00154	0.0469
12	0.00245	0.0746
14	0.00389	0.119
16	0.00619	0.189
18	0.00984	0.300
20	0.0156	0.477
22	0.0249	0.758
24	0.0396	1.21
26	0.0629	1.92
27	0.0793	2.42
28	0.100	3.05
30	0.159	4.85
32	0.253	7.71
34	0.402	12.3
36	0.639	19.5
40	1.62	49.3

Tantalum $\rho = 15.5 \times 10^{-6}$ Ω·cm		
10	0.000295	0.00898
12	0.000468	0.0143
14	0.000745	0.0227
16	0.00118	0.0361
18	0.00188	0.0574
20	0.00299	0.0913
22	0.00476	0.145
24	0.00757	0.231
26	0.0120	0.367
27	0.0152	0.463
28	0.0191	0.583
30	0.0304	0.928
32	0.0484	1.47
34	0.0770	2.35
36	0.122	3.73
40	0.309	9.43

Tin $\rho = 11.5 \times 10^{-6}$ Ω·cm		
10	0.000219	0.00666
12	0.000348	0.0106
14	0.000553	0.0168
16	0.000879	0.0268
18	0.00140	0.0426
20	0.00222	0.0677
22	0.00353	0.108
24	0.00562	0.171
26	0.00893	0.272
27	0.0113	0.343
28	0.0142	0.433
30	0.0226	0.688
32	0.0359	1.09
34	0.0571	1.74
36	0.0908	2.77
40	0.230	7.00

[a]Trademark.

Resistance of Wires (*continued*)

B. & S. No.	Ω/cm	Ω/ft	B. & S. No.	Ω/cm	Ω/ft
Tungsten $\varrho = 5.51 \times 10^{-6}\,\Omega \cdot cm$			Zinc (0°C) $\varrho = 5.75 \times 10^{-6}\,\Omega \cdot cm$		
10	0.000105	0.00319	10	0.000109	0.00333
12	0.000167	0.00508	12	0.000174	0.00530
14	0.000265	0.00807	14	0.000276	0.00842
16	0.000421	0.0128	16	0.000439	0.0134
18	0.000669	0.0204	18	0.000699	0.0213
20	0.00106	0.0324	20	0.00111	0.0339
22	0.00169	0.0516	22	0.00177	0.0538
24	0.00269	0.0820	24	0.00281	0.0856
26	0.00428	0.130	26	0.00446	0.136
27	0.00540	0.164	27	0.00563	0.172
28	0.00680	0.207	28	0.00710	0.216
30	0.0108	0.330	30	0.0113	0.344
32	0.0172	0.524	32	0.0180	0.547
34	0.0274	0.834	34	0.0286	0.870
36	0.0435	1.33	36	0.0454	1.38
40	0.110	3.35	40	0.115	3.50

Defining Terms

Alloy: Metal composed of more than one element.
Bravais lattice: One of the 14 possible arrangements of points in three-dimensional space.
Crystalline: Having constituent atoms stacked together in a regular, repeating pattern.
Density: Mass per unit volume.
Electromotive force series: Systematic listing of half-cell reaction voltages.
Engineering strain: Increase in sample length at a given load divided by the original (stress-free) length.
Engineering stress: Load on a sample divided by the original (stress-free) area.
Fourier's law: Relationship between rate of heat transfer and temperature gradient.
Index of refraction: Ratio of speed of light in vacuum to that in a transparent material.
Linear coefficient of thermal expansion: Material parameter indicating dimensional change as a function of increasing temperature.
Melting point: Temperature of transformation from solid to liquid upon heating.
Ohm's law: Relationship between voltage, current, and resistance in an electrical circuit.
Property: Observable characteristic of a material.
Resistivity: Electrical resistance normalized for sample geometry.
Tensile strength: Maximum engineering stress during a tensile test.
Thermal conductivity: Proportionality constant in Fourier's law.
Young's modulus (modulus of elasticity):   Ratio of engineering stress to engineering strain for relatively small levels of load application.

References

ASM. 1984. *ASM Metals Reference Book,* 2nd ed. American Society for Metals, Metals Park, OH.
ASM. 1988. *Engineered Materials Handbook,* Vol. 2, *Engineering Plastics.* ASM International, Metals Park, OH.
Bolz, R.E. and Tuve, G.L., eds. 1973. *CRC Handbook of Tables for Applied Engineering Science.* CRC Press, Boca Raton, FL.
Lynch, C.T., ed. 1975. *CRC Handbook of Materials Science,* Vol. 3. CRC Press, Boca Raton, FL.
Shackelford, J.F. 1996. *Introduction to Materials Science for Engineers,* 4th ed. Prentice–Hall, Upper Saddle River, NJ.
Shackelford, J.F., Alexander, W., and Park, J.S., eds. 1994. *The CRC Materials Science and Engineering Handbook,* 2nd ed. CRC Press, Boca Raton, FL.

Further Information

A general introduction to the field of materials science and engineering is available from a variety of introductory textbooks. In addition to Shackelford [1996], readily available references include:

Askeland, D.R. 1994. *The Science and Engineering of Materials*, 3rd ed. PWS–Kent, Boston.
Callister, W.D. 1994. *Materials Science and Engineering-An Introduction*, 3rd ed. Wiley, New York.
Flinn, R.A. and Trojan, P.K. 1990. *Engineering Materials and Their Applications*, 4th ed. Houghton Mifflin, Boston.
Smith, W.F. 1990. *Principles of Materials Science and Engineering*, 2nd ed. McGraw–Hill New York.
Van Vlack, L.H. 1989. *Elements of Materials Science and Engineering*, 6th ed. Addison–Wesley Reading, MA.

As noted earlier, *The CRC Materials Science and Engineering Handbook* [Shackelford, Alexander, and Park 1994] is available as a comprehensive source of property data for engineering materials. In addition, ASM International has published between 1982 and 1996 the *ASM Handbook*, a 19-volume set concentrating on metals and alloys. ASM International has also published a 4-volume set entitled the *Engineered Materials Handbook*, covering composites, engineering plastics, adhesives and sealants, and ceramics and glasses.

9
Frequency Bands and Assignments[1]

Robert D. Greenberg
Federal Communications Commission

9.1 U.S. Table of Frequency Allocations.................................... 118
Categories of Services • Format of the U.S. Table

9.1 U.S. Table of Frequency Allocations

The U.S. Table of Frequency Allocations (columns 4–7 of Table 9.1) is based on the International plan for Region 2 because the relevant area of jurisdiction is located primarily in Region 2 (i.e., the 50 States, the District of Columbia, the Caribbean insular areas[2] and some of the Pacific insular areas [3]).[4] Because there is a need to provide radio spectrum for both Federal government and non-Federal government operations, the U.S. Table is divided into the Government Table of Frequency Allocation and the Nongovernment Table of Frequency Allocations. The Government plan, as shown in column 4, is administered by the National Telecommunications and Information Administration (NTIA),[5] whereas the non-Government plan, as shown in column 5, is administered by the Federal Communications Commission (FCC).[6]

In the U.S., radio spectrum may be allocated to either government or nongovernment use exclusively, or for shared use. In the case of shared use, the type of service(s) permitted need not be the same (e.g., Government fixed, nongovernment mobile). The terms used to designate categories of service in columns 4 and 5, correspond to the terms employed by the International Telecommunication Union (ITU) in the international *Radio Regulations* [RR 1982].

Categories of Services

Any segment of the radio spectrum may be allocated to the government and/or nongovernment sectors either on an exclusive or shared basis for use by one or more radio services. In the case where an allocation has been made to more than one service, such services are listed in the following order:

---

[1] Chapter adapted from the Federal Communications Commission, 47 Code of Federal Regulations, Part 2 (10-1-93 edition).

[2] The Caribbean insular areas are: the Commonwealth of Puerto Rico; the unincorporated territory of the U.S. Virgin Islands; and Navassa Island, Quita Sueno Bank, Roncador Bank, Serrana Bank, and Serranilla Bank.

[3] The Pacific insular areas located in region 2 are: Johnston Island and Midway Island.

[4] The operation of stations in the Pacific insular areas located in region 3 are generally governed by the International plan for region 3 (i.e., column 3 of Table 9.1. The Pacific insular areas located in region 3 are: the Commonwealth of the Northern Mariana Islands; the unincorporated territory of American Samoa; the unincorporated territory of Guam; and Baker Island. Howland Island, Jarvis Island, Kingman Reef, Palmyra Island, and Wake Island.

[5] Section 305(a) of the Communications Act of 1934, as amended; Executive Order 12046 (26 March 1978) and Department of Commerce Organization Order 10–10 (9 May 1979).

[6] The Communications Act of 1934, as amended.

Frequency Bands and Assignments

- Services, the names of which are printed in upper case (example: FIXED): these are called *primary* services.
- Services, the names of which are printed in upper case between oblique strokes (example: /RADIOLOCATION): these are called *permitted services.*
- Services, the names of which are printed in lower case (example: Mobile): these are called *secondary* services.

Permitted and primary services have equal rights, except that, in the preparation of frequency plans, the primary services, as compared with the permitted services, shall have prior choice of frequencies.

Stations of a secondary service have limitations. They shall not cause harmful interference to stations of primary or permitted services to which frequencies are already assigned or to which frequencies may be assigned at a later date. They cannot claim protection from harmful interference from stations of a primary or permitted service to which frequencies are already assigned or may be assigned at a later date. They can claim protection, however, from harmful interference from stations of the same or other secondary service(s) to which frequencies may be assigned at a later date.

Format of the U.S. Table

The frequency band referred to in each allocation, column 4 for government and column 5 for nongovernment, is indicated in the left-hand top corner of the column. If there is no service or footnote indicated for a band of frequencies in either column 4 or 5, then the government or the nongovernment sector, respectively, has no access to that band except as provided for by FCC rules. The government allocation plan, given in column 4, is included for informational purposes only. In the case where there is a parenthetical addition to an allocation in the U.S. Table [example: FIXED-SATELLITE (space-to-earth)], that service allocation is restricted to the type of operation so indicated.

The following symbols are used to designate footnotes in the U.S. Table:

Any footnote not prefixed by a letter, denotes an international footnote. Where such a footnote is applicable, without modification, to the U.S. Table, the symbol appears in the U.S. Table (column 4 or 5) and denotes a stipulation affecting both the government and nongovernment plans.

Any footnote consisting of the letters US followed by one or more digits denotes a stipulation affecting both the government and nongovernment plans.

Any footnote consisting of the letters NG followed by one or more digits, for example, NG1, denotes a stipulation applicable only to the nongovernment plan (column 5).

Any footnote consisting of the letter G following by one or more digits, for example, G1, denotes a stipulation applicable only to the government plan (column 4).

Column 6 provides a reference to indicate which Rule part(s) (e.g., private land mobile radio services, domestic public land mobile radio services, etc.) are given assignments within the allocation plan specified in column 5 for any given band of frequencies. The exact use that can be made of any given frequency or frequencies (e.g., channelling plans, allowable emissions, etc.) is given in the Rule part(s) so indicated. The Rule parts in this column are not allocations. They are provided for informational purposes only.

Column 7 is used to denote certain frequencies that have national and/or international significance.

Nomenclature of Frequencies (§ 2.101)

Band No.	Frequency Subdivision	Frequency Range
4	VLF (very low frequency)	Below 30 kHz
5	LF (low frequency)	30–300 kHz
6	MF (medium frequency)	300–3000 kHz
7	HF (high frequency)	3–30 MHz
8	VHF (very high frequency)	30–300 MHz
9	UHF (ultra high frequency)	300–3000 MHz
10	SHF (super high frequency)	3–30 GHz
11	EHF (extremely high frequency)	30–300 GHz
12		300–3000 GHz

TABLE 9.1 Table of Frequency Allocations

International Table			U.S. Table		FCC Use Designators	
Region 1—allocation, kHz (1)	Region 2—allocation, kHz (2)	Region 3—allocation, kHz (3)	Government Allocation, kHz (4)	Nongovernment Allocation, kHz (5)	Rule part(s) (6)	Special-use frequencies (7)
Below 9	Below 9. (Not allocated.) 444 445		Below 9. (Not allocated.) 444 445	Below 9. (Not allocated.) 444 445		
9–14	RADIONAVIGATION.		9–14 RADIONAVIGATION US18 US294	9–14 RADIONAVIGATION. US18 US294		
14–19.95	FIXED. MARITIME MOBILE. 448 446 447		14–19.95 FIXED. MARITIME MOBILE. 448 US294	14–19.95 Fixed 448 US294	INTERNATIONAL FIXED PUBLIC (23).	
19.95–20.05	STANDARD FREQUENCY AND TIME SIGNAL (20 kHz).		19.95–20.05 STANDARD FREQUENCY AND TIME SIGNAL. US294	19.95–20.05 STANDARD FREQUENCY AND TIME SIGNAL. US294		20 kHz Standard Frequency.
20.05–70	FIXED. MARITIME MOBILE. 448 447 449		20.05–59 FIXED. MARITIME MOBILE. 446 US294	20.05–59 FIXED. 448 US294	INTERNATIONAL FIXED PUBLIC (23).	
			59–61 STANDARD FREQUENCY AND TIME SIGNAL. US294	59–61 STANDARD FREQUENCY AND TIME SIGNAL. US294		60 kHz Standard Frequency.
			61–70 FIXED. MARITIME MOBILE. 448 US294	61–70 FIXED. 448 US294	INTERNATIONAL FIXED PUBLIC (23).	
70–72 RADIONAVIGATION 451	70–90 FIXED. MARITIME MOBILE. 448	70–72 RADIONAVIGATION 451 Fixed.	70–90 FIXED. MARITIME MOBILE. Radiolocation.	70–90 FIXED. Radiolocation.	INTERNATIONAL FIXED PUBLIC (23). Private Land Mobile	

Frequency Bands and Assignments

International Table	US Government Table	US Non-Government Table		FCC Use
MARITIME RADIONAVIGATION 451 Radiolocation. 452	Maritime Mobile 448. 450			(90).
72–84 FIXED. MARITIME MOBILE 448 RADIONAVIGATION 451 447	72–84 FIXED. MARITIME MOBILE 448 RADIONAVIGATION 451	US288 US294	US288 US294	
84–86 RADIONAVIGATION 451	84–86 RADIONAVIGATION 451 Fixed. Maritime Mobile 448 450			
86–90 FIXED. MARITIME MOBILE 448 RADIONAVIGATION 447	86–90 FIXED. MARITIME MOBILE 448 RADIONAVIGATION 451	448 451 US294	448 451 US294	
90–110 RADIONAVIGATION 453 Fixed. 453A 454		90–110 RADIONAVIGATION 453 US18 US104 US294	90–110 RADIONAVIGATION 453 US18 US104 US294	Private Land Mobile (90).
110–112 FIXED. MARITIME MOBILE. MARITIME RADIO- NAVIGATION 451 Radiolocation.	110–112 FIXED. MARITIME MOBILE. RADIONAVIGATION 451	110–130 FIXED. MARITIME MOBILE. Radiolocation.	110–130 FIXED. MARITIME MOBILE. Radiolocation.	INTERNATIONAL FIXED PUBLIC (23). MARITIME (80). Private Land Mobile (90).
112–115 RADIONAVIGATION 451	452 454	112–117.6 RADIONAVIGATION 451 Fixed. Maritime Mobile.	454	454 US294
115–117.6 RADIONAVIGATION 451				

TABLE 9.1 Table of Frequency Allocations (*Continued*)

International Table			U.S. Table		FCC Use Designators	
Region 1—allocation, kHz (1)	Region 2—allocation, kHz (2)	Region 3—allocation, kHz (3)	Government Allocation, kHz (4)	Nongovernment Allocation, kHz (5)	Rule part(s) (6)	Special-use frequencies (7)
Fixed. Maritime Mobile. 454 456						
117.6–126.0 FIXED. MARITIME MOBILE. RADIONAVIGATION 451		117.6–126.0 FIXED. MARITIME MOBILE. RADIONAVIGATION 451 454 455				
454		454				
126–129 RADIONAVIGATION 451		126–129 RADIONAVIGATION 451 Fixed. Maritime Mobile. 454 455				
129–130 FIXED. MARITIME MOBILE. RADIONAVIGATION 451		129–130 FIXED. MARITIME MOBILE. RADIONAVIGATION 451				
454		454	451 454 US294	451 454 US294		
130–148.5 MARITIME MOBILE. /FIXED/.	130–160 FIXED. MARITIME MOBILE.	130–160 FIXED. MARITIME MOBILE. RADIONAVIGATION	130–160 FIXED. MARITIME MOBILE.	130–160 FIXED. MARITIME MOBILE.	INTERNATIONAL FIXED PUBLIC (23). MARITIME (80).	
454 457						
148.5–255 BROADCASTING	160–190 FIXED.	160–190 FIXED. Aeronautical Radionavigation.	160–190 FIXED. MARITIME MOBILE. 454 US294	160–190 FIXED. 454 US294	INTERNATIONAL FIXED PUBLIC (23). MARITIME (80).	
	459		459 US294	459 US294		
	190–200 AERONAUTICAL RADIONAVIGATION.		190–200 AERONAUTICAL RADIONAVIGATION. US18 US226 US294	190–200 AERONAUTICAL RADIONAVIGATION. US18 US226 US294		

Frequency Bands and Assignments

460 461 462	200–275 AERONAUTICAL RADIONAVIGATION. Aeronautical Mobile.	200–285 AERONAUTICAL RADIONAVIGATION. Aeronautical Mobile.	200–275 AERONAUTICAL RADIONAVIGATION. Aeronautical Mobile. US18 US294	200–275 AERONAUTICAL RADIONAVIGATION. Aeronautical Mobile. US18 US294	AVIATION (87).
255–283.5 BROADCASTING. /AERONAUTICAL RADIONAVIGATION/ 463	275–285 AERONAUTICAL RADIONAVIGATION. Aeronautical Mobile Maritime Radionavigation (radiobeacons).		275–285 AERONAUTICAL RADIONAVIGATION. Aeronautical Mobile. Maritime Radionavigation (radiobeacons).	275–285 AERONAUTICAL RADIONAVIGATION. Aeronautical Mobile. Maritime Radionavigation (radiobeacons).	
462 464 464A 283.5–315 MARITIME RADIO-NAVIGATION (radiobeacons) 466 /AERONAUTICAL RADIONAVIGATION/.	285–315 MARITIME RADIONAVIGATION (radiobeacons) 466 /AERONAUTICAL RADIONAVIGATION/.		285–325 MARITIME RADIONAVIGATION (radiobeacons) 466 Aeronautical Radionavigation (radiobeacons). US18 US294	285–325 MARITIME RADIONAVIGATION (radiobeacons) 466 Aeronautical Radionavigation (radiobeacons). US18 US294	AVIATION (87).
464A 465 466A 315–325 AERONAUTICAL RADIONAVIGATION. Maritime Radionavigation (radiobeacons) 466	315–325 MARITIME RADIONAVIGATION (radiobeacons) 466 Aeronautical Radionavigation.	315–325 AERONAUTICAL RADIONAVIGATION. MARITIME RADIONAVIGATION (radiobeacons) 466			
465 467 325–405 AERONAUTICAL RADIONAVIGATION.	325–335 AERONAUTICAL RADIONAVIGATION. Aeronautical Mobile Maritime Radionavigation (radiobeacons).	325–405 AERONAUTICAL RADIONAVIGATION. Aeronautical Mobile.	325–335 AERONAUTICAL RADIONAVIGATION. Aeronautical Mobile. Maritime Radionavigation (radiobeacons). US18 US294	325–335 AERONAUTICAL RADIONAVIGATION. Aeronautical Mobile. Maritime Radionavigation (radiobeacons). US18 US294	AVIATION (87).

TABLE 9.1 Table of Frequency Allocations (*Continued*)

International Table			U.S. Table		FCC Use Designators	
Region 1—allocation, kHz (1)	Region 2—allocation, kHz (2)	Region 3—allocation, kHz (3)	Government Allocation, kHz (4)	Nongovernment Allocation, kHz (5)	Rule part(s) (6)	Special-use frequencies (7)
	335–405 AERONAUTICAL RADIONAVIGATION. Aeronautical Mobile.		335–405 AERONAUTICAL RADIONAVIGATION (radiobeacons). Aeronautical Mobile. US18 US294	335–405 AERONAUTICAL RADIONAVIGATION (radiobeacons). Aeronautical Mobile. US18 US294	AVIATION (87).	
465						
405–415 RADIONAVIGATION 468	405–415 RADIONAVIGATION 468 Aeronautical Mobile.		405–415 RADIONAVIGATION 468 Aeronautical Mobile. US18 US294	405–415 RADIONAVIGATION 468 Aeronautical Mobile. US18 US294	AVIATION (87).	
465						
415–435 AERONAUTICAL RADIO-NAVIGATION. /MARITIME MOBILE/ 470	415–495 MARITIME MOBILE 470 AERONAUTICAL RADIONAVIGATION. 470A	415–435 MARITIME MOBILE 470 AERONAUTICAL RADIONAVIGATION 470A	415–435 AERONAUTICAL RADIONAVIGATION. MARITIME MOBILE 470 469A US294	415–435 AERONAUTICAL RADIONAVIGATION. MARITIME MOBILE. 470 469A US294	AVIATION (87) MARITIME (80).	
465						
435–495 MARITIME MOBILE 470 Aeronautical Radionavigation.			435–495 MARITIME MOBILE 470 AERONAUTICAL RADIO-NAVIGATION. 471 US231 472A US294	435–495 MARITIME MOBILE 470 471 472A US231 US294	MARITIME (80).	
465 471 472A		469 469A 471 472A				
495–505			495–505 MOBILE (distress and calling). 472	495–505 MOBILE (distress and calling). 472	MARITIME (80).	500 kHz: Distress and calling frequency.
	MOBILE (distress and calling). 472					
505–526.5 MARITIME MOBILE 470 /AERONAUTICAL RADIONAVIGATION/	505–510 MARITIME MOBILE 470	505–526.5 MARITIME MOBILE 470 474 /AERONAUTICAL RADIONAVIGATION/. Aeronautical Mobile Land Mobile.	505–510 MARITIME MOBILE 470	505–510 MARITIME MOBILE 470	MARITIME (80).	

Frequency Bands and Assignments

		International Table	US Table			Special-Use Frequencies
		471 510-525 MOBILE. 474 AERONAUTICAL RADIONAVIGATION.	471 510-525 AERONAUTICAL RADIONAVIGATION (radiobeacons). MARITIME MOBILE (Ships only). 474 US14 US18 US225	471 510-525 AERONAUTICAL RADIONAVIGATION (radiobeacons). MARITIME MOBILE (Ships only). 474 US14 US18 US225	AVIATION (87). MARITIME (80).	518 kHz: International NAVTEX in the Maritime Mobile Service.
465 471 474 475 476		525-535	525-535	525-535		
526.5-1606.5 BROADCASTING. 478		526.5-535 BROADCASTING. Mobile. 471	MOBILE. AERONAUTICAL RADIONAVIGATION (radiobeacons). US18 US221 US239	MOBILE. AERONAUTICAL RADIONAVIGATION (radiobeacons). US18 US221 US239	AVIATION (87). PRIVATE LAND MOBILE (90)	530 kHz: Travelers information.
535-1705		535-1705 BROAD-CASTING	RADIO BROADCASTING (AM) (73). Alaska Fixed (80). Auxiliary Broadcasting (74). Private Land Mobile (90)	535-1705 kHz: Travelers Information.		
480 US238, US299, US321, 478		480 US238, US299, US321, NG128 1605-1625	1605-1615	1605-1615	MOBILE.	
1606.5-1625 MARITIME MOBILE. 480A/FIXED/. /LAND MOBILE/. 483 484		BROADCASTING 480	1606.5-1800 FIXED. MOBILE. RADIOLOCATION. RADIONAVIGATION. 482	MOBILE.		1610 kHz: Travelers information.
		480A 481		480 US221	480 US221	
				1615-1625 BROADCASTING 480	1615-1625 BROADCASTING 480	AUXILIARY BROADCASTING (74). PRIVATE LAND MOBILE (90).
1625-1635				US237 US299	US237 US299	ALASKA FIXED (80). AUXILIARY BROADCASTING (74). Private Land Mobile (90).
1625-1705				1625-1705	1625-1705	ALASKA FIXED (80).

TABLE 9.1 Table of Frequency Allocations (*Continued*)

International Table		U.S. Table		FCC Use Designators	
Region 1—allocation, kHz (1)	Region 2—allocation, kHz (2)	Region 3—allocation, kHz (3)	Government Allocation, kHz (4)	Nongovernment Allocation, kHz (5)	Rule part(s) (6)	Special-use frequencies (7)
RADIOLOCATION 487	BROADCASTING 480 /FIXED/. /MOBILE/.		Radiolocation.	BROADCASTING 480 Radiolocation	AUXILIARY BROADCASTING (74). Private Land Mobile (90).	
485 486	Radiolocation 480A 481					
1635–1800 MARITIME MOBILE /FIXED/. /LAND MOBILE/. 480A	1705–1800 FIXED. MOBILE. RADIOLOCATION. AERONAUTICAL RADIONAVIGATION.		480 US238 1705–1800 FIXED. MOBILE. RADIOLOCATION.	US238 US299 1705–1800 FIXED. MOBILE. RADIOLOCATION.	DISASTER (99). INTERNATIONAL FIXED PUBLIC (23). MARITIME (80). PRIVATE LAND MOBILE (90).	
483 484 488						
1800–1810 RADIOLOCATION 487	1800–1850 AMATEUR.	1800–2000 AMATEUR. FIXED. MOBILE except aeronautical mobile. RADIONAVIGATION. Radiolocation.	US240 1800–1900	US240 1800–1900 AMATEUR.	AMATEUR (97).	
485 486						
1810–1850 AMATEUR. 490 491 492 493						
1850–2000 FIXED. MOBILE except aeronautical mobile.	1850–2000 AMATEUR. FIXED. MOBILE expect aeronautical mobile. RADIOLOCATION. RADIONAVIGATION. 494		1900–2000 RADIOLOCATION.	1900–2000 RADIOLOCATION.	PRIVATE LAND MOBILE (90). Amateur (97).	

Frequency Bands and Assignments

484 488 495	2000–2025 FIXED. MOBILE.	489	US290 2000–2065 FIXED. MOBILE.	US290 2000–2065 MARITIME MOBILE.	MARITIME (80).
2000–2025 FIXED. MOBILE except aeronautical mobile (R).					
484 495					
2025–2045 FIXED. MOBILE except aeronautical mobile (R). Meteorological Aids 496				NG19	
484 495					
2045–2160 MARITIME MOBILE. /FIXED/. /LAND MOBILE/.	2065–2107		2065–2107	2065–2107	MARITIME (80).
	MARITIME MOBILE 497		MARITIME MOBILE 497	MARITIME MOBILE 497	
483 484	498				
2160–2170 RADIOLOCATION 487	2107–2170 FIXED. MOBILE.		2107–2170 FIXED. MOBILE.	2107–2170 FIXED. MARITIME MOBILE. LAND MOBILE.	AVIATION (87). INTERNATIONAL FIXED PUBLIC (23). MARITIME (80). PRIVATE LAND MOBILE (90).
485 486 499				NG19	
2170.0–2173.5	MARITIME MOBILE.		2170–2173.5 MARITIME MOBILE.	2170–2173.5 MARITIME MOBILE.	MARITIME (80).
2173.5–2190.5	MOBILE (distress and calling). 500 501 500A 500B	MOBILE (distress and calling). 500 501 500A 500B	2173.5–2190.5 MOBILE (distress and calling). 500 501 US279 500A 500B	2173.5–2190.5 MOBILE (distress and calling). 500 501 US279 500A 500B	AVIATION (87). MARITIME (80).	2182 kHz: Distress and calling.
2190.5–2194.0	MARITIME MOBILE.		2190.5–2194 MARITIME MOBILE.	2190.5–2194 MARITIME MOBILE.	MARITIME (80).
2194–2300 FIXED. MOBILE except	2194–2300 FIXED. MOBILE.		2194–2495 FIXED. MOBILE.	2194–2495 FIXED. LAND MOBILE.	AVIATION (87). INTERNATIONAL

TABLE 9.1 Table of Frequency Allocations (*Continued*)

International Table			U.S. Table		FCC Use Designators	
Region 1—allocation, kHz (1)	Region 2—allocation, kHz (2)	Region 3—allocation, kHz (3)	Government Allocation, kHz (4)	Nongovernment Allocation, kHz (5)	Rule part(s) (6)	Special-use frequencies (7)
aeronautical mobile (R).				MARITIME MOBILE.	FIXED PUBLIC (23). MARITIME (80). PRIVATE LAND MOBILE (90).	
484 495 502	502					
2300–2498 FIXED. MOBILE except aeronautical mobile (R). BROADCASTING 503	2300–2495 FIXED. MOBILE. BROADCASTING 503.					
495 2498–2501 STANDARD FREQUENCY AND TIME SIGNAL (2500 kHz).	2495–2501 STANDARD FREQUENCY AND TIME SIGNAL (2500 kHz).		2495–2505 STANDARD FREQUENCY AND TIME SIGNAL.	NG19 2495–2505 STANDARD FREQUENCY AND TIME SIGNAL.		2500 kHz: Standard frequency
STANDARD FREQUENCY AND TIME SIGNAL (2500 kHz). 2501–2502	STANDARD FREQUENCY AND TIME SIGNAL. Space Research.					
2502–2625 FIXED. MOBILE except aeronautical mobile (R).	2502–2505 STANDARD FREQUENCY AND TIME SIGNAL.					
	2505–2850 FIXED. MOBILE.		G106 2505–2850 FIXED. MOBILE.	2505–2850 FIXED. LAND MOBILE. MARITIME MOBILE.	AVIATION (87). INTERNATIONAL FIXED PUBLIC (23). MARITIME (80). PRIVATE LAND MOBILE (90).	
484 495 504						
2625–2650 MARITIME MOBILE.						

Frequency Bands and Assignments

MARITIME RADIO-NAVIGATION. 484				
2650–2850 FIXED. MOBILE except aeronautical mobile (R). 484 495				
2850–3025 AERONAUTICAL MOBILE (R). 501 505		US285 2850–3025 AERONAUTICAL MOBILE (R). 501 505 US283		AVIATION (87).
3025–3155 AERONAUTICAL MOBILE (OR).		3025–3155 AERONAUTICAL MOBILE (OR).		
3155–3200 FIXED. MOBILE except aeronautical mobile (R). 506 507		3155–3230 FIXED. MOBILE except aeronautical mobile (R).		AVIATION (87). INTERNATIONAL FIXED PUBLIC (23). MARITIME (80). PRIVATE LAND MOBILE (90).
3200–3230 FIXED. MOBILE except aeronautical mobile (R). BROADCASTING 503. 506				
3230–3400 FIXED. MOBILE except aeronautical mobile. BROADCASTING 503 506 508		3230–3400 FIXED. MOBILE except aeronautical mobile. Radiolocation.		AVIATION (87). INTERNATIONAL FIXED PUBLIC (23). MARITIME (80). PRIVATE LAND MOBILE (90).
3400–3500 AERONAUTICAL MOBILE (R).		3400–3500 AERONAUTICAL MOBILE (R). US283		AVIATION (87).
3500–3800 3500–3750	3500–3900	3500–4000		

TABLE 9.1 Table of Frequency Allocations (*Continued*)

International Table			U.S. Table		FCC Use Designators	
Region 1—allocation, kHz (1)	Region 2—allocation, kHz (2)	Region 3—allocation, kHz (3)	Government Allocation, kHz (4)	Nongovernment Allocation, kHz (5)	Rule part(s) (6)	Special-use frequencies (7)
AMATEUR 510 FIXED. MOBILE except aeronautical mobile. 484	AMATEUR 510	AMATEUR 510 FIXED. MOBILE.	510	AMATEUR 510	AMATEUR (97).	
	509 511					
	3750–4000 AMATEUR 510 FIXED. MOBILE except aeronautical mobile (R). 511 512 514 515					
3800–3900 FIXED. AERONAUTICAL MOBILE (OR). LAND MOBILE.		3900–3950 AERONAUTICAL MOBILE. BROADCASTING.				
3900–3950 AERONAUTICAL MOBILE (OR). 513						
3950–4000 FIXED. BROADCASTING.		3950–4000 FIXED. BROADCASTING. 516				
4000–4063 FIXED. MARITIME MOBILE 517 516			4000–4438 MARITIME MOBILE 500A 500B 520 520B	4000–4438 MARITIME MOBILE 500A 500B 520 520B	INTERNATIONAL FIXED PUBLIC (23). MARITIME (80).	
4063–4438	MARITIME MOBILE 500A 500B 520 520B 518 519					
4438–4650 FIXED. MOBILE except		4438–4650 FIXED. MOBILE except	US82 US236 US296 4438–4650 FIXED. MOBILE except	US82 US236 US296 4438–4650 FIXED. MOBILE except	AVIATION (87). INTERNATIONAL	

Frequency Bands and Assignments

aeronautical mobile (R)		aeronautical mobile (R)	aeronautical mobile (R)	aeronautical mobile.	
4650–4700	AERONAUTICAL MOBILE (R).		4650–4700 AERONAUTICAL MOBILE (R). US282 US283	4650–4700 AERONAUTICAL MOBILE (R). US282 US283	AVIATION (87).
4700–4750	AERONAUTICAL MOBILE (OR).		4700–4750 AERONAUTICAL MOBILE (OR).	4700–4750 AERONAUTICAL MOBILE (OR).	
4750–4850 FIXED. AERONAUTICAL MOBILE (OR). LAND MOBILE BROADCASTING 503	MOBILE except aeronautical. BROADCASTING 503	4750–4850 FIXED. BROADCASTING 503 Land Mobile.	4750–4850 FIXED. MOBILE except aeronautical mobile.	4750–4850 FIXED. MOBILE except aeronautical mobile (R).	AVIATION (87). INTERNATIONAL FIXED PUBLIC (23). MARITIME (80).
4850–4995	FIXED. LAND MOBILE. BROADCASTING 503.		4850–4995 FIXED. MOBILE.	4850–4995 FIXED.	AVIATION (87). INTERNATIONAL FIXED PUBLIC (23). MARITIME (80).
4995–5003	STANDARD FREQUENCY AND TIME SIGNAL (5000 kHz).		4995–5005 STANDARD FREQUENCY AND TIME SIGNAL	4995–5005 STANDARD FREQUENCY AND TIME SIGNAL.	
5003–5005	STANDARD FREQUENCY AND TIME SIGNAL. Space Research.		G106		5000 kHz: Standard frequency.
5005–5060	FIXED. BROADCASTING 503		5005–5060 FIXED.	5005–5060 FIXED.	
5060–5250	FIXED. Mobile except aeronautical mobile.		5060–5450 FIXED. MOBILE except aeronautical mobile.	5060–5450 FIXED. MOBILE except aeronautical mobile.	AVIATION (87). INTERNATIONAL FIXED PUBLIC (23). MARITIME (80). PRIVATE LAND MOBILE (90).
					AVIATION (87). INTERNATIONAL FIXED PUBLIC (23).

TABLE 9.1 Table of Frequency Allocations (*Continued*)

International Table			U.S. Table		FCC Use Designators	
Region 1—allocation, kHz (1)	Region 2—allocation, kHz (2)	Region 3—allocation, kHz (3)	Government Allocation, kHz (4)	Nongovernment Allocation, kHz (5)	Rule part(s) (6)	Special-use frequencies (7)
5250–5450	521				MARITIME (80). PRIVATE LAND MOBILE (90).	
5450–5480 FIXED. AERONAUTICAL MOBILE (OR). LAND MOBILE.	FIXED. MOBILE except aeronautical mobile.		US212	US212	AVIATION (87).	
5480–5680	5450–5480 AERONAUTICAL MOBILE (R). 501 505	5450–5480 FIXED. AERONAUTICAL MOBILE (OR). LAND MOBILE.	5450–5480 AERONAUTICAL MOBILE (R). 501 505 US283	5450–5480 AERONAUTICAL MOBILE (R). 501 505 US283		
5680–5730	AERONAUTICAL MOBILE (R). 501 505		5680–5730 AERONAUTICAL MOBILE (OR). 501 505	5680–5730 AERONAUTICAL MOBILE (OR). 501 505		
5730–5950 FIXED. LAND MOBILE	5730–5950 FIXED. MOBILE except aeronautical mobile (R).	5730–5950 FIXED. MOBILE except aeronautical mobile (R).	5730–5950 FIXED. MOBILE except aeronautical mobile (R).	5730–5950 FIXED. MOBILE except aeronautical mobile (R).	AVIATION (87). INTERNATIONAL FIXED PUBLIC (23). MARITIME (80).	
5950–6200	BROADCASTING		5950–6200 BROADCASTING US280	5950–6200 BROADCASTING US280	RADIO BROADCAST (HF)(73).	
6200–6525	MARITIME MOBILE 500A 500B 520 520B. 522		6200–6525 MARITIME MOBILE 500A 500B 520 520B. US82 US296	6200–6525 MARITIME MOBILE 500A 500B 520 520B. US82 US296	MARITIME (80).	
6525–6685	AERONAUTICAL		6525–6685 AERONAUTICAL	6525–6685 AERONAUTICAL	AVIATION (87).	

Frequency Bands and Assignments

	MOBILE (R).	MOBILE (R). US283	MOBILE (R). US283	
6685–6765	AERONAUTICAL MOBILE (OR).	6765–7000 AERONAUTICAL MOBILE (OR).	6765–7000 AERONAUTICAL MOBILE (OR).	
6765–7000	FIXED. Land Mobile 525 524	6765–7000 FIXED. Mobile.	6765–7000 FIXED. Mobile. 524	AVIATION (87). INTERNATIONAL FIXED PUBLIC (23).	6780 ± 15 kHz: Industrial, scientific, and medical frequency.
7000–7100	AMATEUR 510 AMATEUR-SATELLITE. 526 527		7000–7100 AMATEUR 510 AMATEUR-SATELLITE.	AMATEUR (97).
7100–7300 BROADCASTING.	7100–7300 AMATEUR 510 528	7100–7300 BROADCASTING.	7100–7300 AMATEUR 510 528	AMATEUR (97).
7300–8100	FIXED. Land Mobile. 529	7300–8100 FIXED. Mobile.	7300–8100 FIXED. Mobile.	AVIATION (87). INTERNATIONAL FIXED PUBLIC (23). MARITIME (80). PRIVATE LAND MOBILE (90).
8100–8195	FIXED. MARITIME MOBILE.	8100–8195 MARITIME MOBILE. 500A 500B 520B 529A	8100–8815 MARITIME mobile. 500A 500B 520B 529A	MARITIME (80).
8195–8815	MARITIME MOBILE. 500A 500B 520B 529A 501	501 US82 US236 US296	501 US82 US236 US296	
8815–8965	AERONAUTICAL MOBILE (R).	8815–8965 AERONAUTICAL MOBILE (R).	8815–8965 AERONAUTICAL MOBILE (R).	Aviation (87).
8965–9040	AERONAUTICAL MOBILE (OR).	8965–9040 AERONAUTICAL MOBILE (OR).	8965–9040 AERONAUTICAL MOBILE (OR).	

TABLE 9.1 Table of Frequency Allocations (*Continued*)

International Table			U.S. Table		FCC Use Designators	
Region 1—allocation, kHz (1)	Region 2—allocation, kHz (2)	Region 3—allocation, kHz (3)	Government Allocation, kHz (4)	Nongovernment Allocation, kHz (5)	Rule part(s) (6)	Special-use frequencies (7)
9040–9500	FIXED.		9040–9500 FIXED.	9040–9050 FIXED.	Aviation (87). INTERNATIONAL FIXED PUBLIC (23). MARITIME (80).	
9500–9900	BROADCASTING.		9500–9900 BROADCASTING.	9500–9900 BROADCASTING.	RADIO BROADCAST (HF) (73). INTERNATIONAL FIXED PUBLIC (23).	
9900–9995	530 531 FIXED.		US235	US235	AVIATION (87). INTERNATIONAL FIXED PUBLIC (23).	
			9900–9995 FIXED.	9900–9995 FIXED.		
9995–10003	STANDARD FREQUENCY AND TIME SIGNAL (10000 kHz). 501		9995–10005 STANDARD FREQUENCY AND TIME SIGNAL.	9995–10005 STANDARD FREQUENCY AND TIME SIGNAL.		10000 kHz: Standard frequency.
10003–10005	STANDARD FREQUENCY AND TIME SIGNAL Space Research. 501					
10005–10100	AERONAUTICAL MOBILE (R). 501		501 G106 10005–10100 AERONAUTICAL MOBILE (R). 501 US283	501 10005–10100 AERONAUTICAL MOBILE (R). 501 US283	AVIATION (87).	
10100–10150	FIXED. Amateur 510.		10100–10150 510 US247	10100–10150 AMATEUR 510. US247	AMATEUR (97).	

Frequency Bands and Assignments

Band	Allocation 1	Allocation 2	Allocation 3
10150–11175	FIXED. MOBILE except aeronautical mobile (R).	10150–11175 FIXED. MOBILE except aeronautical mobile (R).	10150–11175 AVIATION (87). INTERNATIONAL FIXED PUBLIC (23).
11175–11275	AERONAUTICAL MOBILE (OR).	11175–11275 AERONAUTICAL MOBILE (OR).	
11275–11400	AERONAUTICAL MOBILE (R).	11275–11400 AERONAUTICAL MOBILE (R). US283	AVIATION (87).
11400–11650	FIXED.	11400–11650 FIXED.	AVIATION (87). INTERNATIONAL FIXED PUBLIC (23).
11650–12050	BROADCASTING. 530 531 532	11650–12050 BROADCASTING. US235	RADIO BROADCAST (HF) (73). INTERNATIONAL FIXED PUBLIC (23).
12050–12230	FIXED.	12050–12230 FIXED.	AVIATION (87). INTERNATIONAL FIXED PUBLIC (23).
12230–13200	MARITIME MOBILE. 500A 500B 520B 529A	12230–13200 MARITIME MOBILE. 500A 500B 520B 529A US82 US296	INTERNATIONAL FIXED PUBLIC (23).
13200–13260	AERONAUTICAL MOBILE (OR).	13200–13260 AERONAUTICAL MOBILE (OR).	MARITIME (80).
13260–13360	AERONAUTICAL MOBILE (R).	13260–13360 AERONAUTICAL MOBILE (R). US283	AVIATION (87).
13360–13410	FIXED.	13360–13410 RADIO ASTRONOMY.	

TABLE 9.1 Table of Frequency Allocations (*Continued*)

International Table			U.S. Table		FCC Use Designators	
Region 1—allocation, kHz (1)	Region 2—allocation, kHz (2)	Region 3—allocation, kHz (3)	Government Allocation, kHz (4)	Nongovernment Allocation, kHz (5)	Rule part(s) (6)	Special-use frequencies (7)
	RADIO ASTRONOMY. 533		533 G115	533		
13410–13600	FIXED. Mobile except aeronautical mobile (R). 534		13410–13600 FIXED. Mobile except aeronautical mobile (R). 534	13410–13600 FIXED.	AVIATION (87). INTERNATIONAL FIXED PUBLIC (23).	13560 ± 7 kHz: Industrial, scientific, and medical frequency.
13600–13800	BROADCASTING.		13600–13800 BROADCASTING.	13600–13800 BROADCASTING.	RADIO BROADCAST (HF) (73). INTERNATIONAL FIXED PUBLIC (23).	
	531		US235	US235		
13800–14000	FIXED. Mobile except aeronautical mobile (R)		13800–14000 FIXED. Mobile except aeronautical mobile (R).	13800–14000 FIXED.	AVIATION (87). INTERNATIONAL FIXED PUBLIC (23).	
14000–14250	AMATEUR 510 AMATEUR-SATELLITE.		14000–14350 510	14000–14250 AMATEUR 510 AMATEUR-SATELLITE.	AMATEUR (97).	
14250–14350	AMATEUR 510 535			14250–14350 AMATEUR 510	AMATEUR (97).	
14350–14990	FIXED. Mobile except aeronautical mobile (R).		14350–14990 FIXED. Mobile except aeronautical mobile (R).	14350–14990 FIXED.	AVIATION (87). INTERNATIONAL FIXED PUBLIC (23).	
14990–15005	STANDARD FREQUENCY AND TIME SIGNAL (15000 kHz). 501		14990–15010 STANDARD FREQUENCY AND TIME SIGNAL.	14990–15010 STANDARD FREQUENCY AND TIME SIGNAL.		1500 kHz: Standard frequency.
15005–15010	STANDARD FREQUENCY AND TIME					

Frequency Bands and Assignments

	SIGNAL. Space Research.	501 G106	501	
15010–15100	AERONAUTICAL MOBILE (OR).	15010–15100 AERONAUTICAL MOBILE (OR).	15010–15100 AERONAUTICAL MOBILE (OR).	
15100–15600	BROADCASTING.	15100–15600 BROADCASTING.	15100–15600 BROADCASTING.	RADIO BROADCAST (HF) (73). INTERNATIONAL FIXED PUBLIC (23).
	531	US235	US235	
15600–16360	FIXED.	15600–16360 FIXED.	15600–16360 FIXED.	AVIATION (87). INTERNATIONAL FIXED PUBLIC (23).
	536			
16360–17410	MARITIME MOBILE. 500A 500B 520B 529A 532	16360–17410 MARITIME MOBILE. 500A 500B 520B 529A US82 US296	16360–17410 MARITIME MOBILE. 500A 500B 520B 529A US82 US296	MARITIME (80)
17410–17550	FIXED.	17410–17550 FIXED.	17410–17550 FIXED.	AVIATION (87). INTERNATIONAL FIXED PUBLIC (23).
17550–17900	BROADCASTING.	17550–17900 BROADCASTING.	17550–17900 BROADCASTING.	RADIO BROADCAST (HF) (73). INTERNATIONAL FIXED PUBLIC (23).
	531	US235	US235	
17900–17970	AERONAUTICAL MOBILE (R).	17900–17970 AERONAUTICAL MOBILE (R). US283	17900–17970 AERONAUTICAL MOBILE (R). US283	AVIATION (87).
17970–18030	AERONAUTICAL MOBILE (OR).	17970–18030 AERONAUTICAL MOBILE (OR).	17970–18030 AERONAUTICAL MOBILE (OR).	

TABLE 9.1 Table of Frequency Allocations (*Continued*)

International Table			U.S. Table		FCC Use Designators	
Region 1—allocation, kHz (1)	Region 2—allocation, kHz (2)	Region 3—allocation, kHz (3)	Government Allocation, kHz (4)	Nongovernment Allocation, kHz (5)	Rule part(s) (6)	Special-use frequencies (7)
18030–18052	FIXED.		18030–18068 FIXED.	18030–18068 FIXED.	INTERNATIONAL FIXED PUBLIC (23). MARITIME (80).	
18052–18068	FIXED. Space Research.					
18068–18168	AMATEUR 510 AMATEUR-SATELLITE. 537 538		18068–18168 510 US248	18068–18168 AMATEUR 510 AMATEUR-SATELLITE. US248	AMATEUR (97). INTERNATIONAL FIXED PUBLIC (23). MARITIME (80).	
18168–18780	FIXED. Mobile except aeronautical mobile 532		18168–18780 FIXED. Mobile	18168–18780 FIXED. Mobile	AVIATION (87). INTERNATIONAL FIXED PUBLIC (23). MARITIME (80).	
18780–18900	MARITIME MOBILE.		18780–18900 MARITIME MOBILE. US82 US296	18780–18900 MARITIME MOBILE. US82 US296	INTERNATIONAL FIXED PUBLIC (23). MARITIME (80).	
18900–19680	FIXED.		18900–19680 FIXED.	18900–19680 FIXED.	AVIATION (87). INTERNATIONAL FIXED PUBLIC (23).	
19680–19800	MARITIME MOBILE. 520B 532		19680–19800 MARITIME MOBILE. 520B	19680–19800 MARITIME MOBILE. 520B	MARITIME (80).	
19800–19990	FIXED.		19800–19990 FIXED.	19800–19990 FIXED.	AVIATION (87). INTERNATIONAL FIXED PUBLIC (23).	

Frequency Bands and Assignments

		19990–20010 STANDARD FREQUENCY AND TIME SIGNAL	19990–20010 STANDARD FREQUENCY AND TIME SIGNAL		
19990–19995	STANDARD FREQUENCY AND TIME SIGNAL Space Research. 501				
19995–20010	STANDARD FREQUENCY AND TIME SIGNAL (20000 kHz). 501				20000 kHz: Standard frequency.
20010–21000	FIXED. Mobile.	501 G106	20010–21000 FIXED.		
			501		
21000–21450	AMATEUR 510 AMATEUR-SATELLITE.	21000–21450 AMATEUR 510 AMATEUR-SATELLITE.	21000–21450 AMATEUR 510 AMATEUR-SATELLITE.		AMATEUR (97).
		510			
21450–21850	BROADCASTING.	21450–21850 BROADCASTING.	21450–21850 BROADCASTING.		INTERNATIONAL FIXED PUBLIC (23) RADIO BROADCAST (HF) (73).
	531	US235	US235		
21850–21870	FIXED.	21850–21924 FIXED.	21850–21924 FIXED.		AVIATION (87). INTERNATIONAL FIXED PUBLIC (23).
21870–21924	AERONAUTICAL FIXED. AERONAUTICAL MOBILE (R).				
	539				
22000–22855	MARITIME MOBILE.	22000–22855 AERONAUTICAL MOBILE (R).	22000–22855 AERONAUTICAL MOBILE (R).		AVIATION (87).
		MARITIME MOBILE.	MARITIME MOBILE.		INTERNATIONAL FIXED PUBLIC (23). MARITIME (80).
	520B 532 540	520B US82 US296	520B US82 US296		
22855–23000		22855–23000	22855–23000		

TABLE 9.1 Table of Frequency Allocations (*Continued*)

International Table			U.S. Table		FCC Use Designators	
Region 1—allocation, kHz (1)	Region 2—allocation, kHz (2)	Region 3—allocation, kHz (3)	Government Allocation, kHz (4)	Nongovernment Allocation, kHz (5)	Rule part(s) (6)	Special-use frequencies (7)
	FIXED. 540		FIXED.	FIXED.	AVIATION (87). INTERNATIONAL FIXED PUBLIC (23).	
23000–23200	FIXED. Mobile except aeronautical mobile (R). 540		23000–23200 FIXED. Mobile except aeronautical mobile (R).	23000–23200 FIXED.	AVIATION (87). INTERNATIONAL FIXED PUBLIC (23).	
23200–23350	AERONAUTICAL FIXED. AERONAUTICAL MOBILE (OR).		23200–23350 AERONAUTICAL MOBILE (OR).	23200–23350 AERONAUTICAL MOBILE (OR).		
23350–24000	FIXED. Mobile except aeronautical mobile 541. 542		23350–24890 FIXED. Mobile except aeronautical mobile.	23350–24890 FIXED.	AVIATION (87). INTERNATIONAL FIXED PUBLIC (23).	
24000–24890	FIXED. LAND MOBILE. 542					
24890–24990	AMATEUR 510 AMATEUR-SATELLITE. 542 543		24890–24990 510 US248	24890–24990 AMATEUR 510 AMATEUR-SATELLITE. US248	AMATEUR (97).	
24990–25005	STANDARD FREQUENCY AND TIME SIGNAL (25000 kHz).		24990–25010 STANDARD FREQUENCY AND TIME SIGNAL	24990–25010 STANDARD FREQUENCY AND TIME SIGNAL		25000 kHz: Standard frequency.
25005–25010	STANDARD FREQUENCY AND TIME SIGNAL Space Research.		G106			

Frequency Bands and Assignments

25010–25070	FIXED. MOBILE expect aeronautical mobile.	25010–25070	25010–25070 LAND MOBILE.	PRIVATE LAND MOBILE (90).
25070–25210	MARITIME MOBILE. 544	25070–25210 MARITIME MOBILE. US82 US281 US296	25070–25210 MARITIME MOBILE. NG112	MARITIME (80). PRIVATE LAND MOBILE (90).
25210–25550	FIXED. MOBILE except aeronautical mobile.	25210–25330 LAND MOBILE.		PRIVATE LAND MOBILE (90).
		25330–25550 FIXED. MOBILE except aeronautical mobile.		
25550–25670	RADIO ASTRONOMY. 545	25550–25670 RADIO ASTRONOMY. 545 US74		
25670–26100	BROADCASTING. US25	25670–26100 BROADCASTING.		AUXILIARY BROADCASTING (74). RADIO BROADCAST (HF) (73).
26100–26175	MARITIME MOBILE. 520B 544	26100–26175 MARITIME MOBILE. 520B		AUXILIARY BROADCASTING (74). MARITIME (80).
26175–27500	FIXED. MOBILE except aeronautical mobile.	26175–26480 LAND MOBILE.		AUXILIARY BROADCASTING (74).
		26480–26950 FIXED. MOBILE except aeronautical mobile.	26480–26950	

TABLE 9.1 Table of Frequency Allocations (*Continued*)

International Table			U.S. Table		FCC Use Designators	
Region 1—allocation, kHz (1)	Region 2—allocation, kHz (2)	Region 3—allocation, kHz (3)	Government Allocation, kHz (4)	Nongovernment Allocation, kHz (5)	Rule part(s) (6)	Special-use frequencies (7)
			US10	US10		
			26950–27540	26950–26960 FIXED.	INTERNATIONAL FIXED PUBLIC (23).	
				546		
				26960–27230 MOBILE except aeronautical mobile.	PERSONAL (95).	27120 ± 160 kHz: Industrial, scientific, and medical frequency.
				546		
				27230–27410 FIXED. MOBILE except aeronautical mobile.	PERSONAL (95). PRIVATE LAND MOBILE (90).	
				546		
				27410–27540 LAND MOBILE.	PRIVATE LAND MOBILE (90).	
	546					
27500–28000	METEOROLOGICAL AIDS. FIXED. MOBILE.		27540–28000 FIXED. MOBILE.	27540–28000		

International Table			U.S. Table		FCC Use Designators	
Region 1—allocation, MHz (1)	Region 2—allocation, MHz (2)	Region 3—allocation, MHz (3)	Government Allocation, MHz (4)	Nongovernment Allocation, MHz (5)	Rule part(s) (6)	Special-use frequencies (7)
28.0–29.7	AMATEUR. AMATEUR-SATELLITE.		28.0–29.7	28.0–29.7 AMATEUR. AMATEUR-SATELLITE.	AMATEUR (97).	

Frequency Bands and Assignments

29.7–30.005	FIXED. MOBILE.	29.7–29.89	29.7–29.8 LAND MOBILE.	PRIVATE LAND MOBILE (90).
			29.8–29.89 FIXED.	AVIATION (87). INTERNATIONAL FIXED PUBLIC (23).
		29.89–29.91 FIXED. MOBILE.	29.89–29.91	
		29.91–30.0	29.91–30.0 FIXED.	AVIATION (87). INTERNATIONAL FIXED PUBLIC (23).
		30.0–30.56 MOBILE. Fixed.	30.0–30.56	
30.005–30.01	FIXED. MOBILE. SPACE RESEARCH. SPACE OPERATIONS (Satellite identification)			
30.01–37.5	FIXED. MOBILE.	30.56–32.0	30.56–32.0 LAND MOBILE. NG124	PRIVATE LAND MOBILE (90).
		32.0–33.0 FIXED MOBILE.	32.0–33.0	
		33.0–34.0	33.0–34.0 LAND MOBILE. NG124	PRIVATE LAND MOBILE (90).
		34.0–35.0 FIXED. MOBILE.	34.0–35.0	
		35.0–36.0	35.0–35.19 LAND MOBILE. NG124	PRIVATE LAND MOBILE (90).
			35.19–35.69 LAND MOBILE.	DOMESTIC PUBLIC

TABLE 9.1 Table of Frequency Allocations (*Continued*)

International Table			U.S. Table		FCC Use Designators	
Region 1—allocation, MHz (1)	Region 2—allocation, MHz (2)	Region 3—allocation, MHz (3)	Government Allocation, MHz (4)	Nongovernment Allocation, MHz (5)	Rule part(s) (6)	Special-use frequencies (7)
				NG124	LAND MOBILE (22). PRIVATE LAND MOBILE (90).	
				35.69–36.0 LAND MOBILE. NG124	PRIVATE LAND MOBILE (90).	
			36.0–37.0 FIXED. MOBILE. US220	36.0–37.0		
			37.0–37.5	US220 37.0–37.5 LAND MOBILE. NG124	PRIVATE LAND MOBILE (90).	
37.5–38.25	FIXED. MOBILE. Radio Astronomy.		37.5–38.0 Radio Astronomy. 547	37.5–38.0 LAND MOBILE. Radio Astronomy. 547 NG59 NG124	PRIVATE LAND MOBILE (90).	
			38.0–38.25 FIXED. MOBILE. RADIO ASTRONOMY. 547 US81	38.0–38.25 RADIO ASTRONOMY. 547 US81		
38.25–39.986	FIXED. MOBILE.		38.25–39.0 FIXED. MOBILE.	38.25–39.0		
			39.0–40.0	39.0–40.0 LAND MOBILE. NG124	PRIVATE LAND MOBILE (90).	
39.986–40.02	FIXED. MOBILE.					

Frequency Bands and Assignments

40.02–40.98 Space Research.			
		40.0–42.0 FIXED MOBILE.	40.0–42.0
40.98–41.015 FIXED. MOBILE. Space Research. 549 550 551			
			40.68 MHz ± 0.02 MHz: Industrial, scientific and medical frequency.
41.015–44.0 FIXED. MOBILE. 549 550 551			548 US210 US220
	LAND MOBILE.	42.0–46.6 PRIVATE LAND MOBILE (90).	42.0–43.19 NG124 NG141
			43.19–43.69 LAND MOBILE. DOMESTIC PUBLIC LAND MOBILE (22). PRIVATE LAND MOBILE (90).
			43.69–46.6 LAND MOBILE. NG124 NG141 PRIVATE LAND MOBILE (90). NG124
44.0–47.0 FIXED. MOBILE. 551 552		46.6–47.0 FIXED. MOBILE.	46.6–45.0
		47.0–49.6	47.0–49.6 LAND MOBILE. NG124
47.0–68.0 BROADCASTING.	47.0–50.0 FIXED. MOBILE.	47.0–50.0 FIXED. MOBILE. BROADCASTING.	
			PRIVATE LAND MOBILE (90).
		49.6–50.0 FIXED.	49.6–50.0

TABLE 9.1 Table of Frequency Allocations (*Continued*)

International Table			U.S. Table		FCC Use Designators	
Region 1—allocation, MHz (1)	Region 2—allocation, MHz (2)	Region 3—allocation, MHz (3)	Government Allocation, MHz (4)	Nongovernment Allocation, MHz (5)	Rule part(s) (6)	Special-use frequencies (7)
553 554 555 559 561	50.0–54.0 AMATEUR. 556 557 558 560		50.0–54.0 MOBILE.	50.0–54.0 AMATEUR.	AMATEUR (97).	
	54.0–68.0 BROADCASTING. Fixed. Mobile. 562	54.0–68.0 FIXED. MOBILE. BROADCASTING.	54.0–72.0	54.0–72.0 BROADCASTING. NG128 NG149	RADIO BROADCAST (TT) (73). Auxiliary Broadcasting (74).	
68.0–74.8 FIXED. MOBILE except aeronautical mobile.	68.0–72.0 BROADCASTING. Fixed. Mobile. 563	68.0–74.8 FIXED. MOBILE, 566 568 551 572	72.0–73.0	72.0–73.0 FIXED. MOBILE. NG3 NG49 NG56	DOMESTIC PUBLIC LAND MOBILE (22). PERSONAL (96). PRIVATE LAND MOBILE (90).	
564 565 567 568 571 572	72.0–73.0 FIXED. MOBILE.		73.0–74.6 RADIO ASTRONOMY. US74	73.0–74.6 RADIO ASTRONOMY. US74		
	73.0–74.6 RADIO ASTRONOMY. 570		74.6–74.8 FIXED. MOBILE. 572 US273	74.6–74.8 FIXED. MOBILE. 572 US273	PRIVATE LAND MOBILE (90).	
	74.6–74.8 FIXED. MOBILE. 572		74.8–75.2 AERONAUTICAL RADIONAVIGATION. 572	74.8–75.2 AERONAUTICAL RADIONAVIGATION. 572	AVIATION (87).	75 MHz Marker Beacon.
74.8–75.2 AERONAUTICAL RADIONAVIGATION. 572 572A						
75.2–87.5 FIXED. MOBILE except aeronautical	75.2–75.4 FIXED. MOBILE.		75.2–75.4 FIXED. MOBILE.	75.2–75.4 FIXED. MOBILE.	PRIVATE LAND	

Frequency Bands and Assignments

Band	Int'l Region 1	Int'l Region 2/3	Federal Government	Non-Federal Government	FCC Rule Parts
	mobile 565 571 572 575 578	571 572 75.4–76 FIXED. MOBILE.	75.4–87 FIXED. MOBILE. 573 574 577 579	572 US273 75.4–76 FIXED. MOBILE.	572 US273 75.4–76 FIXED. MOBILE. NG3 NG49 NG56	MOBILE (90) DOMESTIC PUBLIC LAND MOBILE (22). PERSONAL (95). PRIVATE LAND MOBILE (90)
		76.0–88.0 BROADCASTING. Fixed. Mobile.		76.0–88.0	76.0–88.0 BROADCASTING. NG128	RADIO BROADCAST (TV) (73). Auxiliary Broadcasting (74).
87.5–100 BROADCASTING. 576			87–100 FIXED. MOBILE. BROADCASTING. 580	NG128	NG129 NG149	
100–108		88–100 BROADCASTING.		88–108	88–108 BROADCASTING.	RADIO BROADCAST (FM) (73). Auxiliary Broadcasting (74).
581 582 100–108		BROADCASTING. 582 584 585 586 587 588 589		US93	US93 NG2 NG128 NG129	
108–117.975		AERONAUTICAL RADIONAVIGATION. 590A		108–117.975 AERONAUTICAL RADIONAVIGATION. US93	108–117.975 AERONAUTICAL RADIONAVIGATION.	
117.975–136.0		AERONAUTICAL MOBILE (R).		117.975–121.9375 AERONAUTICAL MOBILE (R). 501 591 592 593 US26 US28	117.975–121.9375 AERONAUTICAL MOBILE (R). 501 591 592 593 US26 US28	AVIATION (87).

TABLE 9.1 Table of Frequency Allocations (*Continued*)

International Table			U.S. Table		FCC Use Designators	
Region 1—allocation, MHz (1)	Region 2—allocation, MHz (2)	Region 3—allocation, MHz (3)	Government Allocation, MHz (4)	Nongovernment Allocation, MHz (5)	Rule part(s) (6)	Special-use frequencies (7)
			121.9375–123.0875 AERONAUTICAL MOBILE	121.9375–123.0875 AERONAUTICAL MOBILE	AVIATION (87).	
	501 591 592 593 594		591 US30 US31 US33 US80 US102 US213	591 US30 US31 US33 US80 US102 US213		
			123.0875–123.5875 AERONAUTICAL MOBILE	123.0875–123.5875 AERONAUTICAL MOBILE	AVIATION (87).	123.1 MHz for Scene-of-Action Communication (SAR).
			591 593 US32 US33 US112	591 593 US32 US33 US112		
			123.5875–128.8125 AERONAUTICAL MOBILE (R). 591 US26	123.5875–128.8125 AERONAUTICAL MOBILE (R). 591 US26	AVIATION (87).	
			128.8125–132.0125 AERONAUTICAL MOBILE (R). 591	128.8125–132.0125 AERONAUTICAL MOBILE (R). 591	AVIATION (87).	
			132.0125–136.0 AERONAUTICAL MOBILE (R). 591 US26	132.0125–136.0 AERONAUTICAL MOBILE (R). 591 US26	AVIATION (87).	
136–137	AERONAUTICAL MOBILE (R). FIXED. Mobile except aeronautical mobile (R). 591 595 594A		136.0–137.0	136.0–137.0 AERONAUTICAL MOBILE (R).	AVIATION (87). SATELLITE COMMUNICATIONS (25).	
			591 US244	591 US244		
137.0–138.0	SPACE OPERATION (space-to-Earth).		137.0–138.0 SPACE OPERATION (space-to-Earth).	137.0–138.0 SPACE OPERATION (space-to-Earth).	SATELLITE COMMUNICATIONS (25)	

Frequency Bands and Assignments

	METEOROLOGICAL-SATELLITE (space-to-Earth). SPACE RESEARCH (space-to-Earth) Fixed. Mobile except aeronautical mobile (R). 596 597 598 599		METEOROLOGICAL-SATELLITE (space-to-Earth). SPACE RESEARCH (space-to-Earth)	METEOROLOGICAL-SATELLITE (space-to-Earth). SPACE RESEARCH (space-to-Earth)
138.0–143.6 AERONAUTICAL MOBILE (OR). 600 601 604	138.0–143.6 FIXED. MOBILE. /RADIOLOCATION/. Space Research (space-to-Earth).	138.0–143.6 FIXED. MOBILE. Space Research (space-to-Earth)./ 599 603	138.0–144.0 FIXED. MOBILE.	138.0–144.0
143.6–143.65 AERONAUTICAL MOBILE (OR). SPACE RESEARCH (space-to-Earth). 501 602 604	143.6–143.65 FIXED. MOBILE. SPACE RESEARCH (space-to-Earth). /RADIOLOCATION/.	143.6–143.65 FIXED. MOBILE. SPACE RESEARCH (space-to-Earth)./ 599 603		
143.65–144.0 AERONAUTICAL MOBILE (OR). 600 601 602 604	143.65–144.0 FIXED. MOBILE. /RADIOLOCATION/. Space Research (space-to-Earth).	143.65–144.0 FIXED. MOBILE. Space Research (space-to-Earth)./ 599 603		
144.0–146.0 AMATEUR 510 AMATEUR-SATELLITE. 605 606			US10 G30 144.0–146.0 510	US10 144.0–146.0 AMATEUR 510 AMATEUR-SATELLITE.	AMATEUR (97).
146.0–149.9 FIXED. MOBILE except aeronautical Mobile (R). 607	146.0–148.0 AMATEUR. FIXED. MOBILE. 607		146.0–148.0 AMATEUR. FIXED. MOBILE.	146.0–148.0 AMATEUR.	AMATEUR (97).
608	148.0–149.9 FIXED. MOBILE. 608		148.0–149.9 FIXED. MOBILE. 606 US10 G30	148.0–149.9 606 US10	SATELLITE COMMUNICATIONS (25).

TABLE 9.1 Table of Frequency Allocations (*Continued*)

International Table			U.S. Table		FCC Use Designators	
Region 1—allocation, MHz (1)	Region 2—allocation, MHz (2)	Region 3—allocation, MHz (3)	Government Allocation, MHz (4)	Nongovernment Allocation, MHz (5)	Rule part(s) (6)	Special-use frequencies (7)
149.9–150.05	149.9–150.05 RADIONAVIGATION-SATELLITE. 609 609A		149.9–150.05 RADIONAVIGATION SATELLITE. 609A	149.9–150.05 RADIONAVIGATION SATELLITE. 609A		
150.05–153 FIXED. MOBILE except aeronautical mobile. RADIO ASTRONOMY.	150.05–156.7625 FIXED. MOBILE.		150.05–150.8 FIXED. MOBILE. US216 G30 150.8–156.2475	150.05–150.8 US216 150.8–152 LAND MOBILE. NG51 NG112 NG124 152.0–152.255 LAND MOBILE.	PRIVATE LAND MOBILE (90). DOMESTIC PUBLIC LAND MOBILE (22).	
610 612	611 613 613A		613 US216	US216 152.255–152.495 LAND MOBILE. NG124 152.495–152.855 LAND MOBILE. NG4 152.855–156.2475 LAND MOBILE.	PRIVATE LAND MOBILE (90). DOMESTIC PUBLIC LAND MOBILE (22). PRIVATE LAND MOBILE (90). AUXILIARY BROADCASTING (74) MARITIME (80).	
153–154 FIXED. MOBILE except aeronautical mobile (R).						

Frequency Bands and Assignments

Meteorological Aids. 154.0–156.7625 FIXED. MOBILE except aeronautical mobile (R). 613 613A			
	MARITIME MOBILE (distress and calling). 501 613	156.2475–157.0375	156.2475–157.0375 613 613A NG4 NG112 NG117 NG124 NG148
156.7625–156.8375			MARITIME MOBILE. NG117
156.8375–174 FIXED. MOBILE.			
156.8375–174 FIXED. MOBILE except aeronautical mobile.	613 616 617 618	613 613A US77 US106 US107 US266	613 613A US77 US106 US107 US266
		157.0375–157.1875 MARITIME MOBILE. 613 US214 US266 G109	157.0375–157.1875 613 US214 US266	Private Land Mobile (90).
613 613B 614 615		157.1875–157.45 MARITIME MOBILE. 613 US223 US266	157.1875–157.45 MARITIME MOBILE. 613 US223 US266 NG111	MARITIME (80).
		157.45–161.575	157.45–157.755 LAND MOBILE. 613 US266 NG111 NG124	PRIVATE LAND MOBILE (90).
			157.755–158.115 LAND MOBILE. 613 158.115–161.575	DOMESTIC PUBLIC LAND MOBILE (22).

TABLE 9.1 Table of Frequency Allocations (*Continued*)

International Table			U.S. Table		FCC Use Designators	
Region 1—allocation, MHz (1)	Region 2—allocation, MHz (2)	Region 3—allocation, MHz (3)	Government Allocation, MHz (4)	Nongovernment Allocation, MHz (5)	Rule part(s) (6)	Special-use frequencies (7)
				LAND MOBILE NG6	DOMESTIC PUBLIC LAND MOBILE (22). PRIVATE LAND MOBILE (90) MARITIME (80).	
			613 US266	613 NG6 NG28 NG70 NG112 NG124 NG148		
			161.575–161.625 MARITIME MOBILE.	161.575–161.625 MARITIME MOBILE.	DOMESTIC PUBLIC LAND MOBILE (22) MARITIME (80)	
			613 US77	613 US77 NG6 NG17		
			161.625–161.775	161.625–161.775 LAND MOBILE.	AUXILIARY BROADCASTING (74). DOMESTIC PUBLIC LAND MOBILE (22)	
			613	613 NG6		
			161.775–162.0125 MARITIME MOBILE.	161.775–162.0125 MARITIME MOBILE.	DOMESTIC PUBLIC LAND MOBILE (22) MARITIME (80).	
			613 US266	613 US266 NG6		
			162.0125–173.2 FIXED. MOBILE.	162.0125–173.2	Auxiliary Broadcasting (74). Private Land Mobile (90).	
			613 US8 US11 US13 US216 US223 US300 US312 G5	613 US8 US11 US13 US216 US223 US300 US312		

Frequency Bands and Assignments

			173.2–173.4 FIXED. Land Mobile. NG124	Private Land Mobile (90).
			173.4–174.0 FIXED MOBILE. G5	173.4–174.0
174–223 BROADCASTING.	174–216 BROADCASTING. Fixed. Mobile. 620	174–233 FIXED. MOBILE. BROADCASTING.	174–216 BROADCASTING. NG115 NG128 NG149	174–216 RADIO BROADCAST (TV) (73). Auxiliary Broadcasting (74).
	216–220–FIXED. MARITIME MOBILE. Radiolocation 627, 627A.		216–220 MARITIME MOBILE. Aeronautical Mobile. Fixed. Land Mobile, Radiolocation 627, US210 US229 US274 US317 G2.	216–220 MARITIME MOBILE. Aeronautical Mobile. Fixed. Land Mobile, 627, US210 US229 US274, US317 NG121.	MARITIME (80). Private Land Mobile (90). Personal Radio Service (95).
	220–225 AMATEUR. FIXED. MOBILE. Radiolocation 627		220–222 Land mobile Radiolocation: 627 US243, G2	220–222 Land mobile 627, US243	Private land mobile (90).
		223–230 FIXED. 636 637	222–225 Radiolocation: 627 US243, G2	222–225 Amateur	Amateur (97).
	225–235 FIXED. MOBILE.		225.0–328.6 FIXED. MOBILE.	225.0–328.6 627, US243	
622 628 629 631 632 633 634 635. 230–235 FIXED.		230–235 FIXED.	501 592 642 644	501 592 642 644.	

TABLE 9.1 Table of Frequency Allocations (*Continued*)

International Table			U.S. Table		FCC Use Designators	
Region 1—allocation, MHz (1)	Region 2—allocation, MHz (2)	Region 3—allocation, MHz (3)	Government Allocation, MHz (4)	Nongovernment Allocation, MHz (5)	Rule part(s) (6)	Special-use frequencies (7)
MOBILE. 629 632 633 634 635 638 639.		MOBILE. AERONAUTICAL RADIONAVIGATION. 637	G27 G100			
235–267	FIXED. MOBILE. 501 592 635 640 641 642					
267–272	FIXED. MOBILE. Space Operation (space-to-Earth). 641 643					
272–273	SPACE OPERATION (space-to-Earth). FIXED. MOBILE. 641					
273–322	FIXED. MOBILE. 641					
322.0–328.6	FIXED. MOBILE. RADIO ASTRONOMY. 644					
328.6–335.4	AERONAUTICAL RADIONAVIGATION. 645 645A		328.6–335.4 AERONAUTICAL RADIONAVIGATION. 645	328.6–335.4 AERONAUTICAL RADIONAVIGATION. 645		
335.4–399.9	FIXED. MOBILE. 641		335.4–399.9 FIXED. MOBILE. G27 G100	335.4–399.9		

Frequency Bands and Assignments

399.9–400.05	RADIONAVIGATION-SATELLITE. 609 645B	399.9–400.05 RADIONAVIGATION-SATELLITE. 645B	399.9–400.05 RADIONAVIGATION-SATELLITE. 645B		
400.05–400.15	STANDARD FREQUENCY AND TIME SIGNAL-SATELLITE. (400.1 MHz) 646 647	400.05–400.15 STANDARD FREQUENCY AND TIME SIGNAL-SATELLITE. 646	400.05–400.15 STANDARD FREQUENCY AND TIME SIGNAL-SATELLITE. 646		400.1 MHz: Standard frequency.
400.15–401.0	METEOROLOGICAL AIDS. METEOROLOGICAL-SATELLITE (space-to-Earth). SPACE RESEARCH (space-to-Earth). Space Operation (space-to-Earth). 647	400.15–401.0 METEOROLOGICAL AIDS (radiosonde). METEOROLOGICAL-SATELLITE (space-to-Earth). SPACE RESEARCH (space-to-Earth). Space Operation (space-to-Earth). US70	400.15–401.0 METEOROLOGICAL AIDS (radiosonde). SPACE RESEARCH (space-to-Earth). Space Operation (space-to-Earth). US70	SATELLITE COMMUNICATION (25).	
401–402	METEOROLOGICAL AIDS. SPACE OPERATION (space-to-Earth). Earth Exploration-Satellite (Earth-to-space). Fixed. Meteorological-Satellite (Earth-to-space). Mobile except aeronautical mobile.	401–402 METEOROLOGICAL AIDS (radiosonde). SPACE OPERATION (space-to-Earth). Earth Exploration-Satellite (Earth-to-space). Meteorological-Satellite (Earth-to-space). US70	401–402 METEOROLOGICAL AIDS (radiosonde). SPACE OPERATION (space-to-Earth). Earth Exploration-Satellite (Earth-to-space). Meteorological-Satellite (Earth-to-space). US70	SATELLITE COMMUNICATIONS (25).	
402–403	METEOROLOGICAL AIDS. Earth Exploration-Satellite (Earth-to-space). Fixed. Meteorological-Satellite (Earth-to-space).	402–403 METEOROLOGICAL AIDS (radiosonde) Earth Exploration-Satellite (Earth-to-space). Meteorological-Satellite (Earth-to-space).	402–403 METEOROLOGICAL AIDS (radiosonde). Earth Exploration-Satellite (Earth-to-space). Meteorological-Satellite (Earth-to-space).		

TABLE 9.1 Table of Frequency Allocations (*Continued*)

International Table		U.S. Table		FCC Use Designators	
Region 1—allocation, MHz (1)	Region 2—allocation, MHz (2)	Region 3—allocation, MHz (3)	Government Allocation, MHz (4)	Nongovernment Allocation, MHz (5)	Rule part(s) (6)	Special-use frequencies (7)
	(Earth-to-space). Mobile except aeronautical mobile.					
403–406	METEOROLOGICAL AIDS. Fixed. Mobile except aeronautical mobile. 648		403–406 METEOROLOGICAL AIDS (radiosonde). US70 G6	403–406 METEOROLOGICAL AIDS (radiosonde). US70		
406.0–406.1	MOBILE-SATELLITE (Earth-to-space). 649 649A		406.0–406.1 MOBILE-SATELLITE (Earth-to-space). 649 649A	406.0–406.1 MOBILE-SATELLITE (Earth-to-space). 649 649A		
406.1–410.0	FIXED. MOBILE except aeronautical mobile. RADIO ASTRONOMY. 648 650		406.1–410.0 FIXED. MOBILE. RADIO ASTRONOMY. US13 US74 US117 G5 G6	406.1–410.0 RADIO ASTRONOMY. US13 US74 US117		
410–420	FIXED. MOBILE except aeronautical mobile.		410–420 FIXED. MOBILE. US13 G5	410–420 US13		
420–430	FIXED. MOBILE except aeronautical mobile. Radiolocation.		420–450 RADIOLOCATION. 664 668 US7 US87 US217 US228 US230 G2 G8	420–450 Amateur. 664 668 US7 US87 US217 US228 US230 NG135	LAND MOBILE (90). Amateur (97).	
	651 652 653					
430–440						

Frequency Bands and Assignments

AMATEUR. RADIOLOCATION. 653 654 655 656 657 658 659 661 662 663 664 665	RADIOLOCATION. Amateur. 653 658 659 660 660A 663 664			
440–450	FIXED. MOBILE except aeronautical mobile. Radiolocation. 651 652 653 666 667 668			
450–460	FIXED. MOBILE. 653 668 669 670	450–460 668 669 670 US87	450–451 LAND MOBILE.	AUXILIARY BROADCASTING (74). SATELLITE COMMUNICATION (25).
			451–454 LAND MOBILE. NG112 NG124	PRIVATE LAND MOBILE (90).
			454–455 LAND MOBILE. NG12 NG112 NG148	DOMESTIC PUBLIC LAND MOBILE (22). MARITIME (80).
			455–456 LAND MOBILE.	AUXILIARY BROADCASTING (74).
			456–459 LAND MOBILE. 669 670 NG112 NG124	PRIVATE LAND MOBILE (90).
			459–460 LAND MOBILE. NG12 NG112 NG148	DOMESTIC PUBLIC LAND MOBILE (22). MARITIME (80).

TABLE 9.1 Table of Frequency Allocations (*Continued*)

International Table			U.S. Table		FCC Use Designators	
Region 1—allocation, MHz (1)	Region 2—allocation, MHz (2)	Region 3—allocation, MHz (3)	Government Allocation, MHz (4)	Nongovernment Allocation, MHz (5)	Rule part(s) (6)	Special-use frequencies (7)
460–470	FIXED. MOBILE. Meterological-Satellite (space-to-Earth). 669 670 671 672		460–470 Meterological Satellite (space-to-Earth). 669 670 671 US201 US209 US216	460–462.5375 LAND MOBILE. 671 US201 US209 NG124 462.5375–462.7375 LAND MOBILE. 671 US201 462.7375–467.5375 LAND MOBILE. 669 671 US201 US209 US216 NG124 467.5375–467.7375 LAND MOBILE. 669 671 US201 467.7375–470.0 LAND MOBILE. 669 670 671 US201 US216 NG124	PRIVATE LAND MOBILE (90). PERSONAL (95). PRIVATE LAND MOBILE (90). PERSONAL (95). PRIVATE LAND MOBILE (90).	
470–790 BROADCASTING.	470–512 BROADCASTING. Fixed. Mobile.	470–585 FIXED. MOBILE. BROADCASTING.	470–512	470–512 BROADCASTING. LAND MOBILE.	RADIO BROADCAST (TV) (73). DOMESTIC PUBLIC LAND MOBILE (22). PRIVATE LAND MOBILE (90).	

Frequency Bands and Assignments

International Table		US Government Table	US Non-Government Table	FCC Rule Parts
674 675 512–608 BROADCASTING. 667A 676 682 683 684 685 686 686A 687 689 693 694	673 677 679 585–610 FIXED. MOBILE. BROADCASTING. RADIONAVIGATION.	512–608	NG66 NG114 NG127 NG128 NG149 512–608 BROADCASTING. NG128 NG149	Auxiliary Broadcasting (74). RADIO BROADCAST (TV) (73). Auxiliary Broadcasting (74).
678 608–614 RADIO ASTRONOMY. Mobile-Satellite except aeronautical mobile-satellite (Earth-to-space).	688 689 690 610–890 FIXED. MOBILE. BROADCASTING.	608–614	608–614 RADIO ASTRONOMY.	
614–806 BROADCASTING. Fixed. Mobile.		US74 US246 614–806	US74 US246 614–806 BROADCASTING. NG30 NG43 NG63 NG149	RADIO BROADCAST (TV) (73). Auxiliary Broadcasting (74).
675 692 692A 693 806–890 FIXED. MOBILE. BROADCASTING.		806–902	806–821 LAND MOBILE. NG30 NG43 NG63 NG31 821–824	PRIVATE LAND MOBILE (90).
790–862 FIXED. BROADCASTING. 694 695 695A 696 697 700B 702 862–890 FIXED. MOBILE except aeronautical mobile. BROADCASTING 700B 703.				

TABLE 9.1 Table of Frequency Allocations (*Continued*)

International Table			U.S. Table		FCC Use Designators	
Region 1—allocation, MHz (1)	Region 2—allocation, MHz (2)	Region 3—allocation, MHz (3)	Government Allocation, MHz (4)	Nongovernment Allocation, MHz (5)	Rule part(s) (6)	Special-use frequencies (7)
704	692A 700 700A	677 688 689 690 691 693 701	US116 US268 G2	LAND MOBILE. NG30 NG43 NG63	PRIVATE LAND MOBILE (90)	
				824–849 LAND MOBILE. NG30 NG43 NG63 NG151	DOMESTIC PUBLIC LAND MOBILE (22).	
				849–851 AERONAUTICAL MOBILE. NG30 NG63 NG153	PUBLIC MOBILE (22).	
				851–866 LAND MOBILE. NG30 NG63 NG31	PRIVATE LAND MOBILE(90)	
				866–869 LAND MOBILE. NG30 NG63	PRIVATE LAND MOBILE (90)	
				869–894 LAND MOBILE. NG30 NG63 US116 US268 NG151	DOMESTIC PUBLIC LAND MOBILE (22).	
				894–896 AERONAUTICAL MOBILE. US116 US268 NG153	PUBLIC MOBILE (22).	
890–942 FIXED. MOBILE except aeronautical mobile. BROADCASTING 703. Radiolocation. 704	890–902 FIXED. MOBILE except aeronautical mobile. Radiolocation. 700A 704A 705	890–942 FIXED. MOBILE. BROADCASTING. Radiolocation. 706		896–901 LAND MOBILE US116 US268	PRIVATE LAND MOBILE (90)	

Frequency Bands and Assignments

902–928 FIXED. Amateur. Mobile except aeronautical mobile. Radiolocation. 705 707 707A	902–928 RADIOLOCATION. 707 US215 US218 US267 US275 G11 G59	901–902 FIXED MOBILE US116 US268 US330	PERSONAL COMMUNICATIONS SERVICES (99)
		902–928 707 US215 US21 US267 US275	Amateur (97).	915 ± 13 MHz: Industrial, scientific and medical frequency.
928–942 FIXED. MOBILE except aeronautical mobile. Radiolocation. 705	928–932 US116 US215 US268 G2	928–929 FIXED. US116 US215 US268	DOMESTIC PUBLIC LAND MOBILE (22). PRIVATE LAND MOBILE (90). PRIVATE OPERATIONAL FIXED MICROWAVE (94).
		929–930 LAND MOBILE US116 US215 US268	DOMESTIC PUBLIC LAND MOBILE (22). PRIVATE LAND MOBILE (90).
		930–931 FIXED. MOBILE. US116 US215 US268 US330	PERSONAL COMMUNICATIONS SERVICES (99)
		931–932 LAND MOBILE. US116 US215 US268	DOMESTIC PUBLIC LAND MOBILE (22). PRIVATE LAND MOBILE (90).

TABLE 9.1 Table of Frequency Allocations (*Continued*)

International Table			U.S. Table		FCC Use Designators	
Region 1—allocation, MHz (1)	Region 2—allocation, MHz (2)	Region 3—allocation, MHz (3)	Government Allocation, MHz (4)	Nongovernment Allocation, MHz (5)	Rule part(s) (6)	Special-use frequencies (7)
942–960 FIXED. MOBILE except aeronautical mobile. BROADCASTING 703	942–960 FIXED. MOBILE.	942–960 FIXED. MOBILE. BROADCASTING.	932–935 FIXED US215 US268 G2	932–935 FIXED US215 US268		
			935–941 US116 US215 US268 G2	935–940 LAND MOBILE US116 US215 US268	PRIVATE LAND MOBILE (90)	
				940–941 FIXED MOBILE US116 US268 US330	PERSONAL COMMUNICATIONS SERVICES (99)	
			941–944 FIXED US268 US301 US302	941–944 FIXED US268 US301 US302		
			944–960 FIXED	944–960 FIXED NG120	AUXILIARY BROADCASTING (74). DOMESTIC PUBLIC FIXED (22). INTERNATIONAL FIXED PUBLIC (23). PRIVATE OPERATIONAL FIXED MICROWAVE (94).	
704	708	701				
960–1215	AERONAUTICAL RADIONAVIGATION. 709		960–1215 AERONAUTICAL RADIONAVIGATION. 709 US224	960–1215 AERONAUTICAL RADIONAVIGATION. 709 US224	AVIATION (87).	

Frequency Bands and Assignments

1215–1240	1215–1240 RADIOLOCATION. RADIONAVIGATION- SATELLITE (space-to-Earth) 710. 711 712 712A 713	1215–1240	
1240–1260	1240–1260 RADIOLOCATION. RADIONAVIGATION- SATELLITE (space-to-Earth) 710. Amateur. 711 712 712A 713 714		
1260–1300	1260–1300 RADIOLOCATION. Amateur. 664 711 712 712A 713 714	1240–1300 RADIOLOCATION. 713	Amateur (97).
		1240–1300 Amateur.	
		1240–1300 RADIOLOCATION. 664 713 714	
1300–1350	1300–1350 AERONAUTICAL RADIONAVIGATION 717 Radiolocation 715 716 718	1300–1350 AERONAUTICAL RADIONAVIGATION 717 Radiolocation 718 G2	AVIATION (87).
		1300–1350 AERONAUTICAL RADIONAVIGATION 717 718	
1350–1400 FIXED. MOBILE. RADIOLOCATION. 718 719 720	1350–1400 RADIOLOCATION. 714 718 720	1350–1400 RADIOLOCATION. Fixed. Mobile. 714 718 720 G2 G27 G114 US311	
		1350–1400 714 718 720	
1400–1427	1400–1427 EARTH EXPLORATION- SATELLITE (passive). RADIO ASTRONOMY. SPACE RESEARCH (passive). 721 722	1400–1427 EARTH EXPLORATION- SATELLITE (passive). RADIO ASTRONOMY. SPACE RESEARCH (passive). 722 US74 US246	1400–1427 EARTH EXPLORATION- SATELLITE (passive). RADIO ASTRONOMY. SPACE RESEARCH (passive). 722 US74 US246	
1427–1429	1427–1429 SPACE OPERATION (Earth-to-space).	1427–1429 SPACE OPERATION (Earth-to-space).	1427–1429 SPACE OPERATION (Earth-to-space).	Private Land Mobile (90).

TABLE 9.1 Table of Frequency Allocations (*Continued*)

International Table			U.S. Table		FCC Use Designators	
Region 1—allocation, MHz (1)	Region 2—allocation, MHz (2)	Region 3—allocation, MHz (3)	Government Allocation, MHz (4)	Nongovernment Allocation, MHz (5)	Rule part(s) (6)	Special-use frequencies (7)
	FIXED. MOBILE except aeronautical mobile. 722		FIXED. MOBILE except aeronautical mobile. 722 G30	Fixed (telemetering). Land Mobile (telemetering and telecommand). 722	Satellite Communications (25).	
1429–1525 FIXED. MOBILE 723 722	1429–1525 FIXED. MOBILE 723 722		1429–1435 FIXED. MOBILE. 722 G30	1429–1435 Land Mobile (telemetering and telecommand). Fixed (telemetering). 722	Private Land Mobile (90).	
			1435–1530 MOBILE (aeronautical telemetering)	1435–1530 MOBILE (aeronautical telemetering).	AVIATION (87).	
1525–1530 SPACE OPERATION (space-to-Earth). FIXED. MARITIME MOBILE-SATELLITE (space-to-Earth). Land Mobile-Satellite (space-to-Earth) 726B. Earth Exploration-Satellite Mobile except aeronautical mobile 724. 722 723B 725 726A 726D	1525–1530 SPACE OPERATION (space-to-Earth). MOBILE-SATELLITE (space-to-Earth). Earth Exploration-satellite. Fixed. Mobile 723. 722 723A 726A 726D	1525–1530 SPACE OPERATION (space-to-Earth). FIXED. MOBILE-SATELLITE (space-to-Earth). Earth Exploration-satellite. Mobile 723 724. 722 726A 726D	722 US78	722 US78		
1530–1533 SPACE OPERATION (space-to-Earth). MARITIME MOBILE-SATELLITE (space-to-Earth). LAND MOBILE-SATELLITE (space-to-Earth).	1530–1533 SPACE OPERATION (space-to-Earth). MARITIME MOBILE-SATELLITE (space-to-Earth). MOBILE-SATELLITE (space-to-Earth). LAND MOBILE-SATELLITE (space-to-Earth).	1530–1533 SPACE OPERATION (space-to-Earth). MARITIME MOBILE-SATELLITE (space-to-Earth). MOBILE-SATELLITE (space-to-Earth). LAND MOBILE-SATELLITE (space-to-Earth).	1530–1533 MARITIME MOBILE-SATELLITE (space-to-Earth). MOBILE-SATELLITE (space-to-Earth). Mobile (aeronautical telemetering).	1530–1533 MARITIME MOBILE-SATELLITE (space-to-Earth). MOBILE-SATELLITE (space-to-Earth). Mobile (aeronautical telemetering).	SATELLITE COMMUNICATION (25). Aviation (87).	

Frequency Bands and Assignments

Earth Exploration-Satellite. Fixed. Mobile except aeronautical mobile. 722 723B 726A 726D	Earth Exploration-Satellite. Fixed. Mobile 723. 722 726A 726C 726D	Earth Exploration-Satellite. Fixed. Mobile 723. 722 726A 726C 726D			
1533–1535 SPACE OPERATION (space-to-Earth). MARITIME MOBILE-SATELLITE (space-to-Earth). Earth Exploration-Satellite. Fixed. Mobile except aeronautical mobile. Land Mobile-Satellite (space-to-Earth) 726B. 722 726A 726C 726D	1533–1535 SPACE OPERATION (space-to-Earth). MARITIME MOBILE-SATELLITE (space-to-Earth). Earth Exploration-Satellite. Fixed. Mobile 723. Land Mobile-Satellite (space-to-Earth) 726B. 722 726A 726C 726D				
1544–1545 MOBILE-SATELLITE (space-to-Earth). 722 726D 727 727A	1544–1545 MOBILE-SATELLITE (space-to-Earth). 722 726D 727 727A		722 726A US78 US315	722 726A US78 US315	
1545–1555 AERONAUTICAL MOBILE-SATELLITE (R) (space-to-Earth). 722 726A 726D 727 729 729A 730	1545–1555 AERONAUTICAL MOBILE-SATELLITE (R) (space-to-Earth). 722 726A 726D 727 729 729A 730		1544–1545 MOBILE-SATELLITE (space-to-Earth). 722 727A	1544–1545 MOBILE-SATELLITE (space-to-Earth). 722 727A	MARITIME (80). SATELLITE COMMUNICATION (25).
			1545–1549.5 AERONAUTICAL MOBILE-SATELLITE (R) (space-to-Earth). Mobile-satellite (space-to-Earth). 722 726A US308 US309	1545–1549.5 AERONAUTICAL MOBILE-SATELLITE (R) (space-to-Earth). Mobile-satellite (space-to-Earth). 722 726A US308 US309	AVIATION (87).
			1549.5–1558.5 AERONAUTICAL MOBILE-SATELLITE (R) (space-to-Earth). MOBILE-SATELLITE (space-to-Earth).	1549.5–1558.5 AERONAUTICAL MOBILE-SATELLITE (R) (space-to-Earth). MOBILE-SATELLITE (space-to-Earth).	AVIATION (87).

TABLE 9.1 Table of Frequency Allocations (*Continued*)

International Table		U.S. Table		FCC Use Designators	
Region 1—allocation, MHz (1)	Region 2—allocation, MHz (2)	Region 3—allocation, MHz (3)	Government Allocation, MHz (4)	Non-government Allocation, MHz (5)	Rule part(s) (6)	Special-use frequencies (7)
---	---	---	---	---	---	---
1555–1559 LAND MOBILE-SATELLITE (space-to-Earth). 722 726A 726D 727 730 731A 731B 731D 732 733 733A 733B 733E 733F 734	1555–1559 LAND MOBILE-SATELLITE (space-to-Earth). 722 726A 726D 727 730 730A 730B 730C	1555–1559 LAND MOBILE-SATELLITE (space-to-Earth). 722 726A 726D 727 730 730A 730B 730C	722 726A US308 US309	722 726A US308 US309		
			1558.5–1559 AERONAUTICAL MOBILE-SATELLITE (R) (space-to-Earth). 722 726A US308 US309	1558.5–1559 AERONAUTICAL MOBILE-SATELLITE (R) (space-to-Earth). 722 726A US308 US309	AVIATION (87).	
1559–1610 AERONAUTICAL RADIONAVIGATION- SATELLITE (space-to-Earth). 722 727 730 731 731A 731B 731C 731D	722 US28		1559–1610 AERONAUTICAL RADIONAVIGATION- SATELLITE (space-to-Earth). 722 US208 US260 US280	1559–1610 AERONAUTICAL RADIONAVIGATION- SATELLITE (space-to-Earth)	AVIATION (87).	
1610–1626.5 AERONAUTICAL RADIONAVIGATION. Radiodetermination-satellite (Earth-to-space) 733A 733E 722 731B 731C 731D 732 733 733C 733D 734	1610–1626.5 AERONAUTICAL RADIONAVIGATION. Radiodetermination-satellite (Earth-to-space) 733A 733E 722 727 730 731B 731C 732 733 733B 734		1610–1626.5 AERONAUTICAL RADIONAVIGATION. 722 732 733 734 US208 US260 US306	AERONAUTICAL RADIONAVIGATION. 722 732 733 734 US208 US260 US306	AVIATION (87). Satellite communication (25).	
1626.5–1631.5 MARITIME MOBILE-SATELLITE (Earth-to-space). Land Mobile-Satellite (Earth-to-space) 726B. 722 726A 726D 727 730	1625.5–1631.5 MOBILE-SATELLITE (Earth-to-space). 722 726A 726C 726D 727 730	1626.5–1631.5 MOBILE-SATELLITE (Earth-to-space). 722 726A 726C 726D 727 730	1626.5–1645.5 MARITIME MOBILE-SATELLITE (Earth-to-space). MOBILE-SATELLITE (Earth-to-space).	1626.5–1645.5 MARITIME MOBILE-SATELLITE (Earth-to-space). MOBILE-SATELLITE (Earth-to-space).	MARITIME (80). SATELLITE COMMUNICATION (25).	
1631.5–1634.5 MARITIME MOBILE-		1631.5–1634.5 MARITIME MOBILE-				

Frequency Bands and Assignments

SATELLITE (Earth-to-space). LAND MOBILE-SATELLITE (Earth-to-space). 722 726A 726C 726D 727 730 734A	SATELLITE (Earth-to-space). LAND MOBILE-SATELLITE (Earth-to-space). 722 726A 726C 726D 727 730 734A	SATELLITE (Earth-to-space). LAND MOBILE-SATELLITE (Earth-to-space). 722 726A 726C 726D 727 730 734A				
1634.5–1645.5 MARITIME MOBILE-SATELLITE (Earth-to-space). Land Mobile-Satellite (Earth-to-space) 726B. 722 726A 726C 726D 727 730	1634.5–1645.5 MARITIME MOBILE-SATELLITE (Earth-to-space). Land Mobile-Satellite (Earth-to-space). 726B. 722 726A 726C 726D 727 730	1634.5–1645.5 MARITIME MOBILE-SATELLITE (Earth-to-space). Land Mobile-Satellite (Earth-to-space). 726B. 722 726A 726C 726D 727 730		722 726A US315	722 726A US315	
1645.5–1646.5 MOBILE-SATELLITE (Earth-to-space). 722 726D 734B	1645.5–1646.5 MOBILE-SATELLITE (Earth-to-space). 722 726D 734B	1645.5–1646.5 MOBILE-SATELLITE (Earth-to-space). 722 726D 734B		1645.5–1646.5 MOBILE-SATELLITE (Earth-to-space). 722 734B	1645.5–1646.5 MOBILE-SATELLITE (Earth-to-space). 722 734B	MARITIME (80). SATELLITE COMMUNICATION (25).
1646.5–1656.5 AERONAUTICAL MOBILE-SATELLITE (R) (Earth-to-space). 722 726A 726D 727 729A 730 735	1646.5–1656.5 AERONAUTICAL MOBILE-SATELLITE (R) (Earth-to-space). 722 726A 726D 727 729A 730 735	1646.5–1656.5 AERONAUTICAL MOBILE-SATELLITE (R) (Earth-to-space). 722 726A 726D 727 729A 730 735		1646.5–1651 AERONAUTICAL MOBILE-SATELLITE (R) (Earth-to-space). Mobile-Satellite (Earth-to-space). 722 726A US308 US309	1646.5–1651 AERONAUTICAL MOBILE-SATELLITE (R) (Earth-to-space). Mobile-Satellite (Earth-to-space). 722 726A US308 US309	AVIATION (87).
1656.5–1660 LAND MOBILE-SATELLITE (Earth-to-space). 722 726A 726D 727 730 730A 730B 730C 734A	1656.5–1660 LAND MOBILE-SATELLITE (Earth-to-space). 722 726A 726D 727 730 730A 730B 730C 734A	1656.5–1660 LAND MOBILE-SATELLITE (Earth-to-space). 722 726A 726D 727 730 730A 730B 730C 734A		1651–1660 AERONAUTICAL MOBILE-SATELLITE (R) (Earth-to-space). MOBILE-SATELLITE (Earth-to-space). 722 726A US39	1651–1660 AERONAUTICAL MOBILE-SATELLITE (R) (Earth-to-space). MOBILE-SATELLITE (Earth-to-space). 722 726A US39	AVIATION (87).
1660–1660.5 RADIO ASTRONOMY LAND MOBILE-SATELLITE (Earth-to-space).	1660–1660.5 RADIO ASTRONOMY LAND MOBILE-SATELLITE (Earth-to-space).	1660–1660.5 RADIO ASTRONOMY LAND MOBILE-SATELLITE (Earth-to-space).		1660–1660.5 AERONAUTICAL MOBILE-SATELLITE (R) (Earth-to-space). RADIO ASTRONOMY.	1660–1660.5 AERONAUTICAL MOBILE-SATELLITE (R) (Earth-to-space). RADIO ASTRONOMY.	AVIATION (87).

TABLE 9.1 Table of Frequency Allocations (*Continued*)

International Table			U.S. Table		FCC Use Designators	
Region 1—allocation, MHz (1)	Region 2—allocation, MHz (2)	Region 3—allocation, MHz (3)	Government Allocation, MHz (4)	Nongovernment Allocation, MHz (5)	Rule part(s) (6)	Special-use frequencies (7)
722 726A 726D 730A 730B 730C 736	722 726A 726D 730A 730B 730C 736	722 726A 726D 730A 730B 730C 736	722 726A 736 US309	722 726A 736 US309		
1660.5–1668.4 RADIO ASTRONOMY SPACE RESEARCH (passive). Fixed. Mobile except aeronautical mobile. 722 736 737 738 739			1660.5–1668.4 RADIO ASTRONOMY SPACE RESEARCH (passive). 722 US74 US246	1660.5–1668.4 RADIO ASTRONOMY SPACE RESEARCH (passive). 722 US74 US246		
1668.4–1670.0 METEOROLOGICAL AIDS. FIXED. MOBILE except aeronautical mobile. RADIO ASTRONOMY 722 736			1668.4–1670.0 METEOROLOGICAL AIDS (radiosonde). RADIO ASTRONOMY. 722 736 US74 US99	1668.4–1670.0 METEOROLOGICAL AIDS (radiosonde). RADIO ASTRONOMY. 722 736 US74 US99		
1670–1690 METEOROLOGICAL AIDS. FIXED. METEOROLOGICAL-SATELLITE (space-to-Earth). MOBILE except aeronautical mobile. 722			1670–1690 METEOROLOGICAL AIDS (radiosonde). METEOROLOGICAL-SATELLITE (space-to-Earth). 722 US211	1670–1690 METEOROLOGICAL AIDS (radiosonde). METEOROLOGICAL-SATELLITE (space-to-Earth). 722 US211		
1690–1700 METEOROLOGICAL AIDS. METEOROLOGICAL-SATELLITE (space-to-			1690–1700 METEOROLOGICAL AIDS (radiosonde). METEOROLOGICAL-	1690–1700 METEOROLOGICAL AIDS (radiosonde). METEOROLOGICAL-		

Earth). Mobile except aeronautical mobile. 671 722 742	SATELLITE (space-to-Earth). 671 722 740 742		SATELLITE (space-to-Earth). 671 722		
1700–1710 FIXED. METEOROLOGICAL-SATELLITE (space-to-Earth). Mobile except aeronautical mobile. 671 722 743A	1700–1710 FIXED. METEOROLOGICAL SATELLITE (space-to-Earth). MOBILE except aeronautical mobile. 671 722 743		1700–1710 FIXED. METEOROLOGICAL-SATELLITE (space-to-Earth). Fixed. 671 722 G118		
1710–2290 FIXED. Mobile. 722 743A 744 746 747 748 750	1710–2290 FIXED. MOBILE. 722 744 745 746 747 748 749 750	1710–1850 FIXED. MOBILE. 722 US256 G42	1710–1850 722 US256		
		1850–1990	1850–1990 FIXED. NG153	1850–1990 PRIVATE OPERATIONAL-FIXED MICROWAVE (94).	EMERGING TECHNOLOGIES
		1990–2110	1990–2100 Fixed MOBILE. US90 US111 US219 US222 NG23 NG118	AUXILIARY BROAD-CAST (74) CABLE TELEVISION (78) DOMESTIC PUBLIC FIXED (21)	
		US90 US111 US219 US222 2110–2200 US111 US252			
		2200–2290 FIXED. MOBILE. SPACE RESEARCH (space-to-Earth) (space-to-Earth) US303 G101	2100–2290 US303		

TABLE 9.1 Table of Frequency Allocations (*Continued*)

International Table			U.S. Table		FCC Use Designators	
Region 1—allocation, MHz (1)	Region 2—allocation, MHz (2)	Region 3—allocation, MHz (3)	Government Allocation, MHz (4)	Nongovernment Allocation, MHz (5)	Rule part(s) (6)	Special-use frequencies (7)
2290–2300 FIXED. SPACE RESEARCH (space-to-Earth) (deep space). MOBILE except aeronautical mobile. 743A	2290–2300 FIXED. MOBILE except aeronautical mobile. SPACE RESEARCH (space-to-Earth) (deep space).		2290–2300 FIXED. MOBILE except aeronautical mobile. SPACE RESEARCH (space-to-Earth) (deep space only).	2290–2300 SPACE RESEARCH (space-to-Earth) (deep space only).		
2300–2450 FIXED. Amateur. Mobile. Radiolocation.	2300–2450 FIXED. MOBILE. RADIOLOCATION. Amateur.		2300–2310 RADIOLOCATION. Fixed. Mobile. US253 G2	2300–2310 Amateur. US253	Amateur (97).	
			2310–2390 MOBILE. RADIOLOCATION. Fixed. US276 G2	2310–2390 MOBILE US276		
664 743A 752	664 751 752		2390–2450 RADIOLOCATION. 664 752 G2	2390–2450 Amateur. 664 752	Amateur (97).	
2450–2483.5 FIXED. MOBILE. Radiolocation. 752 753	2450–2483.5 FIXED. MOBILE. RADIOLOCATION. 752		2450–2483.5	2450–2483.5 FIXED. MOBILE. Radiolocation. 752 US41		2450 ± 50 MHz: Industrial, scientific and medical frequency.
			752 US41			
2483.5–2500 FIXED MOBILE. Radiolocation.	2483.5–2500 FIXED-MOBILE RADIODETERMINATION SATELLITE. (space-to-Earth). Radiolocation.	2483.5–2500 FIXED-MOBILE RADIOLOCATION RADIODETERMINATION SATELLITE (space-to-earth)	2483.5–2500	2483.5–2500 RADIODETERMINATION SATELLITE (space-to-Earth).	Satellite communication (25).	
733F 752 753A 753B 753C 763E	752 753D	752 753A 753C	752 US41	752 US41 NG147		

Frequency Bands and Assignments

2500–2655 FIXED 762 763 764 MOBILE except aeronautical mobile. BROADCASTING-SATELLITE 757 760	2500–2655 FIXED 762 764 FIXED-SATELLITE (space-to-Earth) 761 MOBILE except aeronautical mobile. BROADCASTING-SATELLITE 757 760	2500–2535 FIXED 762 764 FIXED SATELLITE (space-to-Earth) 761 MOBILE except aeronautical mobile. BROADCASTING-SATELLITE 757 754 754A	2500–2655	2500–2655 FIXED. BROADCASTING-SATELLITE.	AUXILIARY BROADCASTING (74) DOMESTIC PUBLIC FIXED RADIO (21)
		2536–2655 FIXED 762 764 MOBILE except aeronautical mobile. BROADCASTING-SATELLITE 757 760			
720 753 756 758 759	720 755	720	720 US205 US269	720 US205 US269 NG101 NG102	
2655–2690 FIXED 762 763 764 MOBILE except aeronautical mobile. BROADCASTING-SATELLITE 757 760 Earth Exploration-Satellite (passive). Radio Astronomy. Space Research passive).	2655–2690 FIXED 762 764 FIXED-SATELLITE (Earth-to-space) 761 MOBILE except aeronautical mobile. BROADCASTING-SATELLITE 757 760 Earth Exploration-Satellite (passive). Radio Astronomy. Space Research passive).	2655–2690 FIXED 762 764 FIXED-SATELLITE (Earth-to-space) 761 MOBILE except aeronautical mobile. BROADCASTING-SATELLITE 757 760 Earth Exploration-Satellite (passive). Radio Astronomy. Space Research passive).	2655–2690 Earth Exploration-Satellite (passive) Radio Astronomy Space Research passive).	2655–2690 FIXED. BROADCASTING-SATELLITE. Earth Exploration-Satellite (passive). Radio Astronomy. Space Research (passive)	AUXILIARY BROADCASTING (74). PRIVATE OPERATIONAL-FIXED MICROWAVE (94).
758 759 765	765	765 766	US205 US269	US205 US269 NG47 NG101 NG102	
2690–2700	EARTH EXPLORATION-SATELLITE (passive). RADIO ASTRONOMY. SPACE RESEARCH (passive). 767 768 769		2690–2700 EARTH EXPLORATION-SATELLITE(passive). RADIO ASTRONOMY. SPACE RESEARCH (passive). US74 US246	2690–2700 EARTH EXPLORATION SATELLITE (passive). RADIO ASTRONOMY. SPACE RESEARCH (passive). US74 US246	
2700–2900			2700–2900	2700–2900	

TABLE 9.1 Table of Frequency Allocations (*Continued*)

International Table			U.S. Table		FCC Use Designators	
Region 1—allocation, MHz (1)	Region 2—allocation, MHz (2)	Region 3—allocation, MHz (3)	Government Allocation, MHz (4)	Nongovernment Allocation, MHz (5)	Rule part(s) (6)	Special-use frequencies (7)
	AERONAUTICAL RADIONAVIGATION 717. Radiolocation. 770 771		AERONAUTICAL RADIONAVIGATION 717 METEOROLOGICAL AIDS. Radiolocation. 770 US18 G2 G15			
2900–3100	RADIONAVIGATION. Radiolocation. 772 773 775A		2900–3100 MARITIME RADIO-NAVIGATION 775A Radiolocation. US44 US316 G56	2900–3100 MARITIME RADIO-NAVIGATION 775A Radiolocation. US44 US316 717 770 US18	MARITIME (80).	
3100–3300	RADIOLOCATION. 713 777 778		3100–3300 RADIOLOCATION. 713 778 US110 G59	3100–3300 Radiolocation. 713 778 US110		
3300–3400 RADIOLOCATION	3300–3400 RADIOLOCATION. Amateur. Fixed. Mobile. 778 780	3300–3400 RADIOLOCATION. Amateur. 778 779	3300–3500 RADIOLOCATION.	3300–3500 Amateur. Radiolocation.	Amateur (97).	
778 779 780			664 778 US108 G31	664 778 US108		
3400–3600 FIXED. FIXED-SATELLITE (space-to-Earth). Mobile. Radiolocation.	3400–3500 FIXED. FIXED-SATELLITE (space-to-Earth). Amateur. Mobile. Radiolocation 784 664 783					
	3500–3700 FIXED. FIXED-SATELLITE (space-to-Earth) MOBILE except		3500–3600 AERONAUTICAL RADIONAVIGATION (ground-based). RADIOLOCATION	3500–3600 Radiolocation.		

Frequency Bands and Assignments

781 782 785				
3600–4200 FIXED. FIXED-SATELLITE (space-to-Earth). Mobile.	aeronautical mobile Radiolocation 784			
		US110 G59 G110	US110	
	786	3600–3700 AERONAUTICAL RADIONAVIGATION (ground-based). RADIOLOCATION. US110 US245 G59 G110	3600–3700 FIXED-SATELLITE (space-to-Earth). Radiolocation. US110 US245	
	3700–4200 FIXED. FIXED-SATELLITE (space-to-Earth). MOBILE except aeronautical mobile. 787	3700–4200	3700–4200 FIXED. FIXED-SATELLITE (space-to-Earth). NG41	DOMESTIC PUBLIC FIXED (21). SATELLITE COMMUNICATIONS (25) PRIVATE OPERATIONAL FIXED MICROWAVE (94).
4200–4400	AERONAUTICAL RADIONAVIGATION 789 788 790 791	4200–4400 AERONAUTICAL RADIONAVIGATION. 791 US261	4200–4400 AERONAUTICAL RADIONAVIGATION. 791 US261	AVIATION (87).
4400–4500	FIXED. MOBILE.	4400–4500 FIXED. MOBILE.	4400–4500	
4500–4800	FIXED. FIXED-SATELLITE (space-to-Earth). MOBILE. 792A	4500–4800 FIXED. MOBILE. US245	4500–4800 FIXED-SATELLITE (space-to-Earth). 792A US245	
4800–4990	FIXED. MOBILE 793 Radio Astronomy. 720 778 794	4800–4990 FIXED. MOBILE. 720 778 US203 US257	4800–4990 720 778 US203 US257	
4990–5000	FIXED. MOBILE except	4990–5000 RADIO ASTRONOMY. Space Research	4990–5000 RADIO ASTRONOMY. Space Research	

TABLE 9.1 Table of Frequency Allocations (*Continued*)

International Table		U.S. Table		FCC Use Designators	
Region 1—allocation, MHz (1)	Region 2—allocation, MHz (2)	Region 3—allocation, MHz (3)	Government Allocation, MHz (4)	Nongovernment Allocation, MHz (5)	Rule part(s) (6)	Special-use frequencies (7)
	aeronautical mobile. RADIO ASTRONOMY. Space Research (passive). 795		(passive).	(passive).		
5000–5250	AERONAUTICAL RADIONAVIGATION. 733 796 797 797A 797B		US74 US246 5000–5250 AERONAUTICAL RADIONAVIGATION. 733 796 797 797A US211 US260 US307	US74 US246 5000–5250 AERONAUTICAL RADIONAVIGATION. 733 796 797 797A US211 US260 US307	AVIATION (87). Satellite communication (25).	
5250–5255	RADIOLOCATION. Space Research. 713 798		5250–5350 RADIOLOCATION.	5250–5350 Radiolocation.		
5255–5350	RADIOLOCATION. 713 798		713 US110 G59	713 US110		
5350–5460	AERONAUTICAL RADIONAVIGATION 799 Radiolocation.		5350–5460 AERONAUTICAL RADIONAVIGATION 799 RADIOLOCATION. US48 G56	5350–5460 AERONAUTICAL RADIONAVIGATION 799 Radiolocation. US48	AVIATION (87).	
5460–5470	RADIONAVIGATION 799 Radiolocation.		5460–5470 RADIONAVIGATION 799 Radiolocation. US49 US65 G56	5460–5470 RADIONAVIGATION 799 Radiolocation. US49 US65		
5470–5650	MARITIME RADIO-NAVIGATION Radiolocation.		5470–5650 MARITIME RADIO-NAVIGATION. Radiolocation. US50 US65 G56 5600–5650 MARITIME RADIO-NAVIGATION.	5470–5650 MARITIME RADIO-NAVIGATION. Radiolocation. US50 US65 5600–5650 MARITIME RADIO-NAVIGATION.	MARITIME (80). MARITIME (80).	

Frequency Bands and Assignments

				800 801 802		
5650–5725				RADIOLOCATION. Amateur. Space Research (deep space). 664 801 803 804 805		
5725–5850 FIXED-SATELLITE (Earth-to-space). RADIOLOCATION. Amateur. 801 803 805 806 807 808				5725–5850 RADIOLOCATION. Amateur. 803 805 806 808	METEOROLOGICAL AIDS. Radiolocation. 772 802 US51 US65 G56 5650–5850 RADIOLOCATION. 664 806 808 G2	METEOROLOGICAL AIDS. Radiolocation. 772 802 US51 US65 5650–5850 Amateur. 644 806 808	Amateur (97). Industrial, Scientific, and Medical Equipment (18).	58000 ± 75 MHz Industrial, scientific and medical frequency.
5850–5925 FIXED. FIXED-SATELLITE (Earth-to-space) MOBILE. Amateur. Radiolocation. 806				5850–5925 FIXED. FIXED-SATELLITE (Earth-to-space) MOBILE. Radiolocation.	5850–5925 RADIOLOCATION. 806 US245 G2	5850–5925 FIXED-SATELLITE (Earth-to-space). Amateur. 806 US245	Amateur (97).	
5925–7075				FIXED. FIXED-SATELLITE (Earth-to-space). MOBILE. 791 792A 809	5925–7125 791 809	5925–6425 FIXED. FIXED SATELLITE (Earth-to-space). NG41 6425–6525 FIXED-SATELLITE (Earth-to-space). MOBILE.	DOMESTIC PUBLIC FIXED (21) SATELLITE COMMUNICATIONS (25). PRIVATE OPERATIONAL FIXED MICROWAVE (94). AUXILIARY BROADCAST (74) CABLE TELEVISION (78) DOMESTIC PUBLIC	

176   Conversion Factors, Standards, and Constants

TABLE 9.1 Table of Frequency Allocations (*Continued*)

	International Table		U.S. Table		FCC Use Designators	
Region 1—allocation, MHz (1)	Region 2—allocation, MHz (2)	Region 3—allocation, MHz (3)	Government Allocation, MHz (4)	Nongovernment Allocation, MHz (5)	Rule part(s) (6)	Special-use frequencies (7)
7075–7250	FIXED. MOBILE. 809 810 811			791 809 6525–6875 FIXED. FIXED-SATELLITE (Earth-to-space).	FIXED (21) PRIVATE OPERATIONAL-FIXED MICROWAVE (94). DOMESTIC PUBLIC FIXED (21) SATELLITE COMMUNICATIONS (25) PRIVATE OPERATION-FIXED MICROWAVE (94).	
				792A 809 6875–7075 FIXED. FIXED-SATELLITE (Earth-to-space). MOBILE 792A 809 NG118	AUXILIARY BROADCAST (74) CABLE TELEVISION (78) DOMESTIC PUBLIC FIXED (21).	
			7125–7190 FIXED. 809 US252 G116 7190–7235 FIXED. SPACE RESEARCH (Earth-to-space).	7075–7125 FIXED MOBILE. 809 NG118 7125–8450	AUXILIARY BROADCAST (74) CABLE TELEVISION (78) DOMESTIC PUBLIC FIXED (21).	

Frequency Bands and Assignments

		809	
		7235–7250 FIXED. 809	809 US252 US258
7250–7300	FIXED. FIXED-SATELLITE (space-to-Earth). MOBILE. 812	7250–7300 FIXED-SATELLITE (space-to-Earth). MOBILE-SATELLITE (space-to-Earth). Fixed. G117	
7300–7450	FIXED. FIXED-SATELLITE (space-to-Earth) MOBILE except aeronautical mobile. 812	7300–7450 FIXED. FIXED-SATELLITE (space-to-Earth) Mobile-Satellite (space-to-Earth). G117	
7450–7550	FIXED. FIXED-SATELLITE (space-to-Earth). METEOROLOGICAL- SATELLITE (space-to-Earth). MOBILE except aeronautical mobile.	7450–7550 FIXED. FIXED-SATELLITE (space-to-Earth). METEOROLOGICAL- SATELLITE (space-to-Earth). Mobile-Satellite (space-to-Earth). G104 G117	
7550–7750	FIXED. FIXED-SATELLITE (space-to-Earth). MOBILE except aeronautical mobile.	7550–7750 FIXED. FIXED-SATELLITE (space-to-Earth). Mobile-Satellite (space-to-Earth). G117	
7750–7900	FIXED. MOBILE except aeronautical mobile.	7750–7900 FIXED.	
7900–7975	FIXED. FIXED-SATELLITE	7900–8025 FIXED-SATELLITE (Earth-to-space).	

TABLE 9.1 Table of Frequency Allocations (*Continued*)

International Table			U.S. Table		FCC Use Designators	
Region 1—allocation, MHz (1)	Region 2—allocation, MHz (2)	Region 3—allocation, MHz (3)	Government Allocation, MHz (4)	Nongovernment Allocation, MHz (5)	Rule part(s) (6)	Special-use frequencies (7)
	(Earth-to-space). MOBILE. 812		MOBILE-SATELLITE (Earth-to-space). Fixed.			
7975–8025 FIXED. FIXED-SATELLITE (Earth-to-space). MOBILE. 812			G117			
8025–8175 FIXED. FIXED-SATELLITE (Earth-to-space). MOBILE. Earth Exploration-Satellite (space-to-Earth) 813 815	8025–8175 EARTH EXPLORATION-SATELLITE (space-to-Earth) FIXED. FIXED-SATELLITE (Earth-to-space). MOBILE 814	8025–8175 FIXED. FIXED-SATELLITE (Earth-to-space). MOBILE. Earth Exploration-Satellite (space-to-Earth) 813 815	8025–8175 EARTH EXPLORATION-SATELLITE (space-to-Earth). FIXED. FIXED-SATELLITE (Earth-to-space). Mobile-Satellite (Earth-to-space) (no airborne transmission). US258 G117			
8175–8215 FIXED. FIXED-SATELLITE (Earth-to-space). METEOROLOGICAL-SATELLITE (Earth-to-space). MOBILE. Earth Exploration-Satellite (space-to-Earth) 813 815	8175–8215 EARTH EXPLORATION-SATELLITE (space-to-Earth) FIXED. FIXED-SATELLITE (Earth-to-space). METEOROLOGICAL-SATELLITE (Earth-to-space). MOBILE 814	8175–8215 FIXED. FIXED-SATELLITE (Earth-to-space). METEOROLOGICAL-SATELLITE (Earth-to-space). MOBILE. Earth Exploration-Satellite (space-to-Earth) 813 815	8175–8215 EARTH EXPLORATION-SATELLITE (space-to-Earth). FIXED. FIXED-SATELLITE (Earth-to-space). METEOROLOGICAL-SATELLITE (Earth-to-space). Mobile-Satellite (Earth-to-space) (no airborne transmissions). US258 G104 G117			
8215–8400 FIXED.	8215–8400 EARTH EXPLORATION	8215–8400 FIXED	8215–8400 EARTH EXPLORATION			

Frequency Bands and Assignments

Band	Col 2	Col 3	Col 4	Col 5	Col 6
	FIXED-SATELLITE (Earth-to-space). 813 Earth Exploration-Satellite (space-to-Earth) 815	SATELLITE (space-to-Earth). FIXED. FIXED-SATELLITE (Earth-to-space). MOBILE 814	FIXED-SATELLITE (Earth-to-space). MOBILE. Earth Exploration-Satellite (space-to-Earth). 813 815	SATELLITE (space-to-Earth). FIXED. FIXED-SATELLITE (Earth-to-space). Mobile-Satellite (Earth-to-space) (no airborne transmissions). US258 G117	
8400–8500	FIXED. MOBILE except aeronautical mobile. SPACE RESEARCH (space-to-Earth) 816 817 818		8400–8450 FIXED. SPACE RESEARCH (space-to-Earth) (deep space only). FIXED. SPACE RESEARCH (space-to-Earth).		
8500–8750	RADIOLOCATION. 713 819 820		8500–9000 RADIOLOCATION.	8500–9000 Radiolocation.	
8750–8850	RADIOLOCATION. AERONAUTICAL RADIONAVIGATION 821 822				
8850–9000	RADIOLOCATION. MARITIME RADIO-NAVIGATION 823 824				
			713 US53 US110 G59	713 US53 US110	
9000–9200	AERONAUTICAL RADIONAVIGATION. 717 Radiolocation. 822		9000–9200 AERONAUTICAL RADIONAVIGATION 717 Radiolocation. US48 US54	9000–9200 AERONAUTICAL RADIONAVIGATION 717 Radiolocation. US48 US54	AVIATION (87).

TABLE 9.1 Table of Frequency Allocations (*Continued*)

International Table			U.S. Table		FCC Use Designators	
Region 1—allocation, MHz (1)	Region 2—allocation, MHz (2)	Region 3—allocation, MHz (3)	Government Allocation, MHz (4)	Nongovernment Allocation, MHz (5)	Rule part(s) (6)	Special-use frequencies (7)
9200–9300	RADIOLOCATION. MARITIME RADIO-NAVIGATION 823 824 824A		9200–9300 MARITIME RADIO-NAVIGATION. RADIOLOCATION. US110 G59 823 824A G2 G19	9200–9300 MARITIME RADIO-NAVIGATION. Radiolocation. US110 823 824A		
9300–9500	RADIONAVIGATION. RADIOLOCATION. 825A 775A 824A 825		9300–9500 RADIONAVIGATION 825A Meterological Aids. Radiolocation. 775A 824A US51 US56 US67 US71 G56	9300–9500 RADIONAVIGATION 825A Meterological Aids. Radiolocation. 775A 824A US51 US66 US67 US71		
9500–9800	RADIOLOCATION. RADIONAVIGATION. 713		9500–10,000 RADIOLOCATION.	9500–10,000 Radiolocation.		
9800–10,000	RADIOLOCATION. Fixed. 826 827 828		713 828 US110	713 828 US110		

International Table			U.S. Table		FCC Use Designators	
Region 1—allocation, GHz (1)	Region 2—allocation, GHz (2)	Region 3—allocation, GHz (3)	Government Allocation, GHz (4)	Nongovernment Allocation, GHz (5)	Rule part(s) (6)	Special-use frequencies (7)
10.0–10.45 FIXED. MOBILE. RADIOLOCATION. Amateur. 828	10.00–10.45 RADIOLOCATION. Amateur. 828 829	10.00–10.45 FIXED. MOBILE. RADIOLOCATION. Amateur. 828	10.00–10.45 RADIOLOCATION. 828 US58 US108 G32	10.00–10.45 Amateur. Radiolocation. 828 US58 US108 NG42	Amateur (97). Private Land Mobile (90).	
10.45–10.50	RADIOLOCATION.		10.45–10.50 RADIOLOCATION.	10.45–10.50 Amateur.	Amateur (97).	

Frequency Bands and Assignments

				Private Land Mobile (90).
10.50–10.55 FIXED. MOBILE. Radiolocation.	Amateur Amateur-Satellite. 830			
	10.50–10.55 FIXED. MOBILE. RADIOLOCATION.	US58 US108 G32	Amateur-Satellite. Radiolocation. US58 US108 NG42 NG134	
		10.50–10.55 RADIOLOCATION.	10.50–10.55 RADIOLOCATION.	PRIVATE LAND MOBILE (90).
10.55–10.60 FIXED. MOBILE except aeronautical mobile. Radiolocation.		US59	US59	
		10.55–10.60 FIXED.	10.55–10.60 FIXED.	DOMESTIC PUBLIC FIXED (21). PRIVATE OPERATONAL-FIXED MICROWAVE (94).
10.60–10.68 EARTH EXPLORATION-SATELLITE (passive). FIXED. MOBILE except aeronautical mobile. RADIO ASTRONOMY. SPACE RESEARCH (passive). Radiolocation. 831 832		10.60–10.68 EARTH-EXPLORATION-SATELLITE (passive). SPACE RESEARCH (passive). 833 834	10.60–10.68 EARTH EXPLORATION-SATELLITE (passive). FIXED. SPACE RESEARCH (passive).	DOMESTIC PUBLIC FIXED (21). PRIVATE OPERATIONAL FIXED. MICROWAVE (94).
10.68–10.70 EARTH-EXPLORATION-SATELLITE (passive). RADIO ASTRONOMY. SPACE RESEARCH (passive). 833 834		US265 US277	US265 US277	
		10.68–10.70 EARTH-EXPLORATION-SATELLITE (passive). RADIO ASTRONOMY. SPACE RESEARCH (passive). US74 US246	10.68–10.70 EARTH EXPLORATION SATELLITE (passive). RADIO ASTRONOMY. SPACE RESEARCH (passive). US74 US246	
10.7–11.7 FIXED. FIXED-SATELLITE (space-to-Earth). (Earth-to-space) 835 MOBILE except aeronautical mobile.	10.7–11.7 FIXED. FIXED-SATELLITE (space-to-Earth). MOBILE except aeronautical mobile. 792A	10.7–11.7	10.7–11.7 FIXED. FIXED-SATELLITE (space-to-Earth).	DOMESTIC PUBLIC FIXED (21). PRIVATE OPERATIONAL-FIXED MICROWAVE (94).

TABLE 9.1 Table of Frequency Allocations (*Continued*)

International Table			U.S. Table		FCC Use Designators	
Region 1—allocation, GHz (1)	Region 2—allocation, GHz (2)	Region 3—allocation, GHz (3)	Government Allocation, GHz (4)	Nongovernment Allocation, GHz (5)	Rule part(s) (6)	Special-use frequencies (7)
792A			US211	792A US211 NG41 NG104		
11.7–12.5 FIXED. BROADCASTING. BROADCASTING-SATELLITE. Mobile except aeronautical mobile.	11.7–12.1 FIXED 837 FIXED-SATELLITE (space-to-Earth). Mobile except aeronautical mobile. 836 839	11.7–12.2 FIXED. MOBILE except aeronautical mobile. BROADCASTING. BROADCASTING-SATELLITE.	11.7–12.2	11.7–12.2 FIXED-SATELLITE (space-to-Earth). Mobile except aeronautical mobile.	DOMESTIC PUBLIC FIXED (21) SATELLITE COMMUNICATION (25).	
	12.1–12.2 FIXED-SATELLITE (space-to-Earth) 836 839 842.	838 845	839	837 839 NG143 NG145		
838	12.2–12.7 FIXED. MOBILE except aeronautical mobile. BROADCASTING. BROADCASTING-SATELLITE.	12.2–12.5 FIXED. MOBILE except aeronautical mobile. BROADCASTING.	12.2–12.7	12.2–12.7 FIXED. BROADCASTING-SATELLITE.	INTERNATIONAL PUBLIC (23). PRIVATE OPERATIONAL-FIXED MICROWAVE (94). DIRECT BROADCAST SATELLITE SERVICE (100).	
12.5–12.75 FIXED-SATELLITE (space-to-Earth) (Earth-to-space).	839 844 846	12.5–12.75 FIXED. FIXED-SATELLITE (space-to-Earth). MOBILE except aeronautical mobile.	839 843 844	839 843 844 NG139		

Frequency Bands and Assignments

		12.7–12.75	12.7–12.75	
	BROADCASTING-SATELLITE 847		FIXED. FIXED-SATELLITE (Earth-to-space). MOBILE.	AUXILIARY BROADCASTING (74). CABLE TELEVISION RELAY (78). PRIVATE OPERATIONAL FIXED MICROWAVE (94).
12.7–12.75 FIXED. FIXED-SATELLITE. MOBILE except aeronautical mobile.				
848 849 850			NG53 NG118	
12.75–13.25 FIXED. FIXED-SATELLITE (Earth-to-space). MOBILE Space Research (deep space) (space-to-Earth).		12.75–13.25	12.75–13.25 FIXED. FIXED-SATELLITE (Earth-to-space). MOBILE.	AUXILIARY BROADCASTING (74). CABLE TELEVISION RELAY (78). DOMESTIC PUBLIC FIXED (21). PRIVATE OPERATIONAL-FIXED MICROWAVE (94).
792A		US251	792A US251 NG53 NG104 NG118	
13.25–13.40 AERONAUTICAL RADIONAVIGATION. 851 852 853		13.25–13.40 AERONAUTICAL RADIONAVIGATION 851 Space Research (Earth-to-space).	13.25–13.40 AERONAUTICAL RADIONAVIGATION 851 Space Research (Earth-to-space).	AVIATION (87).
13.4–14.0 RADIOLOCATION. Standard Frequency and Time Signal-Satellite (Earth-to-space). Space Research. 713 853 854 855		13.4–14.0 RADIOLOCATION. Standard Frequency and Time Signal-Satellite (Earth-to-space). Space Research. 713 US110 G59	13.4–14.0 Radiolocation. Standard Frequency and Time Signal-Satellite (Earth-to-space). Space Research 713 US110	PRIVATE LAND MOBILE (90).
14.00–14.25 FIXED-SATELLITE (Earth-to-Space) 858 RADIONAVIGATION 856 Space Research.		14.0–14.2 RADIONAVIGATION. Space Research.	14.0–14.2 FIXED-SATELLITE (Earth-to-space). RADIONAVIGATION. Space Research	Aviation (87). MARITIME (80). SATELLITE COMMUNICATION (25).

TABLE 9.1 Table of Frequency Allocations (*Continued*)

International Table			U.S. Table		FCC Use Designators	
Region 1—allocation, GHz (1)	Region 2—allocation, GHz (2)	Region 3—allocation, GHz (3)	Government Allocation, GHz (4)	Nongovernment Allocation, GHz (5)	Rule part(s) (6)	Special-use frequencies (7)
14.25–14.30	857 859		14.2–14.3	14.2–14.3 Fixed-satellite (Earth-to-space) Mobile except aeronautical mobile. US287	Satellite communications (25). Domestic public fixed (21).	
	FIXED-SATELLITE (Earth-to-space) 858 RADIONAVIGATION 856 Space Research. 857 859 860 861		US287			
14.3–14.4 FIXED. FIXED-SATELLITE (Earth-to-space) 858 MOBILE except aeronautical mobile. Radionavigation-Satellite.	14.3–14.4 FIXED-SATELLITE (Earth-to-space) 858 Radionavigation-Satellite.	14.3–14.4 FIXED. FIXED-SATELLITE (Earth-to-space) 858 MOBILE except aeronautical mobile. Radionavigation-Satellite.	14.3–14.4	14.3–14.4 Fixed-satellite (Earth-to-space) Mobile except aeronautical mobile.	Satellite communication (25). Domestic public fixed (21).	
859	859	859	US287	US287		
14.40–14.47	FIXED. FIXED-SATELLITE (Earth-to-space) 858 MOBILE except aeronautical mobile. Space Research (space-to-Earth). 859		14.4–14.5 Fixed. Mobile.	14.4–14.5 FIXED-SATELLITE (Earth-to-space).	SATELLITE COMMUNICATION (25).	
			862 US203 US287	862 US203 US287		
14.47–14.50	FIXED. FIXED-SATELLITE (Earth-to-space) 858					

Frequency Bands and Assignments

Band (GHz)	Col 2	Col 3	Col 4	Col 5
14.5–14.8	MOBILE except aeronautical mobile. Radio Astronomy. 859 862			
	FIXED. FIXED-SATELLITE (Earth-to-space) 863 MOBILE. Space Research.	14.5000–14.7145 FIXED. Mobile. Space Research.	14.50–15.35	
		14.7145–15.1365 MOBILE. Fixed. Space Research. US310 G119		
14.80–15.35	FIXED. MOBILE. Space Research. 720	15.1365–15.35 FIXED. Mobile. Space Research. 720 US211		
15.35–15.40	EARTH EXPLORATION SATELLITE (passive). RADIO ASTRONOMY. SPACE RESEARCH (passive). 864 865	15.35–15.40 EARTH EXPLORATION SATELLITE (passive). RADIO ASTRONOMY. SPACE RESEARCH (passive). US74 US246	15.35–15.40 EARTH EXPLORATION SATELLITE (passive). RADIO ASTRONOMY. SPACE RESEARCH (passive). 720 US211 US310 US74 US246	
15.4–15.7	AERONAUTICAL RADIONAVIGATION. 733 797	15.4–15.7 AERONAUTICAL RADIONAVIGATION. 733 797 US211 US260	15.4–15.7 AERONAUTICAL RADIONAVIGATION. 733 797 US211 US260	AVIATION (87).
15.7–16.6	RADIOLOCATION. 866 867	15.7–16.6 RADIOLOCATION. US110 G59	15.7–17.2 Radiolocation.	Private Land Mobile (90).
16.6–17.1	RADIOLOCATION. Space Research (deep space) (Earth-to-space).	16.6–17.1 RADIOLOCATION. Space Research (deep space) (Earth-to-space).		

TABLE 9.1 Table of Frequency Allocations (*Continued*)

International Table			U.S. Table		FCC Use Designators	
Region 1—allocation, GHz (1)	Region 2—allocation, GHz (2)	Region 3—allocation, GHz (3)	Government Allocation, GHz (4)	Nongovernment Allocation, GHz (5)	Rule part(s) (6)	Special-use frequencies (7)
17.1–17.2	RADIOLOCATION. 866 867		US110 G59	US110		
17.2–17.3	RADIOLOCATION. Earth Exploration-Satellite (active). Space Research (active). 866 867		17.1–17.2 RADIOLOCATION. US110 G59			
			17.2–17.3 RADIOLOCATION. Earth Exploration-Satellite (active). Space Research (active). US110 G59	17.2–17.3 Radiolocation. Earth Exploration-Satellite (active). Space Research (active). US110	Private Land Mobile (90).	
17.3–17.7	FIXED-SATELLITE (Earth-to-space) 869 Radiolocation. 868		17.3–17.7 Radiolocation. US259 US271 G59	17.3–17.7 FIXED-SATELLITE (Earth-to-space). US259 US271 NG140		
17.7–18.1	FIXED. FIXED-SATELLITE (space-to-Earth) (Earth-to-space) 869 MOBILE.		17.7–17.8	17.7–17.8 FIXED. FIXED-SATELLITE (space-to-Earth) (Earth-to-Earth). MOBILE.	AUXILIARY BROADCASTING (74). CABLE TELEVISION RELAY (78). DOMESTIC PUBLIC FIXED (21). PRIVATE OPERATIONAL-FIXED MICROWAVE (94).	
			US271	US271 NG140 NG144		
18.1–18.6	FIXED. FIXED-SATELLITE (space-to-Earth). MOBILE.		17.8–18.6	17.8–18.6 FIXED. FIXED-SATELLITE (space-to-Earth). MOBILE.	AUXILIARY BROADCASTING (74). CABLE TELEVISION RELAY (78). DOMESTIC PUBLIC FIXED (21). PRIVATE OPERATIONAL	

Frequency Bands and Assignments

18.6–18.8 FIXED. FIXED-SATELLITE (space-to-Earth) 872 MOBILE except aeronautical mobile. Earth Exploration-Satellite (passive). Space Research (passive).	18.6–18.8 EARTH EXPLORATION-SATELLITE (passive). FIXED. FIXED-SATELLITE (space-to-Earth) 872 MOBILE except aeronautical mobile. Earth Exploration-Satellite (passive). SPACE RESEARCH (passive). 870 871	18.6–18.8 FIXED. FIXED-SATELLITE (space-to-Earth) 872 MOBILE except aeronautical mobile. Earth-Exploration-Satellite (passive). Space Research (passive). 871	870 18.6–18.8 EARTH EXPLORATION-SATELLITE (passive). SPACE RESEARCH (passive). US254 US255	870 NG144 18.6–18.8 EARTH EXPLORATION-SATELLITE (passive). FIXED. FIXED-SATELLITE (space-to-Earth). MOBILE except aeronautical mobile SPACE RESEARCH (passive). US254 US255	FIXED MICROWAVE (94). AUXILIARY BROADCASTING (74). CABLE TELEVISION RELAY (78). DOMESTIC PUBLIC FIXED (21). PRIVATE OPERATIONAL FIXED MICROWAVE (94).
18.8–19.7	18.8–19.7 FIXED. FIXED-SATELLITE (space-to-Earth). MOBILE.			18.8–19.7 FIXED. FIXED-SATELLITE (space-to-Earth). MOBILE.	AUXILIARY BROADCASTING (74). CABLE TELEVISION RELAY (78). DOMESTIC PUBLIC FIXED (21). PRIVATE OPERATIONAL FIXED MICROWAVE (94).
19.7–20.2	19.7–20.2 FIXED-SATELLITE (space-to-Earth). MOBILE-SATELLITE (space-to-Earth). 873		19.7–20.2	19.7–20.2 FIXED SATELLITE (space-to-Earth). Mobile-Satellite (space-to-Earth).	
20.2–21.2	20.2–21.2 FIXED-SATELLITE (space-to-Earth). MOBILE-SATELLITE (space-to-Earth). Standard Frequency and Time Signal-Satellite (space-to-Earth). 873		20.2–21.2 FIXED-SATELLITE (space-to-Earth). MOBILE-SATELLITE (space-to-Earth). Standard Frequency and Time Signal-Satellite (space-to-Earth). G117	20.2–21.2 Standard Frequency and Time Signal-Satellite (space-to-Earth).	
21.2–21.4	21.2–21.4		21.2–21.4	21.2–21.4	21.2–21.4

TABLE 9.1 Table of Frequency Allocations (*Continued*)

International Table		U.S. Table		FCC Use Designators	
Region 1—allocation, GHz (1)	Region 2—allocation, GHz (2)	Region 3—allocation, GHz (3)	Government Allocation, GHz (4)	Nongovernment Allocation, GHz (5)	Rule part(s) (6)	Special-use frequencies (7)

(1)	(2)	(3)	(4)	(5)	(6)	(7)
	EARTH EXPLORATION-SATELLITE (passive). FIXED. MOBILE. SPACE RESEARCH (passive).		EARTH EXPLORATION-SATELLITE (passive). FIXED. MOBILE. SPACE RESEARCH (passive). US263	EARTH EXPLORATION-SATELLITE (passive). FIXED. MOBILE. SPACE RESEARCH (passive). US263	DOMESTIC PUBLIC FIXED (21). PRIVATE OPERATIONAL-FIXED MICROWAVE (94).	
21.4–22.0	FIXED. MOBILE.		21.4–22.0 FIXED. MOBILE.	21.4–22.0 FIXED. MOBILE.	DOMESTIC PUBLIC FIXED (21). PRIVATE OPERATIONAL-FIXED MICROWAVE (94).	
22.00–22.21	FIXED. MOBILE except aeronautical mobile. 874		22.00–22.21 FIXED. MOBILE except aeronautical mobile. 874	22.00–22.21 FIXED. MOBILE except aeronautical mobile. 874	DOMESTIC PUBLIC MOBILE (22). PRIVATE OPERATIONAL-FIXED MICROWAVE (94).	
22.21–22.50	EARTH EXPLORATION-SATELLITE (passive). FIXED. MOBILE except aeronautical mobile. RADIO ASTRONOMY. SPACE RESEARCH (passive). 875 876		22.21–22.50 EARTH EXPLORATION-SATELLITE (passive). FIXED. MOBILE except aeronautical mobile. RADIO ASTRONOMY. SPACE RESEARCH (passive). 875 US263	22.21–22.50 EARTH EXPLORATION-SATELLITE (passive). FIXED. MOBILE except aeronautical mobile. RADIO ASTRONOMY. SPACE RESEARCH (passive). 875 US263	DOMESTIC PUBLIC FIXED (21). PRIVATE OPERATIONAL-FIXED MICROWAVE (94).	
22.50–22.55 FIXED. MOBILE.	22.50–22.55 BROADCASTING-SATELLITE 877 FIXED. MOBILE.		22.50–22.55 FIXED. MOBILE.	22.50–22.55 BROADCASTING-SATELLITE. FIXED. MOBILE.	DOMESTIC PUBLIC FIXED (21). PRIVATE OPERATIONAL-FIXED	

Frequency Bands and Assignments

22.55–23.00 FIXED. INTER-SATELLITE. MOBILE. 879	878 22.55–23.00 BROADCASTING- SATELLITE 877 FIXED. INTER-SATELLITE. MOBILE. 878 879	US211	US211 22.55–23.00 FIXED. INTER-SATELLITE. MOBILE. 879 US278	MICROWAVE (94). DOMESTIC PUBLIC FIXED (21). PRIVATE OPERATIONAL-FIXED MICROWAVE (94).
23.00–23.55	FIXED. INTER-SATELLITE. MOBILE. 879		23.00–23.55 FIXED. INTER-SATELLITE. MOBILE. 879 US278	DOMESTIC PUBLIC FIXED (21). PRIVATE OPERATIONAL FIXED MICROWAVE (94).
23.55–23.60	FIXED. MOBILE.		23.55–23.60 FIXED. MOBILE.	DOMESTIC PUBLIC FIXED (21). PRIVATE OPERATIONAL-FIXED MICROWAVE (94).
23.6–24.0	EARTH EXPLORATION SATELLITE (passive). RADIO ASTRONOMY. SPACE RESEARCH (passive). 880		23.6–24.0 EARTH EXPLORATION SATELLITE (passive). RADIO ASTRONOMY. SPACE RESEARCH (passive). US74 US246	
24.00–24.05	AMATEUR. AMATEUR-SATELLITE. 881		24.00–24.05 AMATEUR. AMATEUR-SATELLITE. 881 US211	AMATEUR (97).
24.05–24.25	RADIOLOCATION. Amateur. Earth Exploration-Satellite (active). 881	24.05–24.25 RADIOLOCATION. Earth Exploration-Satellite (active). 881 US110 G59	24.05–24.25 Amateur. Radiolocation. Earth Exploration-Satellite (active). 881 US110	Amateur (97). Private Land Mobile (90).	24.125 ± 125 GHz: Industrial scientific and medical frequency.
24.25–25.25	RADIONAVIGATION.		24.25–25.25 RADIONAVIGATION.	
25.25–27.00			25.25–27.00	AVIATION (87).

TABLE 9.1 Table of Frequency Allocations (*Continued*)

International Table			U.S. Table		FCC Use Designators	
Region 1—allocation, GHz (1)	Region 2—allocation, GHz (2)	Region 3—allocation, GHz (3)	Government Allocation, GHz (4)	Nongovernment Allocation, GHz (5)	Rule part(s) (6)	Special-use frequencies (7)
	FIXED. MOBILE. Earth Exploration-Satellite (space-to-space). Standard Frequency and Time Signal-Satellite (Earth-to-space).		FIXED. MOBILE. Earth Exploration-Satellite (space-to-space). Standard Frequency and Time Signal-Satellite (Earth-to-space).	Earth Exploration-Satellite (space-to-space). Standard Frequency and Time Signal-Satellite (Earth-to-space).		
27.0–27.5 FIXED. MOBILE. Earth Exploration-Satellite (space-to-space).	27.0–27.5 FIXED. FIXED-SATELLITE (Earth-to-space). MOBILE. Earth Exploration-Satellite (space-to-space).		27.0–27.5 FIXED. MOBILE. Earth-Exploration-Satellite (space-to-space).	27.0–27.5 Earth Exploration Satellite (space-to-space).		
	27.5–29.5 FIXED. FIXED-SATELLITE (Earth-to-space). MOBILE.		27.5–29.5	27.5–29.5 FIXED. FIXED-SATELLITE (Earth-to-space). MOBILE.	DOMESTIC PUBLIC FIXED (21).	
29.5–30.0	FIXED-SATELLITE (Earth-to-space). Mobile-Satellite (Earth-to-space). 882 883		29.5–30.0 882	29.5–30.0 FIXED-SATELLITE (Earth-to-space). Mobile-Satellite (Earth-to-space). 882		
30.0–31.0	FIXED-SATELLITE (Earth-to-space). MOBILE-SATELLITE (Earth-to-space). Standard Frequency and Time Signal-Satellite (space-to-Earth). 883		30.0–31.0 FIXED-SATELLITE (Earth-to-space). MOBILE-SATELLITE (Earth-to-space). Standard Frequency and Time Signal-Satellite (space-to-Earth). G117	30.0–31.0 Standard Frequency and Time Signal-Satellite (space-to-Earth).		
31.0–31.3			31.0–31.3	31.0–31.3	AUXILIARY	

190   *Conversion Factors, Standards, and Constants*

Frequency Bands and Assignments

Band	Col 2	Col 3	Col 4	Col 5	Col 6
31.3–31.5	FIXED. MOBILE. Standard Frequency and Time Signal-Satellite (space-to-Earth). Space Research. 884 885 886		Standard Frequency and Time Signal-Satellite (space-to-Earth). 886 US211	FIXED. MOBILE. Standard Frequency and Time Signal-Satellite (space-to-Earth). 884 886 US211	BROADCASTING (74). DOMESTIC PUBLIC FIXED (21). CABLE TELEVISION RELAY (78). GENERAL MOBILE RADIO (95). PRIVATE OPERATIONAL-FIXED MICROWAVE (94).
31.5–31.8	EARTH EXPLORATION-SATELLITE (passive). RADIO ASTRONOMY. SPACE RESEARCH (passive). 887	31.5–31.8 EARTH EXPLORATION-SATELLITE (passive). RADIO ASTRONOMY. SPACE RESEARCH (passive). Fixed. Mobile except aeronautical mobile. 888	31.3–31.8 EARTH EXPLORATION-SATELLITE (passive). RADIO ASTRONOMY. SPACE RESEARCH (passive).	31.3–31.8 EARTH EXPLORATION-SATELLITE (passive). RADIO ASTRONOMY. SPACE RESEARCH (passive).	
	EARTH EXPLORATION-SATELLITE (passive). RADIO ASTRONOMY. SPACE RESEARCH (passive). Fixed. Mobile except aeronautical mobile. 888 889				
31.8–32.0	RADIONAVIGATION. Space Research. 890 891 892		31.8–32.0 RADIONAVIGATION. US69 US211 US262	US74 US246 31.8–32.0 RADIONAVIGATION. US69 US211 US262	
32.0–32.3	INTERSATELLITE RADIONAVIGATION. Space Research. 890 891 892 893		32.0–33.0 INTERSATELLITE RADIONAVIGATION.	32.0–33.0 INTERSATELLITE RADIONAVIGATION.	
32.3–33.0	INTERSATELLITE RADIONAVIGATION. 892 893		893 US69 US262	893 US69 US262	

TABLE 9.1 Table of Frequency Allocations (*Continued*)

International Table			U.S. Table		FCC Use Designators	
Region 1—allocation, GHz (1)	Region 2—allocation, GHz (2)	Region 3—allocation, GHz (3)	Government Allocation, GHz (4)	Nongovernment Allocation, GHz (5)	Rule part(s) (6)	Special-use frequencies (7)
33.0–33.4	RADIONAVIGATION. 892		33.0–33.4 RADIONAVIGATION. US69 US278	33.0–33.4 RADIONAVIGATION. US69 US278		
33.4–34.2	RADIOLOCATION. 892 894		33.4–36.0 RADIOLOCATION. 897 US110 US252 G34	33.4–36.0 Radiolocation. 897 US110 US252	Private Land Mobile (90).	
34.2–35.2	RADIOLOCATION. Space Research. 895 896 894					
35.2–36.0	METEOROLOGICAL AIDS. RADIOLOCATION. 894 897					
36.0–37.0	EARTH EXPLORATION-SATELLITE (passive). FIXED. MOBILE. SPACE RESEARCH (passive). 898		36.0–37.0 EARTH EXPLORATION-SATELLITE (passive). FIXED. MOBILE. SPACE RESEARCH (passive). 898 US263	36.0–37.0 EARTH EXPLORATION-SATELLITE (passive). FIXED. MOBILE. SPACE RESEARCH (passive). 898 US263		
37.0–37.5	FIXED. MOBILE. 899		37.0–38.6 FIXED. MOBILE.	37.0–38.6 FIXED. MOBILE.	DOMESTIC PUBLIC FIXED (21). PRIVATE OPERATIONAL-FIXED MICROWAVE (94).	
37.5–39.5	FIXED. FIXED-SATELLITE					

Frequency Bands and Assignments

	(space-to-Earth). MOBILE.		38.6–39.5 FIXED. MOBILE. FIXED-SATELLITE (space-to-Earth). US291		DOMESTIC PUBLIC FIXED (21). PRIVATE OPERATIONAL-FIXED MICROWAVE (90). Auxiliary Broadcasting (74).
39.5–40.5	FIXED. FIXED-SATELLITE (space-to-Earth). MOBILE. MOBILE-SATELLITE (space-to-Earth). 899		39.5–40.5 FIXED. FIXED-SATELLITE (space-to-Earth). MOBILE. MOBILE-SATELLITE (space-to-Earth). US291		DOMESTIC PUBLIC FIXED (21). PRIVATE OPERATIONAL-FIXED MICROWAVE (94). Auxiliary Broadcasting (74).
		40.0–40.5 FIXED-SATELLITE (space-to-Earth). MOBILE-SATELLITE (space-to-Earth). US291 G117	40.0–40.5 FIXED-SATELLITE (space-to-Earth). MOBILE-SATELLITE (space-to-Earth). US291		
40.5–42.5	BROADCASTING-SATELLITE. /BROADCASTING/. Fixed. Mobile.	40.5–42.5	40.5–42.5 BROADCASTING-SATELLITE. /BROADCASTING/. Fixed. Mobile. US211		
42.5–43.5	FIXED. FIXED-SATELLITE (Earth-to-space) 901 MOBILE except aeronautical mobile. RADIO ASTRONOMY. 900	42.5–43.5 FIXED. FIXED-SATELLITE (Earth-to-space). MOBILE except aeronautical mobile. RADIO ASTRONOMY. 900	42.5–43.5 FIXED. FIXED-SATELLITE (Earth-to-space). MOBILE except aeronautical mobile. RADIO ASTRONOMY. 900		
43.5–47.0	MOBILE 902	43.5–45.5 FIXED-SATELLITE	43.5–45.5		

TABLE 9.1 Table of Frequency Allocations (*Continued*)

International Table			U.S. Table		FCC Use Designators	
Region 1—allocation, GHz (1)	Region 2—allocation, GHz (2)	Region 3—allocation, GHz (3)	Government Allocation, GHz (4)	Nongovernment Allocation, GHz (5)	Rule part(s) (6)	Special-use frequencies (7)
	MOBILE-SATELLITE. RADIONAVIGATION. RADIONAVIGATION-SATELLITE. 903		(Earth-to-space). MOBILE-SATELLITE (Earth-to-space). G117			
			45.5–47.0 MOBILE. MOBILE-SATELLITE (Earth-to-space). RADIONAVIGATION. RADIONAVIGATION-SATELLITE. 903	45.5–47.0 MOBILE. MOBILE-SATELLITE (Earth-to-space). RADIONAVIGATION. RADIONAVIGATION-SATELLITE. 903		
47.0–47.2	AMATEUR. AMATEUR-SATELLITE.		47.0–47.2	47.0–47.2 AMATEUR. AMATEUR-SATELLITE.	AMATEUR (97).	
47.2–50.2	FIXED. FIXED-SATELLITE (Earth-to-space) 901. MOBILE 905. 904		47.2–50.2 FIXED. FIXED-SATELLITE (Earth-to-space). MOBILE. 904 US264 US297	47.2–50.2 FIXED. FIXED-SATELLITE (Earth-to-space). MOBILE. 904 US264 US297		
50.2–50.4	EARTH EXPLORATION-SATELLITE (passive). FIXED. MOBILE. SPACE RESEARCH (passive).		50.2–50.4 EARTH EXPLORATION-SATELLITE (passive). FIXED. MOBILE. SPACE RESEARCH (passive). US263	50.2–50.4 EARTH EXPLORATION-SATELLITE (passive). FIXED. MOBILE. SPACE RESEARCH (passive). US263		
50.4–51.4	FIXED. FIXED-SATELLITE (Earth-to-space). MOBILE		50.4–51.4 FIXED. FIXED-SATELLITE (Earth-to-space). MOBILE.	50.4–51.4 FIXED. FIXED-SATELLITE (Earth-to-space). MOBILE.		

Frequency Bands and Assignments

51.4–54.25	Mobile-Satellite (Earth-to-space).	MOBILE-SATELLITE (Earth-to-space). G117		
	51.4–54.25 EARTH EXPLORATION-SATELLITE (passive). SPACE RESEARCH (passive). 906 907	51.4–54.25 EARTH EXPLORATION-SATELLITE (passive). RADIO ASTRONOMY. SPACE RESEARCH (passive). US246		
54.25–58.2	EARTH EXPLORATION-SATELLITE (passive). FIXED. INTERSATELLITE. MOBILE 909. SPACE RESEARCH (passive). 908	54.25–58.2 EARTH EXPLORATION SATELLITE (passive). FIXED. INTERSATELLITE. MOBILE 909. SPACE RESEARCH (passive). US263		
58.2–59.0	EARTH EXPLORATION-SATELLITE (passive). RADIO ASTRONOMY. SPACE RESEARCH (passive). 906 907	58.2–59.0 EARTH EXPLORATION SATELLITE (passive). RADIO ASTRONOMY. SPACE RESEARCH (passive). US246		
59–64	FIXED. INTERSATELLITE. MOBILE 909. RADIOLOCATION 910 911	59–64 FIXED. INTERSATELLITE. MOBILE 909. RADIOLOCATION 910 911		61.25 GHz ± 250 MHz: Industrial, scientific and medical frequency
64–65	EARTH EXPLORATION SATELLITE (passive). SPACE RESEARCH (passive). 906 907	64–65 EARTH EXPLORATION SATELLITE (passive). RADIO ASTRONOMY. SPACE RESEARCH (passive). US246		
65–66	EARTH EXPLORATION-SATELLITE.	65–66 EARTH EXPLORATION SATELLITE.		

TABLE 9.1 Table of Frequency Allocations (*Continued*)

International Table			U.S. Table		FCC Use Designators	
Region 1—allocation, GHz (1)	Region 2—allocation, GHz (2)	Region 3—allocation, GHz (3)	Government Allocation, GHz (4)	Nongovernment Allocation, GHz (5)	Rule part(s) (6)	Special-use frequencies (7)
66–71	SPACE RESEARCH. Fixed. Mobile.		SPACE RESEARCH. Fixed. Mobile.	SPACE RESEARCH. Fixed. Mobile.		
	MOBILE 902 MOBILE-SATELLITE. RADIONAVIGATION. RADIONAVIGATION-SATELLITE. 903		66–71 MOBILE 902 MOBILE-SATELLITE. RADIONAVIGATION. RADIONAVIGATION-SATELLITE. 903	66–71 MOBILE 902 MOBILE-SATELLITE. RADIONAVIGATION. RADIONAVIGATION-SATELLITE. 903		
71–74	FIXED. FIXED-SATELLITE (Earth-to-space). MOBILE. MOBILE-SATELLITE (Earth-to-space). 906		71–74 FIXED. FIXED-SATELLITE (Earth-to-space). MOBILE. MOBILE-SATELLITE (Earth-to-space). US270	71–74 FIXED. FIXED-SATELLITE (Earth-to-space). MOBILE. MOBILE-SATELLITE (Earth-to-space). US270		
74.0–75.5	FIXED. FIXED-SATELLITE (Earth-to-space). MOBILE.		74.0–75.5 FIXED. FIXED-SATELLITE (Earth-to-space). MOBILE. US297	74.0–75.5 FIXED. FIXED-SATELLITE (Earth-to-space). MOBILE. US297		
75.5–76.0	AMATEUR. AMATEUR-SATELLITE.		75.5–76.0	75.5–76.0 AMATEUR. AMATEUR-SATELLITE.	AMATEUR (97).	
76–81	RADIOLOCATION. Amateur. Amateur-Satellite. 912		76–81 RADIOLOCATION. 912	76–81 RADIOLOCATION. Amateur. Amateur-Satellite. 912	Amateur (97).	
81–84	FIXED. FIXED-SATELLITE (space-to-Earth). MOBILE.		81–84 FIXED. FIXED-SATELLITE (space-to-Earth). MOBILE.	81–84 FIXED. FIXED-SATELLITE (space-to-Earth). MOBILE.		

Frequency Bands and Assignments

	MOBILE-SATELLITE (space-to-Earth). FIXED. MOBILE. BROADCASTING. BROADCASTING-SATELLITE. 913	MOBILE-SATELLITE (space-to-Earth). FIXED. MOBILE.	MOBILE-SATELLITE (space-to-Earth). 84–86 FIXED. MOBILE. BROADCASTING. BROADCASTING-SATELLITE. 913 US211	
84–86				
86–92	EARTH EXPLORATION-SATELLITE (passive). RADIO ASTRONOMY. SPACE RESEARCH (passive). 907	86–92 EARTH EXPLORATION-SATELLITE (passive). RADIO ASTRONOMY. SPACE RESEARCH (passive). US74 US246	86–92 EARTH EXPLORATION SATELLITE (passive). RADIO ASTRONOMY. SPACE RESEARCH (passive). US74 US246	
92–95	FIXED. FIXED-SATELLITE (Earth-to-space). MOBILE. RADIOLOCATION. 914	92–95 FIXED. FIXED-SATELLITE (Earth-to-space). MOBILE. RADIOLOCATION. 914	92–95 FIXED. FIXED-SATELLITE (Earth-to-space). MOBILE. RADIOLOCATION. 914	
95–100	MOBILE 902 MOBILE-SATELLITE. RADIONAVIGATION. RADIONAVIGATION-SATELLITE. Radiolocation. 903 904	95–100 MOBILE 902 MOBILE-SATELLITE. RADIONAVIGATION. RADIONAVIGATION-SATELLITE. Radiolocation. 903 904	95–100 MOBILE 902 MOBILE-SATELLITE. RADIONAVIGATION. RADIONAVIGATION-SATELLITE. Radiolocation. 903 904	
100–102	EARTH EXPLORATION SATELLITE (passive). FIXED. MOBILE. SPACE RESEARCH (passive). 722	100–102 EARTH EXPLORATION SATELLITE (passive). SPACE RESEARCH (passive). 722 US246	100–102 EARTH EXPLORATION SATELLITE (passive). SPACE RESEARCH (passive). 722 US246	
102–105	FIXED. FIXED-SATELLITE	102–105 FIXED. FIXED-SATELLITE	102–105 FIXED. FIXED-SATELLITE	

TABLE 9.1 Table of Frequency Allocations (*Continued*)

International Table			U.S. Table		FCC Use Designators	
Region 1—allocation, GHz (1)	Region 2—allocation, GHz (2)	Region 3—allocation, GHz (3)	Government Allocation, GHz (4)	Nongovernment Allocation, GHz (5)	Rule part(s) (6)	Special-use frequencies (7)
(space-to-Earth). 722			(space-to-Earth). 722 US211	(space-to-Earth). 722		
105–116 EARTH EXPLORATION-SATELLITE (passive). RADIO ASTRONOMY. SPACE RESEARCH (passive). 722 907			105–116 EARTH EXPLORATION-SATELLITE (passive). RADIO ASTRONOMY. SPACE RESEARCH (passive). 722 US74 US246	105–116 EARTH EXPLORATION-SATELLITE (passive). RADIO ASTRONOMY. SPACE RESEARCH (passive). 722 US74 US246		
116–126 EARTH EXPLORATION-SATELLITE (passive). FIXED. INTERSATELLITE. MOBILE 909. SPACE RESEARCH (passive). 722 915 916			116–126 EARTH EXPLORATION-SATELLITE (passive). FIXED. INTERSATELLITE. MOBILE 909. SPACE RESEARCH (passive). 722 915 916 US211 US263	116–126 EARTH EXPLORATION-SATELLITE (passive). FIXED. INTERSATELLITE. MOBILE 909. SPACE RESEARCH (passive). 722 915 916 US211 US263		122.5 ± 5GHz Industrial scientific and medical frequency
126–134 FIXED. INTERSATELLITE. MOBILE 909. RADIOLOCATION. 910			126–134 FIXED. INTERSATELLITE. MOBILE 909. RADIOLOCATION. 910	126–134 FIXED. INTERSATELLITE. MOBILE 909. RADIOLOCATION. 910		
134–142 MOBILE 902 MOBILE-SATELLITE. RADIONAVIGATION. RADIONAVIGATION-SATELLITE. Radiolocation. 903 917 918			134–142 MOBILE 902 MOBILE-SATELLITE. RADIONAVIGATION. RADIONAVIGATION-SATELLITE. Radiolocation. 903 917 918	134–142 MOBILE 902 MOBILE-SATELLITE. RADIONAVIGATION. RADIONAVIGATION-SATELLITE. Radiolocation. 903 917 818		
142–144 AMATEUR. AMATEUR-SATELLITE.			142–144	142–144 AMATEUR. AMATEUR-SATELLITE.	AMATEUR (97).	

Frequency Bands and Assignments

Band			
144–149	RADIOLOCATION. Amateur. Amateur-Satellite. 918	144–149 RADIOLOCATION. Amateur. Amateur-Satellite. 918	Amateur (97).
149–150	FIXED. FIXED-SATELLITE (space-to-Earth). MOBILE.	149–150 FIXED. FIXED-SATELLITE (space-to-Earth). MOBILE.	
150–151	EARTH EXPLORATION-SATELLITE (passive). FIXED. FIXED-SATELLITE (space-to-Earth). MOBILE. SPACE RESEARCH (passive). 919	150–151 EARTH EXPLORATION-SATELLITE (passive). FIXED. FIXED-SATELLITE (space-to-Earth). MOBILE. SPACE RESEARCH (passive). 919 US263	
151–164	FIXED. FIXED-SATELLITE (space-to-Earth). 211	151–164 FIXED. FIXED-SATELLITE. 211	
164–168	EARTH EXPLORATION-SATELLITE (passive). RADIO ASTRONOMY. SPACE RESEARCH (passive).	164–168 EARTH EXPLORATION-SATELLITE (passive). RADIO ASTRONOMY. SPACE RESEARCH (passive). US246	
168–170	FIXED. MOBILE.	168–170 FIXED. MOBILE.	
170.0–174.5	FIXED. INTERSATELLITE. MOBILE. 909. 919	170.0–174.5 FIXED. INTERSATELLITE. MOBILE. 909. 919	
174.5–176.5	EARTH EXPLORATION-	174.5–176.5 EARTH EXPLORATION-	

TABLE 9.1 Table of Frequency Allocations (*Continued*)

International Table			U.S. Table		FCC Use Designators	
Region 1—allocation, GHz (1)	Region 2—allocation, GHz (2)	Region 3—allocation, GHz (3)	Government Allocation, GHz (4)	Nongovernment Allocation, GHz (5)	Rule part(s) (6)	Special-use frequencies (7)
	SATELLITE (passive). FIXED. INTERSATELLITE. MOBILE. 909. SPACE RESEARCH (passive). 919		SATELLITE (passive). FIXED. INTERSATELLITE. MOBILE. 909. SPACE RESEARCH (passive). 919 US263	SATELLITE (passive). FIXED. INTERSATELLITE. MOBILE. 909. SPACE RESEARCH (passive). 919 US263		
176.5–182.0	FIXED. INTERSATELLITE. MOBILE. 909. 919		176.5–182.0 FIXED. INTERSATELLITE. MOBILE. 909. 919 US211	176.5–182.0 FIXED. INTERSATELLITE. MOBILE. 909. 919 US211		
182–185	EARTH EXPLORATION-SATELLITE (passive). RADIO ASTRONOMY. SPACE RESEARCH (passive). 920 921		182–185 EARTH EXPLORATION-SATELLITE (passive). RADIO ASTRONOMY. SPACE RESEARCH (passive). US246	182–185 EARTH EXPLORATION-SATELLITE (passive). RADIO ASTRONOMY. SPACE RESEARCH (passive). US246		
185–190	FIXED. INTERSATELLITE. MOBILE 909 919		185–190 FIXED. INTERSATELLITE. MOBILE 909 919 US211	185–190 FIXED. INTERSATELLITE. MOBILE 909 919 US211		
190–200	MOBILE 902 MOBILE-SATELLITE. RADIONAVIGATION. RADIONAVIGATION-SATELLITE. 722 903		190–200 MOBILE 902 MOBILE-SATELLITE. RADIONAVIGATION. RADIONAVIGATION-SATELLITE. 722 903	190–200 MOBILE 902 MOBILE-SATELLITE. RADIONAVIGATION. RADIONAVIGATION-SATELLITE. 722 903		
200–202	EARTH EXPLORATION-SATELLITE (passive). FIXED. MOBILE. SPACE RESEARCH		200–202 EARTH EXPLORATION-SATELLITE (passive). FIXED. MOBILE. SPACE RESEARCH	200–202 EARTH EXPLORATION-SATELLITE (passive). FIXED. MOBILE. SPACE RESEARCH		

	(passive). 722	(passive). 722 US263	(passive). 722 US263	
202–217	FIXED. FIXED-SATELLITE (Earth-to-space). MOBILE. 722	202–217 FIXED. FIXED-SATELLITE (Earth-to-space). MOBILE. 722	202–217 FIXED. FIXED-SATELLITE (Earth-to-space). MOBILE. 722	
217–231	EARTH EXPLORATION-SATELLITE (passive). RADIO ASTRONOMY. SPACE RESEARCH (passive). 722 907	217–231 EARTH EXPLORATION-SATELLITE (passive). RADIO ASTRONOMY. SPACE RESEARCH (passive). 722 US74 US246	217–231 EARTH EXPLORATION-SATELLITE (passive). RADIO ASTRONOMY. SPACE RESEARCH (passive). 722 US74 US246	
231–235	FIXED. FIXED-SATELLITE (space-to-Earth). MOBILE. Radiolocation.	213–235 FIXED. FIXED-SATELLITE (space-to-Earth). MOBILE. Radiolocation. US211	231–235 FIXED. FIXED-SATELLITE (space-to-Earth). MOBILE. Radiolocation. US211	
235–238	EARTH EXPLORATION-SATELLITE (passive). FIXED. FIXED-SATELLITE (space-to-Earth). MOBILE. SPACE RESEARCH (passive).	235–238 EARTH EXPLORATION-SATELLITE (passive). FIXED. FIXED-SATELLITE (space-to-Earth). MOBILE. SPACE RESEARCH (passive). US263	235–238 EARTH EXPLORATION-SATELLITE (passive). FIXED. FIXED-SATELLITE (space-to-Earth). MOBILE. SPACE RESEARCH (passive). US263	
238–241	FIXED. FIXED-SATELLITE (space-to-Earth). MOBILE. Radiolocation.	238–241 FIXED. FIXED-SATELLITE (space-to-Earth). MOBILE. Radiolocation.	238–241 FIXED. FIXED-SATELLITE (space-to-Earth). MOBILE. Radiolocation.	
241–248	RADIOLOCATION.	241–248 RADIOLOCATION.	241–248 RADIOLOCATION.	Amateur (97).	245 ± 1 GHz:

TABLE 9.1 Table of Frequency Allocations (*Continued*)

International Table			U.S. Table		FCC Use Designators	
Region 1—allocation, GHz (1)	Region 2—allocation, GHz (2)	Region 3—allocation, GHz (3)	Government Allocation, GHz (4)	Nongovernment Allocation, GHz (5)	Rule part(s) (6)	Special-use frequencies (7)
	Amateur. Amateur-Satellite. 922					Industrial, scientific and medical frequency.
248–250	AMATEUR. AMATEUR-SATELLITE.		248–250 AMATEUR. AMATEUR-SATELLITE.	248–250 AMATEUR. AMATEUR-SATELLITE. 922	AMATEUR (97).	
250–252	EARTH EXPLORATION-SATELLITE (passive). SPACE RESEARCH (passive). 923		250–252 EARTH EXPLORATION-SATELLITE (passive). SPACE RESEARCH (passive). 923	250–252 EARTH EXPLORATION-SATELLITE (passive). SPACE RESEARCH (passive). 923		
252–265	MOBILE 902 MOBILE-SATELLITE. RADIONAVIGATION. RADIONAVIGATION-SATELLITE. 903 923 924 925		252–265 MOBILE 902 MOBILE-SATELLITE. RADIONAVIGATION. RADIONAVIGATION-SATELLITE. 903 923 924 US211	252–265 MOBILE 902 MOBILE-SATELLITE. RADIONAVIGATION. RADIONAVIGATION-SATELLITE. 903 923 924 US211		
265–275	FIXED. FIXED-SATELLITE (Earth-to-space). MOBILE. RADIO ASTRONOMY. 926		265–275 FIXED. FIXED-SATELLITE (Earth-to-space). MOBILE. RADIO ASTRONOMY. 926	265–275 FIXED. FIXED-SATELLITE (Earth-to-space). MOBILE. RADIO ASTRONOMY. 926		
275–400	(Not allocated) 927		275–300 FIXED. MOBILE. 927 Above 300 (Not allocated). 927	275–300 FIXED. MOBILE. 927 Above 300. (Not allocated). 927	Amateur (97).	

Defining Terms

Accepted interference:[7] Interference at a higher level than defined as permissible interference and which has been agreed upon between two or more administrations without prejudice to other administrations [RR 1982].

Active satellite: A satellite carrying a station intended to transmit or retransmit radiocommunication signals [RR 1982].

Active sensor: A measuring instrument in the Earth exploration-satellite service or in the space research service by means of which information is obtained by transmission and reception of radio waves [RR 1982].

Administration: Any governmental department or service responsible for discharging the obligations undertaken in the Convention of the International Telecommunication Union and the Regulations [CONV 1973].

Aeronautical Earth station: An Earth station in the fixed-satellite service, or, in some cases, in the aeronautical mobile-satellite service, located at a specified fixed point on land to provide a feeder link for the aeronautical mobile-satellite service [RR 1982].

Aeronautical fixed service: A radiocommunication service between specified fixed points provided primarily for the safety of air navigation and for the regular, efficient and economical operation of air transport [RR 1982].

Aeronautical fixed station: A station in the aeronautical fixed service [RR 1982].

Aeronautical mobile off-route (OR) service: An aeronautical mobile service intended for communications, including those relating to flight coordination, primarily outside national or international civil air routes [RR 1982].

Aeronautical mobile route (R) service: An aeronautical mobile service reserved for communications relating to safety and regularity of flight, primarily along national or international civil air routes [RR 1982].

Aeronautical mobile-satellite off-route (OR) service: An aeronautical mobile-satellite service intended for communications, including those relating to flight coordination, primarily outside nautical and international civil air routes [RR 1982].

Aeronautical mobile-satellite route (R) service: An aeronautical mobile-satellite service reserved for communications relating to safety and regularity of flights primarily along national or international civil air routes. [RR 1982].

Aeronautical mobile-satellite service: A mobile-satellite service in which mobile Earth stations are located onboard aircraft; survival craft stations and emergency position-indicating radiobeacon stations may also participate in this service [RR 1982].

Aeronautical mobile service: A mobile service between aeronautical stations and aircraft stations, or between aircraft stations, in which survival craft stations may participate; emergency position-indicating radiobeacon stations may also participate in this service on designated distress and emergency frequencies [RR 1982].

Aeronautical radionavigation-satellite service: A radionavigation-satellite service in which Earth stations are located onboard aircraft [RR 1982].

Aeronautical radionavigation service: A radio-navigation service intended for the benefit and for the safe operation of aircraft [RR 1982].

Aeronautical station: A land station in the aeronautical mobile service. Note: in certain instances, an aeronautical station may be located, for example, onboard ship or on a platform at sea [RR 1982].

Aircraft Earth station: A mobile Earth station in the aeronautical mobile-satellite service located onboard an aircraft [RR 1982].

Aircraft station: A mobile station in the aeronautical mobile service, other than a survival craft station, located onboard an aircraft [RR 1982].

Allocation (of a frequency band): Entry in the Table of Frequency Allocations of a given frequency band for the purpose of its use by one or more terrestrial or space radiocommunication services or the radio astronomy

---

[7]The terms *permissible interference* and *accepted interference* are used in the coordination of frequency assignments between administrations.

service under specified conditions. This term shall also be applied to the frequency band concerned [RR 1982].

Allotment (of a radio frequency or radio frequency channel): Entry of a designated frequency channel in an agreed plan, adopted by a competent conference, for use by one or more administrations for a terrestrial or space radiocommunication service in one or more identified countries or geographical area and under specified conditions [RR 1982].

Altitude of the apogee or perigee: The altitude of the apogee or perigee above a specified reference surface serving to represent the surface of the Earth [RR 1982].

Amateur-satellite service: A radiocommunication service using space stations on Earth satellites for the same purposes as those of the amateur service [RR 1982].

Amateur service: A radiocommunication service for the purpose of self-training, intercommunication and technical investigations carried out by amateurs, that is, by duly authorized persons interested in radio technique solely with a personal aim and without pecuniary interest [RR 1982].

Amateur station: A station in the amateur service [RR 1982].

Assigned frequency: The center of the frequency band assigned to a station [RR 1982].

Assigned frequency band: The frequency band within which the emission of a station is authorized; the width of the band equals the necessary bandwidth plus twice the absolute value of the frequency tolerance. Where space stations are concerned, the assigned frequency band includes twice the maximum Doppler shift that may occur in relation to any point of the Earth's surface [RR 1982].

Assignment (of a radio frequency or radio frequency channel): Authorization given by an administration for a radio station to use a radio frequency or radio frequency channel under specified conditions [RR 1982].

Base Earth station: An Earth station in the fixed-satellite service or, in some cases, in the land mobile-satellite service, located at a specified fixed point or within a specified area on land to provide a feeder link for the land mobile-satellite service [RR 1982].

Base station: A land station in the land mobile service [RR 1982].

Broadcasting-satellite service: A radiocommunication service in which signals transmitted or retransmitted by space stations are intended for direct reception by the general public. Note: in the broadcasting-satellite service, the term *direct reception* shall encompass both individual reception and community reception [RR 1982].

Broadcasting service: A radiocommunication service in which the transmissions are intended for direct reception by the general public. This service may include sound transmissions, television transmissions or other types of transmission [CONV 1973].

Broadcasting station: A station in the broadcasting service [RR 1982].

Carrier power (of a radio transmitter): The average power supplied to the antenna transmission line by a transmitter during one radio frequency cycle taken under the condition of no modulation [RR 1982].

Characteristic frequency: A frequency that can be easily identified and measured in a given emission. Note: a carrier frequency may, for example, be designated as the characteristic frequency [RR 1982].

Class of emission: The set of characteristics of an emission, designated by standard symbols, for example, type of modulation, modulating signal, type of information to be transmitted, and also if appropriate, any additional signal characteristics [RR 1982].

Coast Earth station: An Earth station in the fixed-satellite service or, in some cases, in the maritime mobile-satellite service, located at a specified fixed point on land to provide a feeder link for the maritime mobile-satellite service [RR 1982].

Coast station: A land station in the maritime mobile service [RR 1982].

Community reception (in the broadcasting-satellite service): The reception of emissions from a space station in the broadcasting-satellite service by receiving equipment, which in some cases may be complex and have antennae larger than those for individual reception, and intended for use: (1) by a group of the general public at one location; or (2) through a distribution system covering a limited area [RR 1982].

Coordinated universal time (UTC): Time scale, based on the second (S.I.), as defined and recommended by the CCIR,[8] and maintained by the Bureau International de l'Heure (BIH). Note: for most practical purposes associated with the Radio Regulations, UTC is equivalent to mean solar time at the prime meridian (0° longitude), formerly expressed in GMT [RR 1982].

Coordination area: The area associated with an Earth station outside of which a terrestrial station sharing the same frequency band neither causes nor is subject to interfering emissions greater than a permissible level [RR 1982].

Coordination contour: The line enclosing the coordination area [RR 1982].

Coordination distance: Distance on a given azimuth from an Earth station beyond which a terrestrial causes nor is subject to interfering emissions greater than a permissible level [RR 1982].

Deep space: Space at distance from the Earth equal to or greater than 2×10^6 km [RR 1982].

Direct sequence systems: A direct sequence system is a spread spectrum system in which the incoming information is usually digitized, if it is not already in a binary format, and modulo 2 added to a higher speed code sequence. The combined information and code are then used to modulate a RF carrier. Since the high-speed code sequence dominates the modulating function, it is the direct cause of the wide spreading of the transmitted signal.

Duplex operation: Operating method in which transmission is possible simultaneously in both directions of a telecommunication channel[9] [RR 1982].

Earth exploration-satellite service: A radiocommunication service between Earth stations and one or more space stations, which may include links between space stations in which:

1. Information relating to the characteristics of the Earth and its natural phenomena is obtained from active sensors or passive sensors on earth satellites
2. Similar information is collected from air-borne or Earth-based platforms
3. Such information may be distributed to Earth stations within the system concerned
4. Platform interrogation may be included

[Note: this service may also include feeder links necessary for its operation [RR 1982].

Earth station: A station located either on the Earth's surface or within the major portion of Earth's atmosphere and intended for communication: (1) with one or more space stations; or (2) with one or more stations of the same kind by means of one or more reflecting satellites or other objects in space [RR 1982].

Effective radiated power (erp) (in a given direction): The product of the power supplied to the antenna and its gain relative to a half-wave dipole in a given direction [RR 1982].

Emergency position-indicating radiobeacon station: A station in the mobile service the emissions of which are intended to facilitate search and rescue operations [RR 1982].

Emission: Radiation produced, or the production of radiation, by a radio transmitting station. Note: for example, the energy radiated by the local oscillator of a radio receiver would not be an emission but a radiation [RR 1982].

Equivalent isotropically radiated power (eirp): The product of the power supplied to the antenna and the antenna gain in a given direction relative to an isotropic antenna [RR 1982].

Equivalent monopole radiated power (emrp) (in a given direction): The product of the power supplied to the antenna and its gain relative to a short vertical antenna in a given direction [RR 1982].

Equivalent satellite link noise temperature: The noise temperature referred to the output of the receiving antenna of the Earth station corresponding to the radio-frequency noise power, which produces the total observed noise at the output of the satellite link excluding the noise due to interference coming from satellite links using other satellites and from terrestrial systems [RR 1982].

Experimental station: A station utilizing radio waves in experiments with a view to the development of science or technique. Note: this definition does not include amateur stations [RR 1982].

Facsimile: A form of telegraphy for the transmission of fixed images, with or without half-tones, with a view to

---

[8]The full definition is contained in International Radio Consultive Committee (CCRI) Recommendation 460–2.

[9]In general, duplex operation and semi-duplex operation require two frequencies in radiocommunication; simplex operation may use either one or two.

their reproduction in a permanent form. Note: in this definition the term telegraphy has the same general meaning as defined in the Convention [RR 1982].

Feeder link: A radio link from an Earth station at a given location to a space station, or vice versa, conveying information for a space radiocommunication service other than for the fixed-satellite service. The given location may be at a specified fixed point, or at any fixed point within specified areas [RR 1982].

Fixed-satellite service: A radiocommunication service between Earth stations at given positions, when one or more satellites are used; the given position may be a specified fixed point or any fixed point within specified areas; in some cases this service includes satellite-to-satellite links, which may also be operated in the intersatellite service; the fixed-satellite service may also include feeder links for other space radiocommunication services [RR 1982].

Fixed service: A radiocommunication service between specified fixed points [RR 1982].

Fixed station: A station in the fixed service [RR 1982].

Frequency hopping systems: A frequency hopping system is a spread spectrum system in which the carrier is modulated with the coded information in a conventional manner causing a conventional spreading of the RF energy about the carrier frequency. However, the frequency of the carrier is not fixed but changes at fixed intervals under the direction of a pseudorandom coded sequence. The wide RF bandwidth needed by such a system is not required by a spreading of the RF energy about the carrier but rather to accommodate the range of frequencies to which the carrier frequency can hop.

Frequency-shift telegraphy: Telegraphy by frequency modulation in which the telegraph signal shifts the frequency of the carrier between predetermined values [RR 1982].

Frequency tolerance: The maximum permissible departure by the center frequency of the frequency band occupied by an emission from the assigned frequency or, by the characteristic frequency of an emission from the reference frequency. Note: the frequency tolerance is expressed in parts in 10^6 or in hertz [RR 1982].

Full carrier single-sideband emission: A single-sideband emission without suppression of the carrier [RR 1982].

Gain of an antenna: The ratio, usually expressed in decibels, of the power required at the input of a loss free reference antenna to the power supplied to the input of the given antenna to produce, in a given direction, the same field strength or the same power flux-density at the same distance. When not specified otherwise, the gain refers to the direction of maximum radiation. The gain may be considered for a specified polarization. Note: depending on the choice of the reference antenna a distinction is made between:
1. Absolute or isotropic gain (Gi), when the reference antenna is an isotropic antenna isolated in space
2. Gain relative to a half-wave dipole (Gd), when the reference antenna is a half-wave dipole isolated in space whose equatorial plane contains the given direction
3. Gain relative to a short vertical antenna (Gv), when the reference antenna is a linear conductor, much shorter than one-quarter of the wavelength, normal to the surface of a perfectly conducting plane, which contains the given direction [RR 1982].

General purpose mobile service: A mobile service that includes all mobile communications uses including those within the aeronautical mobile, land mobile, or the maritime mobile services.

Geostationary satellite: A geosynchronous satellite whose circular and direct orbit lies in the plane of the Earth's equator and which, thus, remains fixed relative to the Earth; by extension, a satellite which remains approximately fixed relative to the Earth [RR 1982].

Geostationary satellite orbit: The orbit in which a satellite must be placed to be a geostationary satellite [RR 1982].

Geosynchronous satellite: An Earth satellite whose period of revolution is equal to the period of rotation of the Earth about its axis [RR 1982].

Harmful interference: [10] Interference that endangers the functioning of a radionavigation service or of other safety services or seriously degrades, obstructs, or repeatedly interrupts a radiocommunication service operating in accordance with these (international) Radio Regulations [RR 1982].

---

[10] See Resolution 68 of the *Radio Regulations*.

Frequency Bands and Assignments 207

Hybrid spread spectrum systems: Hybrid spread spectrum systems are those that use combinations of two or more types of direct sequence, frequency hopping, time hopping and pulsed FM modulation in order to achieve their wide occupied bandwidths.

Inclination of an orbit (of an Earth satellite): The angle determined by the plane containing the orbit and the plane of the Earth's equator [RR 1982].

Individual reception (in the broadcasting-satellite service): The reception of emissions from a space station in the broadcasting-satellite service by simple domestic installations and, in particular, those possessing small antennas [RR 1982].

Industrial, scientific and medical (ISM) (of radio frequency energy) applications: Operation of equipment or appliances designed to generate and use locally radio-frequency energy for industrial, scientific, medical, domestic, or similar purposes, excluding applications in the field of telecommunications [RR 1982].

Instrument landing system (ILS): A radionavigation system that provides aircraft with horizontal and vertical guidance just before and during landing and, at certain fixed points, indicates the distance to the reference point of landing [RR 1982].

Instrument landing system glide path: A system of vertical guidance embodied in the instrument landing system that indicates the vertical deviation of the aircraft from its optimum path of descent [RR 1982].

Instrument landing system localizer: A system of horizontal guidance embodied in the instrument landing system that indicates the horizontal deviation of the aircraft from its optimum path of descent along the axis of the runway [RR 1982].

Interference: The effect of unwanted energy due to one or a combination of emissions, radiations, or inductions upon reception in a radiocommunication system, manifested by any performance degradation, misinterpretation, or loss of information, which could be extracted in the absence of such unwanted energy [RR 1982].

Intersatellite service: A radiocommunication service providing links between artificial Earth satellites [RR 1982].

Ionospheric scatter: The propagation of radio waves by scattering as a result of irregularities or discontinuities in the ionization of the ionosphere [RR 1982].

Land Earth station: An Earth station in the fixed-satellite service or, in some cases, in the mobile-satellite service, located at a specified fixed point or within a specified area on land to provide a feeder link for the mobile-satellite service [RR 1982].

Land mobile Earth station: A mobile Earth station in the land mobile-satellite service capable of surface movement within the geographical limits of a country or continent [RR 1982].

Land mobile-satellite service: A mobile-satellite service in which mobile Earth stations are located on land [RR 1982].

Land mobile service: A mobile service between base stations and land mobile stations, or between land mobile stations [RR 1982].

Land mobile station: A mobile station in the land mobile service capable of surface movement within the geographical limits of a country or continent.

Land station: A station in the mobile service not intended to be used while in motion [RR 1982].

Left-hand (or anticlockwise) polarized wave: An elliptically or circularly-polarized wave, in fixed plane, normal to the direction of propagation, while looking in the direction of propagation, rotates with time in a left-hand or anticlockwise direction [RR 1982].

Line A: Begins at Aberdeen, Washington, running by great circle arc to the intersection of 48° N, 120° W, then along parallel 48° N, to the intersection of 95° W, then by great circle arc through the southernmost point of Duluth, Minnesota, then by great circle arc to 45° N, 85° W, then southward along meridian 85° W, to its intersection with parallel 41° N, then along parallel 41° N, to its intersection with meridian 82° W, then by great circle arc through the southernmost point of Bangor, Maine, then by great circle arc through the southernmost point of Searsport, Maine, at which point it terminates [FCC].

Line B: Begins at Tofino, British Columbia, running by great circle arc to the intersection of 50° N, 125° W, then along parallel 50° N, to the intersection of 90° W, then by great circle arc to the intersection of 45° N, 79° 30′ W, then by great circle arc through the northernmost point of Drummondville, Quebec (lat.

45°52′ N., long. 72°30′ W), then by great circle arc to 48°30′ N, 70° W, then by great circle arc through the northernmost point of Compbellton, New Brunswick, then by great circle arc through the northernmost point of Liverpool, Nova Scotia, at which point it terminates [FCC].

Line C: Begins at the intersection of 70° N, 144° W, then by great circle arc to the intersection of 60° N, 143° W, then by great circle arc so as to include all of the Alaskan Panhandle [FCC].

Line D: Begins at the intersection of 70° N, 138° W, then by great circle arc to the intersection of 61°20′ N, 139° W (Burwash Landing), then by great circle arc to the intersection of 60°45′ N, 135° W, then by great circle arc to the intersection of 56° N, 128° W, then south along 128° meridian to lat. 55° N, then by great circle arc to the intersection of 54° N, 130° W, then by great circle arc to Port Clements, then to the Pacific Ocean where it ends [FCC].

Maritime mobile-satellite service: A mobile-satellite service in which mobile Earth stations are located onboard ships; survival craft stations and emergency position-indicating radiobeacon stations may also participate in this service [RR 1982].

Maritime mobile service: A mobile service between coast stations and ship stations, or between ship stations, or between associated onboard communication stations; survival craft stations and emergency position-indicating radiobeacon stations may also participate in this service [RR 1982].

Maritime radionavigation-satellite service: A radionavigation-satellite service in which Earth stations are located onboard ships [RR 1982].

Maritime radionavigation service: A radionavigation service intended for the benefit and for the safe operation of ships [RR 1982].

Marker beacon: A transmitter in the aeronautical radionavigation service that radiates vertically a distinctive pattern for providing position information to aircraft [RR 1982].

Mean power (of a radio transmitter): The average power supplied to the antenna transmission line by a transmitter during an interval of time sufficiently long compared with the lowest frequency encountered in the modulation taken under normal operating conditions [RR 1982].

Meteorological aids service: A radiocommunication service used for meteorological, including hydrological, observation and exploration [RR 1982].

Meteorological-satellite service: An Earth exploration-satellite service for meteorological purposes [RR 1982].

Mobile Earth station: An Earth station in the mobile-satellite service intended to be used while in motion or during halts at unspecified points [RR 1982].

Mobile-satellite service: A radiocommunication service (1) between mobile Earth stations and one or more space stations, or between space stations used by this service; or (2) between mobile Earth stations by means of one or more space stations. Note: this service may also include feeder links necessary for its operation [RR 1982].

Mobile service: A radiocommunication service between mobile and land stations, or between mobile stations [CONV 1973].

Mobile station: A station in the mobile service intended to be used while in motion or during halts at unspecified points [RR 1982].

Multisatellite link: A radio link between a transmitting Earth station and a receiving Earth station through two or more satellites, without any intermediate Earth station. Note: a multisatellite link comprises one uplink, one or more satellite-to-satellite links, and one downlink [RR 1982].

Necessary bandwidth: For a given class of emission, the width of the frequency band that is just sufficient to ensure the transmission of information at the rate and with the quality required under specified conditions [RR 1982].

Occupied bandwidth: The width of a frequency band such that, below the lower and above the upper frequency limits, the mean powers emitted are each equal to a specified percentage beta/2 of the total mean power of a given emission. Note: unless otherwise specified by the CCIR for the appropriate class of emission, the value of beta/2 should be taken as 0.5% [RR 1982].

Onboard communication station: A low-powered mobile station in the maritime mobile service intended for use for internal communications onboard a ship, or between a ship and its lifeboats and life-rafts during lifeboat drills or operations, or for communication within a group of vessels being towed or pushed, as well as for line handling and mooring instructions [RR 1982].

Orbit: The path, relative to a specified frame of reference, described by the center of mass of a satellite or other object in space subjected primarily to natural forces, mainly the force of gravity [RR 1982].

Out-of-band emission: Emission on a frequency or frequencies immediately outside the necessary bandwidth, which results from the modulation process, but excluding spurious emissions [RR 1982].

Passive sensor: A measuring instrument in the Earth exploration-satellite service or in the space research service by means of which information is obtained by reception of radio waves of natural origin [RR 1982].

Peak envelope power (of a radio transmitter): The average power supplied to the antenna transmission line by a transmitter during one radio frequency cycle at the crest of the modulation envelope taken under normal operating conditions [RR 1982].

Period (of a satellite): The time elapsing between two consecutive passages of a satellite through a characteristic point on its orbit [RR 1982].

Permissible interference: Observed or predicted interference, which complies with quantitative interference and sharing criteria contained in these [international] Radio Regulations or in CCIR Recommendations or in special agreements as provided for in these regulations [RR 1982].

Port operations service: A maritime mobile service in or near a port, between coast stations and ship stations, or between ship stations, in which messages are restricted to those relating to the operational handling, the movement and the safety of ships and, in emergency, to the safety of persons. Note: messages that are of a public correspondence nature shall be excluded from this service [RR 1982].

Port station: A coast station in the port operations service [RR 1982].

Power: Whenever the power of a radio transmitter, etc., is referred to it shall be expressed in one of the following forms, according to the class of emission, using the arbitrary symbols indicated:

1. Peak envelope power (PX or pX)
2. Mean power (PY or pY)
3. Carrier power (PZ or pZ)

Note: for different classes of emission, the relationships between peak envelope power, mean power and carrier power, under the conditions of normal operation and of no modulation, are contained in CCIR Recommendations, which may be used as a guide. Also note for use in formulas, the symbol p denotes power expressed in watts and the symbol p denotes power expressed in decibels relative to the reference level [RR 1982].

Primary radar: A radiodetermination system based on the comparison of reference signals with radio signals reflected from the position to be determined [RR 1982].

Protection ratio: The minimum value of the wanted-to-unwanted signal ratio, usually expressed in decibels, at the receiver input determined under specified conditions such that a specified reception quality of the wanted signal is achieved at the receiver output [RR 1982].

Pseudorandom sequence: A sequence of binary data that has some of the characteristics of a random sequence but also has some characteristics that are not random. It resembles a true random sequence in that the one bits and zero bits of the sequence are distributed randomly throughout every length N of the sequence and the total numbers of the one and zero bits in that length are approximately equal. It is not a true random sequence, however, because it consists of a fixed number (or length) of coded bits, which repeats itself exactly whenever that length is exceeded, and because it is generated by a fixed algorithm from some fixed initial state.

Public correspondence: Any telecommunication that the offices and stations must, by reason of their being at the disposal of the public, accept for transmission [CONV 1973].

Pulsed FM systems: A pulsed FM system is a spread spectrum system in which a RF carrier is modulated with a fixed period and fixed duty cycle sequence. At the beginning of each transmitted pulse, the carrier frequency is frequency modulated causing an additional spreading of the carrier. The pattern of the frequency modulation will depend on the spreading function that is chosen. In some systems the spreading function is a linear FM chirp sweep, sweeping either up or down in frequency.

Radar: A radiodetermination system based on the comparison of reference signals with radio signals reflected, or retransmitted, from the position to be determined [RR 1982].

Radar beacon (RACON): A transmitter–receiver associated with a fixed navigational mark, which, when triggered by a radar, automatically returns a distinctive signal that can appear on the display of the triggering radar, providing range, bearing and identification information [RR 1982].

Radiation: The outward flow of energy from any source in the form of radio waves [RR 1982].

Radio: A general term applied to the use of radio waves [CONV 1973].

Radio altimeter: Radionavigation equipment, onboard an aircraft or spacecraft or the spacecraft above the Earth's surface or another surface [RR 1982].

Radio astronomy: Astronomy based on the reception of radio waves of cosmic origin [RR 1982].

Radio astronomy service: A service involving the use of radio astronomy [RR 1982].

Radio astronomy station: A station in the radio astronomy service [RR 1982].

Radiobeacon station: A station in the radionavigation service the emissions of which are intended to enable a mobile station to determine its bearing or direction in relation to radiobeacon station [RR 1982].

Radiocommunication: Telecommunication by means of radio waves [CONV 1973].

Radiocommunication service: A service as defined in this section involving the transmission, emission and/or reception of radio waves for specific telecommunication purposes. Note: in these [international] Radio Regulations, unless otherwise stated, any radiocommunication service relates to terrestrial radiocommunication [RR 1982].

Radiodetermination: The determination of the position, velocity, and/or other characteristics of an object, or the obtaining of information relating to these parameters, by means of the propagation properties of radio waves [RR 1982].

Radiodetermination-satellite service: A radiocommunication service for the purpose of radiodetermination involving the use of one or more space stations. This service may also include feeder links necessary for its own operation [RR 1982].

Radiodetermination service: A radiocommunication service for the purpose of radiodetermination [RR 1982].

Radiodetermination station: A station in the radiodetermination service [RR 1982].

Radio direction-finding: Radio-determination using the reception of radio waves for the purpose of determining the direction of a station or object [RR 1982].

Radio direction-finding station: A radiodetermination station using radio direction-finding [RR 1982].

Radiolocation: Radiodetermination used for purposes other than those of radionavigation [RR 1982].

Radiolocation land station: A station in the radiolocation service not intended to be used while in motion [RR 1982].

Radiolocation mobile station: A station in the radiolocation service intended to be used while in motion or during halts at unspecified points [RR 1982].

Radiolocation service: A radiodetermination service for the purpose of radiolocation [RR 1982].

Radionavigation: Radiodetermination used for the purposes of navigation, including obstruction warning.

Radionavigation land station: A station in the radionavigation service not intended to be used while in motion [RR 1982].

Radionavigation mobile station: A station in the radionavigation service intended to be used while in motion or during halts at unspecified points [RR 1982].

Radionavigation-satellite service: A radiodetermination-satellite service used for the purpose of radionavigation. This service may also include feeder links necessary for its operation [RR 1982].

Radionavigation service: A radiodetermination service for the purpose of radionavigation [RR 1982].

Radiosonde: An automatic radio transmitter in the meteorological aids service usually carried on an aircraft, free balloon, kite, or parachute, and which transmits meteorological data [RR 1982].

Radiotelegram: A telegram, originating in or intended for a mobile station or a mobile Earth station transmitted on all or part of its route over the radiocommunication channels of the mobile service or of the mobile-satellite service [RR 1982].

Radiotelemetry: Telemetry by means of radio waves [RR 1982].

Radiotelephone call: A telephone call, originating in or intended for a mobile station or a mobile Earth station, transmitted on all or part of its route over the radiocommunication channels of the mobile service or of the mobile-satellite service [RR 1982].

Frequency Bands and Assignments

Radiotelex call: A telex call, originating in or intended for a mobile station or a mobile Earth station, transmitted on all or part of its route over the radiocommunication channels of the mobile service or the mobile-satellite service [RR 1982].

Radio waves or hertzian waves: Electromagnetic waves of frequencies arbitrarily lower than 3000 GHz, propagated in space without artificial guide [RR 1982].

Reduced carrier single-sideband emission: A single-sideband emission in which the degree of carrier suppression enables the carrier to be reconstituted and to be used for demodulation [RR 1982].

Reference frequency: A frequency having a fixed and specified position with respect to the assigned frequency. The displacement of this frequency with respect to the assigned frequency has the same absolute value and sign that the displacement of the characteristic frequency has with respect to the center of the frequency band occupied by the emission [RR 1982].

Reflecting satellite: A satellite intended to reflect radiocommunication signals [RR 1982].

Right-hand (or clockwise) polarized wave: An elliptically or circularly-polarized wave, in which the electric field vector, observed in any fixed plane, normal to the direction of propagation, while looking in the direction of propagation, rotates with time in a right-hand or clockwise direction [RR 1982].

Safety service: Any radiocommunication service used permanently or temporarily for the safe-guarding of human life and property [CONV 1973].

Satellite: A body that revolves around another body of preponderant mass and that has a motion primarily and permanently determined by the force of attraction of that other body [RR 1982].

Satellite link: A radio link between a transmitting Earth station and a receiving Earth station through one satellite. A satellite link comprises one uplink and one downlink [RR 1982].

Satellite network: A satellite system or a part of a satellite system, consisting of only one satellite and the cooperating Earth stations [RR 1982].

Satellite system: A space system using one or more artificial Earth satellites [RR 1982].

Secondary radar: A radiodetermination system based on the comparison of reference signals with radio signals retransmitted from the position to be determined [RR 1982].

Semi-duplex operation: A method that is simplex operation at one end of the circuit and duplex operation at the other. See footnote 9 [RR 1982].

Ship Earth station: A mobile Earth station in the maritime mobile-satellite service located onboard ship [RR 1982].

Ship movement service: A safety service in the maritime mobile service other than a port operations service, between coast stations and ship stations, or between ship stations, in which messages are restricted to those relating to the movement of ships. Messages of a public correspondence nature shall be excluded from this service [RR 1982].

Ship's emergency transmitter: A ship's transmitter to be used exclusively on a distress frequency for distress, urgency, or safety purposes [RR 1982].

Ship station: A mobile station in the maritime mobile service located onboard a vessel which is not permanently moored, other than a survival craft station [RR 1982].

Simplex operation: Operating method in which transmission is made possible alternatively in each direction of a telecommunication channel, for example, by means of manual control. See footnote 9 [RR 1982].

Single-sideband emission: An amplitude modulated emission with one sideband only [RR 1982].

Spacecraft: A man-made vehicle, which is intended to go beyond the major portion of the Earth's atmosphere [RR 1982].

Space operation service: A radiocommunication service concerned exclusively with the operation of spacecraft, in particular space tracking, space telemetry, and space telecommand. Note: these functions will normally be provided within the service in which the space station is operating [RR 1982].

Space radiocommunication: Any radiocommunication involving the use of one or more space stations or the use of one or more reflecting satellites or other objects in space [RR 1982].

Space research service: A radiocommunication service in which spacecraft or other objects in space are used for scientific or technological research purposes [RR 1982].

Space station: A station located on an object that is beyond, is intended to go beyond, or has been beyond, the major portion of the Earth's atmosphere [RR 1982].

Space system: Any group of cooperating Earth stations and/or space stations employing space radiocommunication for specific purposes [RR 1982].

Space telecommand: The use of radiocommunication for the transmission of signals to a space station to initiate, modify or terminate functions of equipment on a space object, including the space station [RR 1982].

Space telemetry: The use of telemetry for transmission for a space station of results of measurements made in a spacecraft, including those relating to the functioning of the spacecraft [RR 1982].

Space tracking: Determination of the orbit, velocity, or instanteneous position of an object in space by means of radiodetermination, excluding primary radar, for the purpose of following the movement of the object [RR 1982].

Special service: A radiocommunication service, not otherwise defined in this section, carried on exclusively for specific needs of general utility, and not open to public correspondence [RR 1982].

Spread spectrum systems: A spread spectrum system is an information bearing communications system in which: (1) information is conveyed by modulation of a carrier by some conventional means, (2) the bandwidth is deliberately widened by means of a spreading function over that which would be needed to transmit the information alone. (In some spread spectrum systems, a portion of the information being conveyed by the system may be contained in the spreading function.)

Spurious emission: Emission on a frequency or frequencies that are outside the necessary bandwidth and the level of which may be reduced without affecting the corresponding transmission of information. Spurious emissions include harmonic emissions, parasitic emissions, intermodulation products and frequency conversion products, but exclude out-of-band emissions [RR 1982].

Standard frequency and time signal-satellite service: A radiocommunication service using space stations on Earth satellites for the same purposes as those of the standard frequency and time signal service. Note: this service may also include feeder links necessary for its operation [RR 1982].

Standard frequency and time signal service: A radiocommunication service for scientific, technical, and other purposes, providing the transmission of specified frequencies, time signals, or both, of stated high precision, intended for general reception [RR 1982].

Standard frequency and time signal station: A station in the standard frequency and time signal service [RR 1982].

Station: One or more transmitters or receivers or a combination of transmitters and receivers, including the accessory equipment, necessary at one location for carrying on a radiocommunication service, or the radio astronomy service. Note: each station shall be classified by the service in which it operates permanently or temporarily [RR 1982].

Suppressed carrier single-sideband emission: A single-sideband emission in which the carrier is virtually suppressed and not intended to be used for demodulation [RR 1982].

Survival craft station: A mobile station in the maritime mobile service or the aeronautical mobile service intended solely for survival purposes and located on any lifeboat, life-raft, or other survival equipment [RR 1982].

Telecommand: The use of telecommunication for the transmission of signals to initiate, modify, or terminate functions of equipment at a distance [RR 1982].

Telecommunication: Any transmission, emission or reception of signs, signals, writing, images, and sounds or intelligence of any nature by wire, radio, optical or other electromagnetic systems [CONV 1973].

Telegram: Written matter intended to be transmitted by telegraphy for delivery to the addressee. This term also includes radiotelegrams unless otherwise specified. Note: in this definition the term telegraphy has the same general meaning as defined in the Convention [CONV 1973].

Telegraphy: A form of telecommunication that is concerned in any process providing transmission and reproduction at a distance of documentary matter, such as written or printed matter or fixed images, or the reproduction at a distance of any kind of information in such a form. For the purposes of the [international] Radio Regulations, unless otherwise specified therein, telegraphy shall mean a form of telecommunication for the transmission of written matter by the use of a signal code. See footnote 10 [RR 1982].

Telemetry: The use of telecommunication for automatical indicating or recording measurements at a distance from the measuring instrument [RR 1982].

Telephony: A form of telecommunication set up for the transmission of speech or, in some cases, other sounds. See footnote 10 [RR 1982].

Television: A form of telecommunication for the transmission of transient images of fixed or moving objects [RR 1982].

Terrestrial radiocommunication: Any radiocommunication other than space radiocommunication or radio astronomy [RR 1982].

Terrestrial station: A station effecting terrestrial radiocommunication. Note: in these [international] Radio Regulations, unless otherwise stated, any station is a terrestrial station [RR 1982].

Time hopping systems: A time hopping system is a spread spectrum system in which the period and duty cycle of a pulsed RF carrier are varied in a pseudorandom manner under the control of a coded sequence. Time hopping is often used effectively with frequency hopping to form a hybrid time-division, multiple access (TDMA) spread spectrum system.

Transponder: A transmitter–receiver facility the function of which is to transmit signals automatically when the proper interrogation is received [FCC].

Tropospheric scatter: The propagation of radio waves by scattering as a result of irregularities or discontinuities in the physical properties of the troposphere [RR 1982].

Unwanted emissions: Consist of spurious emissions and out-of-band emissions [RR 1982].

References

Code of Federal Regulations: 49 FR 2373, Jan. 19, 1964, as amended at 49 FR 44101, Nov. 2, 1964; 49 FR 2368, June 19, 1984, as amended at 50 FR 25239, June 18, 1985; 51 FR 37399, Oct. 22, 1986; 52 FR 7417, Mar. 11, 1987; 54 FR 49980, Dec. 4, 1990; 55 FR 28761, July 13, 1990; 56 FR 42703, Aug. 29, 1991.

CONV. 1973. International Telecommunication Conference, Malaga-Torremolinos.

FCC. 1934. Federal Communications Commission.

RR. 1982. Radio Regulations, International Telecommunications Union, Geneva, Switzerland.

10

International Standards and Constants[1]

10.1	International System of Units (SI)....................................	214
	Definitions of SI Base Units • Units in Use Together with the SI	
10.2	Physical Constants...	216

10.1 International System of Units (SI)

The International System of units (SI) was adopted by the 11th General Conference on Weights and Measures (CGPM) in 1960. It is a coherent system of units built from seven *SI base units*, one for each of the seven dimensionally independent base quantities: they are the meter, kilogram, second, ampere, kelvin, mole, and candela, for the dimensions length, mass, time, electric current, thermodynamic temperature, amount of substance, and luminous intensity, respectively. The definitions of the SI base units are given subsequently. The *SI derived units* are expressed as products of powers of the base units, analogous to the corresponding relations between physical quantities but with numerical factors equal to unity.

In the International System there is only one SI unit for each physical quantity. This is either the appropriate SI base unit itself or the appropriate SI derived unit. However, any of the approved decimal prefixes, called *SI prefixes*, may be used to construct decimal multiples or submultiples of SI units.

It is recommended that only SI units be used in science and technology (with SI prefixes where appropriate). Where there are special reasons for making an exception to this rule, it is recommended always to define the units used in terms of SI units. This section is based on information supplied by IUPAC.

Definitions of SI Base Units

Meter: The meter is the length of path traveled by light in vacuum during a time interval of 1/299 792 458 of a second (17th CGPM, 1983).

Kilogram: The kilogram is the unit of mass; it is equal to the mass of the international prototype of the kilogram (3rd CGPM, 1901).

Second: The second is the duration of 9 192 631 770 periods of the radiation corresponding the transition between the two hyperfine levels of the ground state of the cesium-133 atom (13th CGPM, 1967).

Ampere: The ampere is that constant current which, if maintained in two straight parallel conductors of infinite length, of negligible circular cross section, and placed 1 m apart in vacuum, would produce between these conductors a force equal to 2×10^{-7} newton per meter of length (9th CGPM, 1948).

Kelvin: The kelvin, unit of thermodynamic temperature, is the fraction 1/273.16 of the thermodynamic temperature of the triple point of water (13th CGPM, 1967).

---

[1] Material herein was reprinted from the following sources:

Lide, D.R., ed. 1992. *CRC Handbook of Chemistry and Physics*, 76th ed. CRC Press, Boca Raton, FL: International System of Units (SI), symbols and terminology for physical and chemical quantities, classification of electromagnetic radiation.

Zwillinger, D., ed. 1996. *CRC Standard Mathematical Tables and Formulae*, 30th ed. CRC Press, Boca Raton, FL: Greek alphabet, physical constants.

International Standards and Constants

Mole: The mole is the amount of substance of a system that contains as many elementary entities as there are atoms in 0.012 kilogram of carbon-12. When the mole is used, the elementary entities must be specified and may be atoms, molecules, ions, electrons, or other particles, or specified groups of such particles (14th CGPM, 1971).

Examples of the use of the mole:

1 mol of H_2 contains about 6.022×10^{23} H_2 molecules, or 12.044×10^{23} H atoms

1 mol of HgCl has a mass of 236.04 g

1 mol of Hg_2Cl_2 has a mass of 472.08 g

1 mol of Hg_2^{2+} has a mass of 401.18 g and a charge of 192.97 kC

1 mol of $Fe_{0.91}S$ has a mass of 82.88 g

1 mol of e^- has a mass of 548.60 μg and a charge of -96.49 kC

1 mol of photons whose frequency is 10^{14} Hz has energy of about 39.90 kJ

Candela: The candela is the luminous intensity, in a given direction, of a source that emits monochromatic radiation of frequency 540×10^{12} hertz and that has a radiant intensity in that direction of (1/683) watt per steradian (16th CGPM, 1979).

Names and Symbols for the SI Base Units

Physical quantity	Name of SI unit	Symbol for SI unit
length	meter	m
mass	kilogram	kg
time	second	s
electric current	ampere	A
thermodynamic temperature	kelvin	K
amount of substance	mole	mol
luminous intensity	candela	cd

SI Derived Units with Special Names and Symbols

Physical quantity	Name of SI unit	Symbol for SI unit	Expression in terms of SI base units
frequency[a]	hertz	Hz	s^{-1}
force	newton	N	$m\ kg\ s^{-2}$
pressure, stress	pascal	Pa	$N\ m^{-2} = m^{-1}\ kg\ s^{-2}$
energy, work, heat	joule	J	$N\ m = m^2\ kg\ s^{-2}$
power, radiant flux	watt	W	$J\ s^{-1} = m^2\ kg\ s^{-3}$
electric charge	coulomb	C	$A\ s$
electric potential, electromotive force	volt	V	$J\ C^{-1} = m^2\ kg\ s^{-3}\ A^{-1}$
electric resistance	ohm	Ω	$V\ A^{-1} = m^2\ kg\ s^{-3}\ A^{-2}$
electric conductance	siemens	S	$\Omega^{-1} = m^{-2}\ kg^{-1}\ s^3\ A^2$
electric capacitance	farad	F	$C\ V^{-1} = m^{-2}\ kg^{-1}\ s^4\ A^2$
magnetic flux density	tesla	T	$V\ s\ m^{-2} = kg\ s^{-2}\ A^{-1}$
magnetic flux	weber	Wb	$V\ s = m^2\ kg\ s^{-2}\ A^{-1}$
inductance	henry	H	$V\ A^{-1}\ s = m^2\ kg\ s^{-2}\ A^{-2}$
Celsius temperature[b]	degree Celsius	°C	K
luminous flux	lumen	lm	cd sr
illuminance	lux	lx	$cd\ sr\ m^{-2}$
activity (radioactive)	becquerel	Bq	s^{-1}
absorbed dose (of radiation)	gray	Gy	$J\ kg^{-1} = m^2 s^{-2}$
dose equivalent (dose equivalent index)	sievert	Sv	$J\ kg^{-1} = m^2\ s^{-2}$
plane angle	radian	rad	$1 = m\ m^{-1}$
solid angle	steradian	sr	$1 = m^2\ m^{-2}$

[a] For radial (circular) frequency and for angular velocity the unit rad s^{-1}, or simply s^{-1}, should be used, and this may not be simplified to Hz. The unit Hz should be used only for frequency in the sense of cycles per second.

[b] The Celsius temperature θ is defined by the equation:

$$\theta/°C = T/K - 273.15$$

The SI unit of Celsius temperature interval is the degree Celsius, °C, which is equal to the kelvin, K. °C should be treated as a single symbol, with no space between the ° sign and the letter C. (The symbol °K and the symbol °, should no longer be used.)

Units in Use Together with the SI

These units are not part of the SI, but it is recognized that they will continue to be used in appropriate contexts. SI prefixes may be attached to some of these units, such as milliliter, ml; millibar, mbar; megaelectronvolt, MeV; kilotonne, ktonne.

Physical quantity	Name of unit	Symbol for unit	Value in SI units
time	minute	min	60 s
time	hour	h	3600 s
time	day	d	86 400 s
plane angle	degree	°	$(\pi/180)$ rad
plane angle	minute	′	$(\pi/10\,800)$ rad
plane angle	second	″	$(\pi/648\,000)$ rad
length	ångstrom[a]	Å	10^{-10} m
area	barn	b	10^{-28} m^2
volume	litre	l, L	dm^3 = 10^{-3}m^3
mass	tonne	t	Mg = 10^3 kg
pressure	bar[a]	bar	10^5 Pa = 10^5 N m^{-2}
energy	electronvolt[b]	eV (= e × V)	≈ 1.60218×10^{-19} J
mass	unified atomic mass unit[b,c]	u (= $m_a(^{12}\text{C})/12$)	≈ 1.66054×10^{-27} kg

[a]The ångstrom and the bar are approved by CIPM for temporary use with SI units, until CIPM makes a further recommendation. However, they should not be introduced where they are not used at present.

[b]The values of these units in terms of the corresponding SI units are not exact, since they depend on the values of the physical constants e (for the electronvolt) and N_A (for the unified atomic mass unit), which are determined by experiment.

[c]The unified atomic mass unit is also sometimes called the dalton, with symbol Da, although the name and symbol have not been approved by CGPM.

Greek Alphabet

Greek letter		Greek name	English equivalent	Greek letter		Greek name	English equivalent
A	α	Alpha	a	N	ν	Nu	n
B	β	Beta	b	Ξ	ξ	Xi	x
Γ	γ	Gamma	g	O	o	Omicron	ŏ
Δ	δ	Delta	d	Π	π	Pi	p
E	ϵ	Epsilon	ĕ	P	ρ	Rho	r
Z	ζ	Zeta	z	Σ	σ ς	Sigma	s
H	η	Eta	ē	T	τ	Tau	t
Θ	θ ϑ	Theta	th	Υ	υ	Upsilon	u
I	ι	Iota	i	Φ	ϕ φ	Phi	ph
K	κ	Kappa	k	X	χ	Chi	ch
Λ	λ	Lambda	l	Ψ	ψ	Psi	ps
M	μ	Mu	m	Ω	ω	Omega	ō

10.2 Physical Constants

General

Equatorial radius of the Earth = 6378.388 km = 3963.34 miles (statute).
Polar radius of the Earth, 6356.912 km = 3949.99 miles (statute).
1 degree of latitude at 40° = 69 miles.
1 international nautical mile = 1.15078 miles (statute) = 1852 m = 6076.115 ft.
Mean density of the Earth = 5.522 g/cm^3 = 344.7 lb/ft^3
Constant of gravitation $(6.673 \pm 0.003) \times 10^{-8}$/cm^{3}gm^{-1}s^{-2}.
Acceleration due to gravity at sea level, latitude 45° = 980.6194 cm/s^2 = 32.1726 ft/s^2.

International Standards and Constants

Length of seconds pendulum at sea level, latitude 45° = 99.3575 cm = 39.1171 in.
1 knot (international) = 101.269 ft/min = 1.6878 ft/s = 1.1508 miles (statute)/h.
1 micron = 10^{-4} cm.
1 ångstrom = 10^{-8} cm.
Mass of hydrogen atom = $(1.67339 \pm 0.0031) \times 10^{-24}$ g.
Density of mercury at 0°C = 13.5955 g/ml.
Density of water at 3.98°C = 1.000000 g/ml.
Density, maximum, of water, at 3.98°C = 0.999973 g/cm^3.
Density of dry air at 0°C, 760 mm = 1.2929 g/l.
Velocity of sound in dry air at 0°C = 331.36 m/s − 1087.1 ft/s.
Velocity of light in vacuum = $(2.997925 \pm 0.000002) \times 10^{10}$ cm/s.
Heat of fusion of water 0°C = 79.71 cal/g.
Heat of vaporization of water 100°C = 539.55 cal/g.
Electrochemical equivalent of silver 0.001118 g/s international amp.
Absolute wavelength of red cadmium light in air at 15°C, 760 mm pressure = 6438.4696 Å.
Wavelength of orange-red line of krypton 86 = 6057.802 Å.

π Constants

π =	3.14159	26535	89793	23846	26433	83279	50288	41971	69399	37511
$1/\pi$ =	0.31830	98861	83790	67153	77675	26745	02872	40689	19291	48091
π^2 =	9.8690	44010	89358	61883	44909	99876	15113	53136	99407	24079
$\log_e \pi$ =	1.14472	98858	49400	17414	34273	51353	05871	16472	94812	91531
$\log_{10} \pi$ =	0.49714	98726	94133	85435	12682	88290	89887	36516	78324	38044
$\log_{10} \sqrt{2\pi}$ =	0.39908	99341	79057	52478	25035	91507	69595	02099	34102	92128

Constants Involving e

e =	2.71828	18284	59045	23536	02874	71352	66249	77572	47093	69996
$1/e$ =	0.36787	94411	71442	32159	55237	70161	46086	74458	11131	03177
e^2 =	7.38905	60989	30650	22723	04274	60575	00781	31803	15570	55185
$M = \log_{10} e$ =	0.43429	44819	03251	82765	11289	18916	60508	22943	97005	80367
$1/M = \log_e 10$ =	2.30258	50929	94045	68401	79914	54684	36420	76011	01488	62877
$\log_{10} M$ =	9.63778	43113	00536	78912	29674	98645	−10			

Numerical Constants

$\sqrt{2}$ =	1.41421	35623	73095	04880	16887	24209	69807	85696	71875	37695
$\sqrt[3]{2}$ =	1.25992	10498	94873	16476	72106	07278	22835	05702	51464	70151
$\log_e 2$ =	0.69314	71805	59945	30941	72321	21458	17656	80755	00134	36026
$\log_{10} 2$ =	0.30102	99956	63981	19521	37388	94724	49302	67881	89881	46211
$\sqrt{3}$ =	1.73205	08075	68877	29352	74463	41505	87236	69428	05253	81039
$\sqrt[3]{3}$ =	1.44224	95703	07408	38232	16383	10780	10958	83918	69253	49935
$\log_e 3$ =	1.09861	22886	68109	69139	52452	36922	52570	46474	90557	82275
$\log_{10} 3$ =	0.47712	12547	19662	43729	50279	03255	11530	92001	28864	19070

Symbols and Terminology for Physical and Chemical Quantities: Classical Mechanics

Name	Symbol	Definition	SI unit
mass	m		kg
reduced mass	μ	$\mu = m_1 m_2 / (m_1 + m_2)$	kg
density, mass density	ρ	$\rho = m/V$	kg m^{-3}
relative density	d	$d = \rho / \rho^\theta$	1
surface density	ρ_A, ρ_S	$\rho_A = m/A$	kg m^{-2}
specific volume	v	$v = V/M = 1/\rho$	m^3 kg^{-1}
momentum	$\boldsymbol{p}$	$\boldsymbol{p} = mv$	kg ms^{-1}

Symbols and Terminology for Physical and Chemical Quantities: Classical Mechanics (*Continued*)

Name	Symbol	Definition	SI unit
angular momentum, action	$\boldsymbol{L}$	$\boldsymbol{L} = \boldsymbol{r} \times \boldsymbol{p}$	J s
moment of inertia	I, J	$I = \sum m_i r_i^2$	kg m^2
force	$\boldsymbol{F}$	$\boldsymbol{F} = d\boldsymbol{p}/dt = m\boldsymbol{a}$	N
torque, moment of a force	$\boldsymbol{T}, (\boldsymbol{M})$	$\boldsymbol{T} = \boldsymbol{r} \times \boldsymbol{F}$	N m
energy	E		J
potential energy	E_p, V, Φ	$E_p = -\int \boldsymbol{F} \cdot d\boldsymbol{s}$	J
kinetic energy	E_k, T, K	$E_k = (1/2)mv^2$	J
work	W, w	$W = \int \boldsymbol{F} \cdot d\boldsymbol{s}$	J
Hamilton function	H	$H(q, p)$ $= T(q, p) + V(q)$	J
Lagrange function	L	$L(q, \dot{q})$ $= T(q, \dot{q}) - V(q)$	J
pressure	p, P	$p = F/A$	Pa, N m^{-2}
surface tension	γ, σ	$\gamma = dW/dA$	N m^{-1}, J m^{-2}
weight	$G, (W, P)$	$G = mg$	N
gravitational constant	G	$F = Gm_1 m_2/r^2$	N m^2 kg^{-2}
normal stress	σ	$\sigma = F/A$	Pa
shear stress	τ	$\tau = F/A$	Pa
linear strain, relative elongation	ε, e	$\varepsilon - \Delta l/l$	1
modulus of elasticity, Young's modulus	E	$E = \sigma/\varepsilon$	Pa
shear strain	γ	$\gamma = \Delta x/d$	1
shear modulus	G	$G = \tau/\gamma$	Pa
volume strain, bulk strain	θ	$\theta = \Delta V/V_0$	1
bulk modulus, compression modulus	K	$K = -V_0(dp/dV)$	Pa
viscosity, dynamic viscosity	η, μ	$\tau_{x,z} = \eta(dv_x/dz)$	Pa s
fluidity	ϕ	$\phi = 1/\eta$	m kg^{-1} s
kinematic viscosity	v	$v = \eta/\rho$	m^2 s^{-1}
friction coefficient	$\mu, (f)$	$F_{\text{frict}} = \mu F_{\text{norm}}$	1
power	P	$P = dW/dt$	W
sound energy flux	P, P_a	$P = dE/dt$	W
acoustic factors			
reflection factor	ρ	$\rho = P_r/P_0$	1
acoustic absorption factor	$\alpha_a, (\alpha)$	$\alpha_a = 1 - \rho$	1
transmission factor	τ	$\tau = P_{\text{tr}}/P_0$	1
dissipation factor	δ	$\delta = \alpha_a - \tau$	1

Symbols and Terminology for Physical and Chemical Quantities: Electricity and Magnetism

Name	Symbol	Definition	SI unit
quantity of electricity, electric charge	Q		C
charge density	ρ	$\rho = Q/V$	C m^{-3}
surface charge density	σ	$\sigma = Q/A$	C m^{-2}
electric potential	V, ϕ	$V = dW/dQ$	V, J C^{-1}
electric potential difference	$U, \Delta V, \Delta \phi$	$U = V_2 - V_1$	V
electromotive force	E	$E = \int (F/Q) \cdot d\boldsymbol{s}$	V
electric field strength	$\boldsymbol{E}$	$\boldsymbol{E} = \boldsymbol{F}/Q = -\text{grad } V$	V m^{-1}
electric flux	Ψ	$\Psi = \int \boldsymbol{D} \cdot d\boldsymbol{A}$	C
electric displacement	$\boldsymbol{D}$	$\boldsymbol{D} = \varepsilon \boldsymbol{E}$	C m^{-2}
capacitance	C	$C = Q/U$	F, C V^{-1}
permittivity	ε	$\boldsymbol{D} = \varepsilon \boldsymbol{E}$	F m^{-1}
permittivity of vacuum	ε_0	$\varepsilon_0 = \mu_0^{-1} c_0^{-2}$	F m^{-1}
relative permittivity	ε_r	$\varepsilon_r = \varepsilon/\varepsilon_0$	1

Symbols and Terminology for Physical and Chemical Quantities: Electricity and Magnetism (*Continued*)

Name	Symbol	Definition	SI unit
dielectric polarization (dipole moment per volume)	$\boldsymbol{P}$	$\boldsymbol{P} = \boldsymbol{D} - \varepsilon_0 \boldsymbol{E}$	$C\,m^{-2}$
electric susceptibility	χ_e	$\chi_e = \varepsilon_r - 1$	1
electric dipole moment	$\boldsymbol{p}, \mu$	$\boldsymbol{p} = Q\boldsymbol{r}$	$C\,m$
electric current	I	$I = dQ/dt$	A
electric current density	$\boldsymbol{j}, \boldsymbol{J}$	$I = \int \boldsymbol{j} \cdot d\boldsymbol{A}$	$A\,m^{-2}$
magnetic flux density, magnetic induction	$\boldsymbol{B}$	$\boldsymbol{F} = Q\boldsymbol{v} \times \boldsymbol{B}$	T
magnetic flux	$\boldsymbol{\Phi}$	$\boldsymbol{\Phi} = \int \boldsymbol{B} \cdot d\boldsymbol{A}$	Wb
magnetic field strength	$\boldsymbol{H}$	$\boldsymbol{B} = \mu \boldsymbol{H}$	$A\,M^{-1}$
permeability	μ	$\boldsymbol{B} = \mu \boldsymbol{H}$	$N\,A^{-2}, H\,m^{-1}$
permeability of vacuum	μ_0		$H\,m^{-1}$
relative permeability	μ_r	$\mu_r = \mu/\mu_0$	1
magnetization (magnetic dipole moment per volume)	$\boldsymbol{M}$	$\boldsymbol{M} = \boldsymbol{B}/\mu_0 - \boldsymbol{H}$	$A\,m^{-1}$
magnetic susceptibility	$\chi, \kappa, (\chi_m)$	$\chi = \mu_r - 1$	1
molar magnetic susceptibility	χ_m	$\chi_m = V_m \chi$	$m^3\,mol^{-1}$
magnetic dipole moment	$\boldsymbol{m}, \mu$	$E_p = -\boldsymbol{m} \cdot \boldsymbol{B}$	$A\,m^2, J\,T^{-1}$
electrical resistance	R	$R = U/I$	Ω
conductance	G	$G = 1/R$	S
loss angle	δ	$\delta = (\pi/2) + \phi_I - \phi_U$	1, rad
reactance	X	$X = (U/I) \sin \delta$	Ω
impedance (complex impedance)	Z	$Z = R + iX$	Ω
admittance (complex admittance)	Y	$Y = 1/Z$	S
susceptance	B	$Y = G + iB$	S
resistivity	ρ	$\rho = E/j$	$\Omega\,m$
conductivity	κ, γ, σ	$\kappa = 1/\rho$	$S\,m^{-1}$
self-inductance	L	$E = -L(dI/dt)$	H
mutual inductance	M, L_{12}	$E_1 = L_{12}(dI_2/dt)$	H
magnetic vector potential	$\boldsymbol{A}$	$\boldsymbol{B} = \nabla \times \boldsymbol{A}$	$Wb\,m^{-1}$
Poynting vector	$\boldsymbol{S}$	$\boldsymbol{S} = \boldsymbol{E} \times \boldsymbol{H}$	$W\,m^{-2}$

Symbols and Terminology for Physical and Chemical Quantities: Electromagnetic Radiation

Name	Symbol	Definition	SI unit
wavelength	λ		m
speed of light in vacuum	c_0		$m\,s^{-1}$
in a medium	c	$c = c_0/n$	$m\,s^{-1}$
wavenumber in vacuum	$\tilde{\nu}$	$\tilde{\nu} = \nu/c_0 = 1/n\lambda$	m^{-1}
wavenumber (in a medium)	σ	$\sigma = 1/\lambda$	m^{-1}
frequency	ν	$\nu = c/\lambda$	Hz
circular frequency, pulsatance	ω	$\omega = 2\pi\nu$	$s^{-1}, rad\,s^{-1}$
refractive index	n	$n = c_0/c$	1
Planck constant	h		$J\,s$
Planck constant/2π	$\hbar$	$\hbar = h/2\pi$	$J\,s$
radiant energy	Q, W		J
radiant energy density	ρ, w	$\rho = Q/V$	$J\,m^{-3}$
spectral radiant energy density			
in terms of frequency	ρ_ν, w_ν	$\rho_\nu = d\rho/d\nu$	$J\,m^{-3}Hz^{-1}$
in terms of wavenumber	$\rho_{\tilde{\nu}}, w_{\tilde{\nu}}$	$\rho_{\tilde{\nu}} = d\rho/d\tilde{\nu}$	$J\,m^{-2}$
in terms of wavelength	ρ_λ, w_λ	$\rho_\lambda = d\rho/d\lambda$	$J\,m^{-4}$

Symbols and Terminology for Physical and Chemical Quantities: Electromagnetic Radiation (*Continued*)

Name	Symbol	Definition	SI unit
Einstein transition probabilities			
spontaneous emission	A_{nm}	$dN_n/dt = -A_{nm}N_n$	s^{-1}
stimulated emission	B_{nm}	$dN_n/dt = -\rho_{\tilde{v}}(\tilde{v}_{nm}) \times B_{nm}N_n$	s kg^{-1}
stimulated absorption	B_{mn}	$dN_n/dt = \rho_{\tilde{v}}(\tilde{v}_{nm})B_{mn}N_m$	s kg^{-1}
radiant power, radiant energy per time	Φ, P	$\Phi = dQ/dt$	W
radiant intensity	I	$I = d\Phi/d\Omega$	W sr^{-1}
radiant exitance (emitted radiant flux)	M	$M = d\Phi/dA_{\text{source}}$	W m^{-2}
irradiance (radiant flux received)	$E, (I)$	$E = d\Phi/dA$	W m^{-2}
emittance	ε	$\varepsilon = M/M_{\text{bb}}$	1
Stefan-Boltzman constant	σ	$M_{\text{bb}} = \sigma T^4$	W m^{-2} K^{-4}
first radiation constant	c_1	$c_1 = 2\pi h c_0^2$	W m^2
second radiation constant	c_2	$c_2 = hc_0/k$	K m
transmittance, transmission factor	τ, T	$\tau = \Phi_{\text{tr}}/\Phi_0$	1
absorptance, absorption factor	α	$\alpha = \Phi_{\text{abs}}/\Phi_0$	1
reflectance, reflection factor	ρ	$\rho = \Phi_{\text{refl}}/\Phi_0$	1
(decadic) absorbance	A	$A = \lg(1 - \alpha_i)$	1
napierian absorbance	B	$B = \ln(1 - \alpha_i)$	1
absorption coefficient			
(linear) decadic	a, K	$a = A/l$	m^{-1}
(linear) napierian	α	$\alpha = B/l$	m^{-1}
molar (decadic)	ε	$\varepsilon = a/c = A/cl$	m^2 mol^{-1}
molar napierian	κ	$\kappa = \alpha/c = B/cl$	m^2 mol^{-1}
absorption index	k	$k = \alpha/4\pi\tilde{v}$	1
complex refractive index	$\hat{n}$	$\hat{n} = n + ik$	1
molar refraction	R, R_m	$R = \dfrac{n^2 - 1}{n^2 + 2} V_m$	m^3 mol^{-1}
angle of optical rotation	α		1, rad

Symbols and Terminology for Physical and Chemical Quantities: Solid State

Name	Symbol	Definition	SI unit
lattice vector	$\mathbf{R}, \mathbf{R}_0$		m
fundamental translation vectors for the crystal lattice	$\mathbf{a}_1; \mathbf{a}_2; \mathbf{a}_3,$ $\mathbf{a}; \mathbf{b}; \mathbf{c}$	$\mathbf{R} = n_1\mathbf{a}_1 + n_2\mathbf{a}_2 + n_3\mathbf{a}_3$	m
(circular) reciprocal lattice vector	$\mathbf{G}$	$\mathbf{G} \cdot \mathbf{R} = 2\pi m$	m^{-1}
(circular) fundamental translation vectors for the reciprocal lattice	$\mathbf{b}_1; \mathbf{b}_2; \mathbf{b}_3,$ $\mathbf{a}^*; \mathbf{b}^*; \mathbf{c}^*$	$\mathbf{a}_i \cdot \mathbf{b}_k = 2\pi\delta_{ik}$	m^{-1}
lattice plane spacing	d		m
Bragg angle	θ	$n\lambda = 2d\sin\theta$	1, rad
order of reflection	n		1
order parameters			
short range	σ		1
long range	s		1
Burgers vector	$\mathbf{b}$		m
particle position vector	$\mathbf{r}, \mathbf{R}_j$		m
equilibrium position vector of an ion	$\mathbf{R}_0$		m

Symbols and Terminology for Physical and Chemical Quantities: Solid State (*Continued*)

Name	Symbol	Definition	SI unit
equilibrium position vector of an ion	$\mathbf{R}_0$		m
displacement vector of an ion	$\mathbf{u}$	$\mathbf{u} = \mathbf{R} - \mathbf{R}_0$	m
Debye–Waller factor	B, D		1
Debye circular wavenumber	q_D		m^{-1}
Debye circular frequency	ω_D		s^{-1}
Grüneisen parameter	γ, Γ	$\gamma = \alpha V / \kappa C_V$	1
Madelung constant	$\alpha, \mathcal{M}$	$E_{\text{coul}} = \dfrac{\alpha N_A z_+ z_- e^2}{4\pi\varepsilon_0 R_0}$	1
density of states	N_E	$N_E = dN(E)/dE$	J^{-1} m^{-3}
(spectral) density of vibrational modes	N_ω, g	$N_\omega = dN(\omega)/d\omega$	s m^{-3}
resistivity tensor	ρ_{ik}	$E = \rho \cdot j$	Ω m
conductivity tensor	σ_{ik}	$\sigma = \rho^{-1}$	S m^{-1}
thermal conductivity tensor	λ_{ik}	$J_q = -\lambda \cdot \operatorname{grad} T$	W m^{-1} K^{-1}
residual resistivity	ρ_R		Ω m
relaxation time	τ	$\tau = l/v_F$	s
Lorenz coefficient	L	$L = \lambda/\sigma T$	V^2 K^{-2}
Hall coefficient	A_H, R_H	$E = \rho \cdot j + R_H(\mathbf{B} \times j)$	m^3 C^{-1}
thermoelectric force	E		V
Peltier coefficient	Π		V
Thomson coefficient	$\mu, (\tau)$		V K^{-1}
work function	Φ	$\Phi = E_\infty - E_F$	J
number density, number concentration	$n, (p)$		m^{-3}
gap energy	E_g		J
donor ionization energy	E_d		J
acceptor ionization energy	E_a		J
Fermi energy	E_F, ε_F		J
circular wave vector, propagation vector	$\mathbf{k}, \mathbf{q}$	$k = 2\pi/\lambda$	m^{-1}
Bloch function	$u_k(\mathbf{r})$	$\psi(\mathbf{r}) = u_k(\mathbf{r})\exp(i\mathbf{k} \cdot \mathbf{r})$	m$^{-3/2}$
charge density of electrons	ρ	$\rho(\mathbf{r}) = -e\psi^*(\mathbf{r})\psi(\mathbf{r})$	C m^{-3}
effective mass	m^*		kg
mobility	μ	$\mu = v_{\text{drift}}/E$	m^2 V^{-1} s^{-1}
mobility ratio	b	$b = \mu_n/\mu_p$	1
diffusion coefficient	D	$dN/dt = -DA(dn/dx)$	m^2 s^{-1}
diffusion length	L	$L = \sqrt{D\tau}$	m
characteristic (Weiss) temperature	ϕ, ϕ_W		K
Curie temperature	T_C		K
Néel temperature	T_N		K

Letter Designations of Microwave Bands

Frequency, GHz	Wavelength, cm	Wavenumber, cm^{-1}	Band
1–2	30–15	0.033–0.067	L-band
1–4	15–7.5	0.067–0.133	S-band
4–8	7.5–3.7	0.133–0.267	C-band
8–12	3.7–2.5	0.267–0.4	X-band
12–18	2.5–1.7	0.4–0.6	Ku-band
18–27	1.7–1.1	0.6–0.9	K-band
27–40	1.1–0.75	0.9–1.33	Ka-band

11
Conversion Factors[1]

Jerry C. Whitaker
Editor-in-Chief

11.1 Introduction 222
11.2 Conversion Constants and Multipliers 238
  Recommended Decimal Multiples and Submultiples • Conversion Factors—Metric to English • Conversion Factors—English to Metric • Conversion Factors—General • Temperature Factors • Conversion of Temperatures

11.1 Introduction

Engineers often find it necessary to convert from one unit of measurement to another. The following table of conversion factors provides a convenient method of accomplishing the task. In Chapter 153, certain conversion factors were listed, grouped by function. In the following table, a more complete listing of conversion factors is presented, grouped in alphabetical order.

To Convert	Into	Multiply by
	A	
abcoulomb	statcoulombs	2.998×10^{10}
acre	square chain (Gunters)	10
acre	rods	160
acre	square links (Gunters)	1×10^5
acre	hectare or square hectometer	0.4047
acres	square feet	43,560.0
acres	square meters	4,047
acres	square miles	1.562×10^{-3}
acres	square yards	4,840
acre-feet	cubic feet	43,560.0
acre-feet	gallons	3.259×10^5
amperes per square centimeter	ampere per square inch	6.452
amperes per square centimeter	ampere per square meter	10^4
amperes per square inch	ampere per square centimeter	0.1550
amperes per square inch	ampere per square meter	1,550.0
amperes per square meter	ampere per square centimeter	10^{-4}
amperes per square meter	ampere per square inch	6.452×10^{-4}
ampere-hours	coulombs	3,600.0
ampere-hours	faradays	0.03731
ampere-turns	gilberts	1.257
ampere-turns per centimeter	ampere-turns per inch	2.540
ampere-turns per centimeter	ampere-turns per meter	100.0
ampere-turns per centimeter	gilberts per centimeter	1.257
ampere-turns per inch	ampere-turns per centimeter	0.3937
ampere-turns per inch	ampere-turns per meter	39.37
ampere-turns per inch	gilberts per centimeter	0.4950

[1] Information contained in this chapter was adapted from Whitaker, J.C. 1991. *Maintaining Electronic Systems.* CRC Press, Boca Raton, FL.

Conversion Factors

To Convert	Into	Multiply by
ampere-turns per meter	ampere-turns per centimeter	0.01
ampere-turns per meter	ampere-turns per inch	0.0254
ampere-turns per meter	gilberts per centimeter	0.01257
Angstrom unit	inch	3937×10^{-9}
Angstrom unit	meter	1×10^{-10}
Angstrom unit	micron or mu(μ)	1×10^{-4}
are	acre (U.S.)	.02471
ares	square yards	119.60
ares	acres	0.02471
ares	square meters	100.0
astronomical unit	kilometers	1.495×10^8
atmospheres	ton per square inch	0.007348
atmospheres	centimeter of mercury	76.0
atmospheres	foot of water (at 4°C)	33.90
atmospheres	inch of mercury (at 0°C)	29.92
atmospheres	kilogram per square centimeter	1.0333
atmospheres	kilogram per square meter	10,332
atmospheres	pounds per square inch	14.70
atmospheres	tons per square foot	1.058

B

To Convert	Into	Multiply by
barrels (U.S., dry)	cubic inch	7056
barrels (U.S., dry)	quarts (dry)	105.0
barrels (U.S., liquid)	gallons	31.5
barrels (oil)	gallons (oil)	42.0
bars	atmospheres	0.9869
bars	dyne per square centimeter	10^4
bars	kilogram per square meter	1.020×10^4
bars	pound per square foot	2,089
bars	pound per square inch	14.50
Baryl	dyne per square centimeter	1.000
bolt (U.S., cloth)	meter	36.576
British thermal unit	liter-atmosphere	10.409
British thermal unit	erg	1.0550×10^{10}
British thermal unit	foot-pound	778.3
British thermal unit	gram-calorie	252.0
British thermal unit	horsepower-hour	3.931×10^{-4}
British thermal unit	joule	1,054.8
British thermal unit	kilogram-calorie	0.2520
British thermal unit	kilogram-meter	107.5
British thermal unit	kilowatt-hour	2.928×10^{-4}
British thermal unit per hour	foot-pound per second	0.2162
British thermal unit per hour	gram-calorie per second	0.0700
British thermal unit per hour	horsepower-hour	3.929×10^{-4}
British thermal unit per hour	watts	0.2931
British thermal unit per minute	foot-pound per second	12.96
British thermal unit per minute	horsepower	0.02356
British thermal unit per minute	kilowatts	0.01757
British thermal unit per minute	watts	17.57
British thermal unit per square foot per minute	watts per square inch	0.1221
bucket (British, dry)	cubic centimeter	1.818×10^4
bushels	cubic foot	1.2445
bushels	cubic inch	2,150.4
bushels	cubic meter	0.03524
bushels	liters	35.24
bushels	pecks	4.0
bushels	pints (dry)	64.0
bushels	quarts (dry)	32.0

To Convert	Into	Multiply by
	C	
calories, gram (mean)	British thermal unit (mean)	3.9685×10^{-3}
candle per square centimeter	lamberts	3.142
candle per square inch	lamberts	.4870
centares (centiares)	square meters	1.0
centigrade	Fahrenheit	$(C° \times 9/5) + 32$
centigrams	grams	0.01
centiliter	ounce fluid (U.S.)	0.3382
centiliter	cubic inch	0.6103
centiliter	drams	2.705
centiliters	liters	0.01
centimeters	feet	3.281×10^{-2}
centimeters	inches	0.3937
centimeters	kilometers	10^{-5}
centimeters	meters	0.01
centimeters	miles	6.214×10^{-6}
centimeters	millimeter	10.0
centimeters	mils	393.7
centimeters	yards	1.094×10^{-2}
centimeter-dynes	centimeter-grams	1.020×10^{-3}
centimeter-dynes	meter-kilogram	1.020×10^{-8}
centimeter-dynes	pound-feet	7.376×10^{-8}
centimeter-grams	centimeter-dynes	980.7
centimeter-grams	meter-kilogram	10^{-5}
centimeter-grams	pound-feet	7.233×10^{-5}
centimeters of mercury	atmospheres	0.01316
centimeters of mercury	feet of water	0.4461
centimeters of mercury	kilogram per square meter	136.0
centimeters of mercury	pounds per square foot	27.85
centimeters of mercury	pounds per square inch	0.1934
centimeters per second	feet per minute	1.9686
centimeters per second	feet per second	0.03281
centimeters per second	kilometer per hour	0.036
centimeters per second	knot	0.1943
centimeters per second	meter per minute	0.6
centimeters per second	mile per hour	0.02237
centimeters per second	mile per minute	3.728×10^{-4}
centimeters per second per second	feet per second per second	0.03281
centimeters per second per second	kilometer per hour per second	0.036
centimeters per second per second	meters per second per second	0.01
centimeters per second per second	miles per hour per second	0.02237
chain	inches	792.00
chain	meters	20.12
chain (surveyors' or Gunter's)	yards	22.00
circular mils	square centimeter	5.067×10^{-6}
circular mils	square mils	0.7854
circular mils	square inches	7.854×10^{-7}
circumference	radians	6.283
cord	cord feet	8
cord feet	cubic feet	16
coulomb	statcoulombs	2.998×10^9
coulombs	faradays	1.036×10^{-5}
coulombs per square centimeter	coulombs per square inch	64.52
coulombs per square centimeter	coulombs per square meter	10^4
coulombs per square inch	coulombs per square centimeter	0.1550
coulombs per square inch	coulombs per square meter	1,550
coulombs per square meter	coulombs per square centimeter	10^{-4}
coulombs per square meter	coulombs per square inch	6.452×10^{-4}
cubic centimeters	cubic feet	3.531×10^{-5}

Conversion Factors

To Convert	Into	Multiply by
cubic centimeters	cubic inches	0.06102
cubic centimeters	cubic meters	10^{-6}
cubic centimeters	cubic yards	1.308×10^{-6}
cubic centimeters	gallons (U.S. liquid)	2.642×10^{-4}
cubic centimeters	liters	0.001
cubic centimeters	pints (U.S. liquid)	2.113×10^{-3}
cubic centimeters	quarts (U.S. liquid)	1.057×10^{-3}
cubic feet	bushels (dry)	0.8036
cubic feet	cubic centimeter	28,320.0
cubic feet	cubic inches	1,728.0
cubic feet	cubic meters	0.02832
cubic feet	cubic yards	0.03704
cubic feet	gallons (U.S. liquid)	7.48052
cubic feet	liters	28.32
cubic feet	pints (U.S. liquid)	59.84
cubic feet	quarts (U.S. liquid)	29.92
cubic feet per minute	cubic centimeter per second	472.0
cubic feet per minute	gallons per second	0.1247
cubic feet per minute	liters per second	0.4720
cubic feet per minute	pounds of water per minute	62.43
cubic feet per second	million galllons per day	0.646317
cubic feet per second	gallons per minute	448.831
cubic inches	cubic centimeters	16.39
cubic inches	cubic feet	5.787×10^{-4}
cubic inches	cubic meter	1.639×10^{-5}
cubic inches	cubic yards	2.143×10^{-5}
cubic inches	gallons	4.329×10^{-3}
cubic inches	liters	0.01639
cubic inches	mil-feet	1.061×10^{5}
cubic inches	pints (U.S. liquid)	0.03463
cubic inches	quarts (U.S. liquid)	0.01732
cubic meters	bushels (dry)	28.38
cubic meters	cubic centimeter	10^{6}
cubic meters	cubic feet	35.31
cubic meters	cubic inches	61,023.0
cubic meters	cubic yards	1.308
cubic meters	gallons (U.S. liquid)	264.2
cubic meters	liters	1,000.0
cubic meters	pints (U.S. liquid)	2,113.0
cubic meters	quarts (U.S. liquid)	1,057
cubic yards	cubic centimeter	7.646×10^{5}
cubic yards	cubic feet	27.0
cubic yards	cubic inches	46,656.0
cubic yards	cubic meters	0.7646
cubic yards	gallons (U.S. liquid)	202.0
cubic yards	liters	764.6
cubic yards	pints (U.S. liquid)	1,615.9
cubic yards	quarts (U.S. liquid)	807.9
cubic yards per minute	cubic feet per second	0.45
cubic yards per minute	gallons per second	3.367
cubic yards per minute	liters per second	12.74

D

To Convert	Into	Multiply by
Dalton	gram	1.650×10^{-24}
days	seconds	86,400.0
decigrams	grams	0.1
deciliters	liters	0.1
decimeters	meters	0.1
degrees (angle)	quadrants	0.01111
degrees (angle)	radians	0.01745

To Convert	Into	Multiply by
degrees (angle)	seconds	3,600.0
degrees per second	radians per second	0.01745
degrees per second	revolutions per minute	0.1667
degrees per second	revolutions per second	2.778×10^{-3}
dekagrams	grams	10.0
dekaliters	liters	10.0
dekameters	meters	10.0
drams (apothecaries' or troy)	ounces (avoirdupois)	0.1371429
drams (apothecaries' or troy)	ounces (troy)	0.125
drams (U.S., fluid or apothecaries')	cubic centimeter	3.6967
drams	grams	1.7718
drams	grains	27.3437
drams	ounces	0.0625
dyne per centimeter	erg per square millimeter	0.01
dyne per square centimeter	atmospheres	9.869×10^{-7}
dyne per square centimeter	inch of mercury at 0°C	2.953×10^{-5}
dyne per square centimeter	inch of water at 4°C	4.015×10^{-4}
dynes	grams	1.020×10^{-3}
dynes	joules per centimeter	10^{-7}
dynes	joules per meter (newtons)	10^{-5}
dynes	kilograms	1.020×10^{-6}
dynes	poundals	7.233×10^{-5}
dynes	pounds	2.248×10^{-6}
dynes per square centimeter	bars	10^{-6}

E

To Convert	Into	Multiply by
ell	centimeter	114.30
ell	inches	45
em, pica	inch	0.167
em, pica	centimeter	0.4233
erg per second	dyne–centimeter per second	1.000
ergs	British thermal unit	9.480×10^{-11}
ergs	dyne-centimeters	1.0
ergs	foot-pounds	7.367×10^{-8}
ergs	gram-calories	0.2389×10^{-7}
ergs	gram-centimeter	1.020×10^{-3}
ergs	horsepower-hour	3.7250×10^{-14}
ergs	joules	10^{-7}
ergs	kilogram-calories	2.389×10^{-11}
ergs	kilogram-meters	1.020×10^{-8}
ergs	kilowatt-hour	0.2778×10^{-13}
ergs	watt-hours	0.2778×10^{-10}
ergs per second	British thermal unit per minute	$5,688 \times 10^{-9}$
ergs per second	foot-pound per minute	4.427×10^{-6}
ergs per second	foot-pound per second	7.3756×10^{-8}
ergs per second	horsepower	1.341×10^{-10}
ergs per second	kilogram-calories per minute	1.433×10^{-9}
ergs per second	kilowatts	10^{-10}

F

To Convert	Into	Multiply by
farads	microfarads	10^6
faraday per second	ampere (absolute)	9.6500×10^4
faradays	ampere-hours	26.80
faradays	coulombs	9.649×10^4
fathom	meter	1.828804
fathoms	feet	6.0
feet	centimeters	30.48
feet	kilometers	3.048×10^{-4}
feet	meters	0.3048
feet	miles (nautical)	1.645×10^{-4}
feet	miles (statute)	1.894×10^{-4}

Conversion Factors

To Convert	Into	Multiply by
feet	millimeters	304.8
feet	mils	1.2×10^4
feet of water	atmospheres	0.02950
feet of water	inch of mercury	0.8826
feet of water	kilogram per square centimeter	0.03048
feet of water	kilogram per square meter	304.8
feet of water	pounds per square foot	62.43
feet of water	pounds per square inch	0.4335
feet per minute	centimeter per second	0.5080
feet per minute	feet per second	0.01667
feet per minute	kilometer per hour	0.01829
feet per minute	meters per minute	0.3048
feet per minute	miles per hour	0.01136
feet per second	centimeter per second	30.48
feet per second	kilometer per hour	1.097
feet per second	knots	0.5921
feet per second	meters per minute	18.29
feet per second	miles per hour	0.6818
feet per second	miles per minute	0.01136
feet per second per second	centimeter per second per second	30.48
feet per second per second	kilometer per hour per second	1.097
feet per second per second	meters per second per second	0.3048
feet per second per second	miles per hour per second	0.6818
feet per 100 feet	per centigrade	1.0
foot-candle	Lumen per square meter	10.764
foot-pounds	British thermal unit	1.286×10^{-3}
foot-pounds	ergs	1.356×10^7
foot-pounds	gram-calories	0.3238
foot-pounds	horsepower per hour	5.050×10^{-7}
foot-pounds	joules	1.356
foot-pounds	kilogram-calories	3.24×10^{-4}
foot-pounds	kilogram-meters	0.1383
foot-pounds	kilowatt-hour	3.766×10^{-7}
foot-pounds per minute	British thermal unit per minute	1.286×10^{-3}
foot-pounds per minute	foot-pounds per second	0.01667
foot-pounds per minute	horsepower	3.030×10^{-5}
foot-pounds per minute	kilogram-calories per minute	3.24×10^{-4}
foot-pounds per minute	kilowatts	2.260×10^{-5}
foot-pounds per second	British thermal unit per hour	4.6263
foot-pounds per second	British thermal unit per minute	0.07717
foot-pounds per second	horsepower	1.818×10^{-3}
foot-pounds per second	kilogram-calories per minute	0.01945
foot-pounds per second	kilowatts	1.356×10^{-3}
furlongs	miles (U.S.)	0.125
furlongs	rods	40.0
furlongs	feet	660.0

G

To Convert	Into	Multiply by
gallons	cubic centimeter	3,785.0
gallons	cubic feet	0.1337
gallons	cubic inches	231.0
gallons	cubic meters	3.785×10^{-3}
gallons	cubic yards	4.951×10^{-3}
gallons	liters	3.785
gallons (liquid British Imperial)	gallons (U.S. liquid)	1.20095
gallons (U.S.)	gallons (Imperial)	0.83267
gallons of water	pounds of water	8.3453
gallons per minute	cubic foot per second	2.228×10^{-3}
gallons per minute	liters per second	0.06308
gallons per minute	cubic foot per hour	8.0208

To Convert	Into	Multiply by
gausses	lines per square inch	6.452
gausses	webers per square centimeter	10^{-8}
gausses	webers per square inch	6.452×10^{-8}
gausses	webers per square meter	10^{-4}
gilberts	ampere-turns	0.7958
gilberts per centimeter	ampere-turns per centimeter	0.7958
gilberts per centimeter	ampere-turns per inch	2.021
gilberts per centimeter	ampere-turns per meter	79.58
gills (British)	cubic centimeter	142.07
gills	liters	0.1183
gills	pints (liquid)	0.25
grade	radian	0.01571
grains	drams (avoirdupois)	0.03657143
grains (troy)	grains (avoirdupois)	1.0
grains (troy)	grams	0.06480
grains (troy)	ounces (avoirdupois)	2.0833×10^{-3}
grains (troy)	pennyweight (troy)	0.04167
grains/U.S. gallon	parts per million	17.118
grains/U.S. gallon	pounds per million gallon	142.86
grains/Imp. gallon	parts per million	14.286
grams	dynes	980.7
grams	grains	15.43
grams	joules per centimeter	9.807×10^{-5}
grams	joules per meter (newtons)	9.807×10^{-3}
grams	kilograms	0.001
grams	milligrams	1,000
grams	ounces (avoirdupois)	0.03527
grams	ounces (troy)	0.03215
grams	poundals	0.07093
grams	pounds	2.205×10^{-3}
grams per centimeter	pounds per inch	5.600×10^{-3}
grams per cubic centimeter	pounds per cubic foot	62.43
grams per cubic centimeter	pounds per cubic inch	0.03613
grams per cubic centimeter	pounds per mil-foot	3.405×10^{-7}
grams per liter	grains per gallon	58.417
grams per liter	pounds per 1000 gallon	8.345
grams per liter	pounds per cubic foot	0.062427
grams per liter	parts per million	1,000.0
grams per square centimeter	pounds per square foot	2.0481
gram-calories	British thermal unit	3.9683×10^{-3}
gram-calories	ergs	4.1868×10^{7}
gram-calories	foot-pounds	3.0880
gram-calories	horsepower-hour	1.5596×10^{-6}
gram-calories	kilowatt-hour	1.1630×10^{-6}
gram-calories	watt-hour	1.1630×10^{-3}
gram-calories per second	British thermal unit per hour	14.286
gram-centimeters	British thermal unit	9.297×10^{-8}
gram-centimeters	ergs	980.7
gram-centimeters	joules	9.807×10^{-5}
gram-centimeters	kilogram-calorie	2.343×10^{-8}
gram-centimeters	kilogram-meters	10^{-5}

H

To Convert	Into	Multiply by
hand	centimeter	10.16
hectares	acres	2.471
hectares	square feet	1.076×10^{5}
hectograms	grams	100.0
hectoliters	liters	100.0
hectometers	meters	100.0
hectowatts	watts	100.0

Conversion Factors

To Convert	Into	Multiply by
henries	millihenries	1,000.0
hogsheads (British)	cubic foot	10.114
hogsheads (U.S.)	cubic foot	8.42184
hogsheads (U.S.)	gallons (U.S.)	63
horsepower	British thermal unit per minute	42.44
horsepower	foot-pound per minute	33,000
horsepower	foot-pound per second	550.0
horsepower (metric) (542.5 foot-pound per second)	horsepower (550 foot-pound per second)	0.9863
horsepower (550 foot-pound per second)	horsepower (metric) (542.5 foot-pound per second)	1.014
horsepower	kilogram-calories per minute	10.68
horsepower	kilowatts	0.7457
horsepower	watts	745.7
horsepower (boiler)	British thermal unit per hour	33.479
horsepower (boiler)	kilowatts	9.803
horsepower-hour	British thermal unit	2,547
horsepower-hour	ergs	2.6845×10^{13}
horsepower-hour	foot-pound	1.98×10^6
horsepower-hour	gram-calories	641,190
horsepower-hour	joules	2.684×10^6
horsepower-hour	kilogram-calories	641.1
horsepower-hour	kilogram-meters	2.737×10^5
horsepower-hour	kilowatt-hour	0.7457
hours	days	4.167×10^{-2}
hours	weeks	5.952×10^{-3}
hundredweights (long)	pounds	112
hundredweights (long)	tons (long)	0.05
hundredweights (short)	ounces (avoirdupois)	1,600
hundredweights (short)	pounds	100
hundredweights (short)	tons (metric)	0.0453592
hundredweights (short)	tons (long)	0.0446429
	I	
inches	centimeters	2.540
inches	meters	2.540×10^{-2}
inches	miles	1.578×10^{-5}
inches	millimeters	25.40
inches	mils	1,000.0
inches	yards	2.778×10^{-2}
inches of mercury	atmospheres	0.03342
inches of mercury	feet of water	1.133
inches of mercury	kilogram per square centimeter	0.03453
inches of mercury	kilogram per square meter	345.3
inches of mercury	pounds per square foot	70.73
inches of mercury	pounds per square inch	0.4912
inches of water (at 4°C)	atmospheres	2.458×10^{-3}
inches of water (at 4°C)	inches of mercury	0.07355
inches of water (at 4°C)	kilogram per square centimeter	2.540×10^{-3}
inches of water (at 4°C)	ounces per square inch	0.5781
inches of water (at 4°C)	pounds per square foot	5.204
inches of water (at 4°C)	pounds per square inch	0.03613
International ampere	ampere (absolute)	.9998
International volt	volts (absolute)	1.0003
International volt	joules (absolute)	1.593×10^{-19}
International volt	joules	9.654×10^4
	J	
joules	British thermal unit	9.480×10^{-4}
joules	ergs	10^7

To Convert	Into	Multiply by
joules	foot-pounds	0.7376
joules	kilogram-calories	2.389×10^{-4}
joules	kilogram-meters	0.1020
joules	watt-hour	2.778×10^{-4}
joules per centimeter	grams	1.020×10^4
joules per centimeter	dynes	10^7
joules per centimeter	joules per meter (newtons)	100.0
joules per centimeter	poundals	723.3
joules per centimeter	pounds	22.48
	K	
kilograms	dynes	980,665
kilograms	grams	1,000.0
kilograms	joules per centimeter	0.09807
kilograms	joules per meter (newtons)	9.807
kilograms	poundals	70.93
kilograms	pounds	2.205
kilograms	tons (long)	9.842×10^{-4}
kilograms	tons (short)	1.102×10^{-3}
kilograms per cubic meter	grams per cubic centimeter	0.001
kilograms per cubic meter	pounds per cubic foot	0.06243
kilograms per cubic meter	pounds per cubic inch	3.613×10^{-5}
kilograms per cubic meter	pounds per mil-foot	3.405×10^{-10}
kilograms per meter	pounds per foot	0.6720
Kilogram per square centimeter	dynes	980,665
kilograms per square centimeter	atmospheres	0.9678
kilograms per square centimeter	feet of water	32.81
kilograms per square centimeter	inches of mercury	28.96
kilograms per square centimeter	pounds per square foot	2,048
kilograms per square centimeter	pounds per square inch	14.22
kilograms per square meter	atmospheres	9.678×10^{-5}
kilograms per square meter	bars	98.07×10^{-6}
kilograms per square meter	feet of water	3.281×10^{-3}
kilograms per square meter	inches of mercury	2.896×10^{-3}
kilograms per square meter	pounds per square foot	0.2048
kilograms per square meter	pounds per square inch	1.422×10^{-3}
kilograms per square millimeter	kilogram per square meter	10^6
kilogram-calories	British thermal unit	3.968
kilogram-calories	foot-pounds	3,088
kilogram-calories	horsepower-hour	1.560×10^{-3}
kilogram-calories	joules	4,186
kilogram-calories	kilogram-meters	426.9
kilogram-calories	kilojoules	4.186
kilogram-calories	kilowatt-hour	1.163×10^{-3}
kilogram meters	British thermal unit	9.294×10^{-3}
kilogram meters	ergs	9.804×10^7
kilogram meters	foot-pounds	7.233
kilogram meters	joules	9.804
kilogram meters	kilogram-calories	2.342×10^{-3}
kilogram meters	kilowatt-hour	2.723×10^{-6}
kilolines	maxwells	1,000.0
kiloliters	liters	1,000.0
kilometers	centimeters	10^5
kilometers	feet	3,281
kilometers	inches	3.937×10^4
kilometers	meters	1,000.0
kilometers	miles	0.6214
kilometers	millimeters	10^4
kilometers	yards	1,094
kilometers per hour	centimeter per second	27.78

Conversion Factors

To Convert	Into	Multiply by
kilometers per hour	feet per minute	54.68
kilometers per hour	feet per second	0.9113
kilometers per hour	knots	0.5396
kilometers per hour	meters per minute	16.67
kilometers per hour	miles per hour	0.6214
kilometers per second	centimeter per second per second	27.78
kilometers per second	feet per second per second	0.9113
kilometers per second	meters per second per second	0.2778
kilometers per second	miles per hour per second	0.6214
kilowatts	British thermal unit per minute	56.92
kilowatts	foot-pounds per minute	4.426×10^4
kilowatts	foot-pounds per second	737.6
kilowatts	horsepower	1.341
kilowatts	kilogram-calories per minute	14.34
kilowatts	watts	1,000.0
kilowatt-hour	British thermal unit	3,413
kilowatt-hour	ergs	3.600×10^{13}
kilowatt-hour	foot-pound	2.655×10^6
kilowatt-hour	gram-calories	859,850
kilowatt-hour	horsepower-hour	1.341
kilowatt-hour	joules	3.6×10^6
kilowatt-hour	kilogram-calories	860.5
kilowatt-hour	kilogram-meters	3.671×10^5
kilowatt-hour	pounds of water evaporated from and at 212°F	8.53
kilowatt-hours	pounds of water raised from 62° to 212°F	22.75
knots	feet per hour	6,080
knots	kilometers per hour	1.8532
knots	nautical miles per hour	1.0
knots	statute miles per hour	1.151
knots	yards per hour	2,027
knots	feet per second	1.689

L

To Convert	Into	Multiply by
league	miles (approximate)	3.0
Light year	miles	5.9×10^{12}
Light year	kilometers	9.4637×10^{12}
lines per square centimeter	gausses	1.0
lines per square inch	gausses	0.1550
lines per square inch	webers per square centimeter	1.550×10^{-9}
lines per square inch	webers per square inch	10^{-8}
lines per square inch	webers per square meter	1.550×10^{-5}
links (engineer's)	inches	12.0
links (surveyor's)	inches	7.92
liters	bushels (U.S. dry)	0.02838
liters	cubic centimeter	1,000.0
liters	cubic feet	0.03531
liters	cubic inches	61.02
liters	cubic meters	0.001
liters	cubic yards	1.308×10^{-3}
liters	gallons (U.S. liquid)	0.2642
liters	pints (U.S. liquid)	2.113
liters	quarts (U.S. liquid)	1.057
liters per minute	cubic foot per second	5.886×10^{-4}
liters per minute	gallons per second	4.403×10^{-3}
Lumen	spherical candle power	.07958
Lumen	watt	.001496
lumens per square foot	foot-candles	1.0
lumen per square foot	lumen per square meter	10.76
lux	foot-candles	0.0929

To Convert	Into	Multiply by
	M	
maxwells	kilolines	0.001
maxwells	webers	10^{-8}
megalines	maxwells	10^6
megohms	microhms	10^{12}
megohms	ohms	10^6
meters	centimeters	100.0
meters	feet	3.281
meters	inches	39.37
meters	kilometers	0.001
meters	miles (nautical)	5.396×10^{-4}
meters	miles (statute)	6.214×10^{-4}
meters	millimeters	1,000.0
meters	yards	1.094
meters	varas	1.179
meters per minute	centimeter per second	1,667
meters per minute	feet per minute	3.281
meters per minute	feet per second	0.05468
meters per minute	kilometer per hour	0.06
meters per minute	knots	0.03238
meters per minute	miles per hour	0.03728
meters per second	feet per minute	196.8
meters per second	feet per second	3.281
meters per second	kilometers per hour	3.6
meters per second	kilometers per minute	0.06
meters per second	miles per hour	2.237
meters per second	miles per minute	0.03728
meters per second per second	centimeter per second per second	100.0
meters per second per second	foot per second per second	3.281
meters per second per second	kilometer per hour per second	3.6
meters per second per second	miles per hour per second	2.237
meter-kilograms	centimeter-dynes	9.807×10^7
meter-kilograms	centimeter-grams	10^5
meter-kilograms	pound-feet	7.233
microfarad	farads	10^{-6}
micrograms	grams	10^{-6}
microhms	megohms	10^{-12}
microhms	ohms	10^{-6}
microliters	liters	10^{-6}
Microns	meters	1×10^{-6}
miles (nautical)	feet	6,080.27
miles (nautical)	kilometers	1.853
miles (nautical)	meters	1,853
miles (nautical)	miles (statute)	1.1516
miles (nautical)	yards	2,027
miles (statute)	centimeters	1.609×10^5
miles (statute)	feet	5,280
miles (statute)	inches	6.336×10^4
miles (statute)	kilometers	1.609
miles (statute)	meters	1,609
miles (statute)	miles (nautical)	0.8684
miles (statute)	yards	1,760
miles per hour	centimeter per second	44.70
miles per hour	feet per minute	88
miles per hour	feet per second	1.467
miles per hour	kilometer per hour	1.609
miles per hour	kilometer per minute	0.02682
miles per hour	knots	0.8684
miles per hour	meters per minute	26.82

Conversion Factors

To Convert	Into	Multiply by
miles per hour	miles per minute	0.1667
miles per hour per second	centimeter per second per second	44.70
miles per hour per second	feet per second per second	1.467
miles per hour per second	kilometer per hour per second	1.609
miles per hour per second	meters per second per second	0.4470
miles per minute	centimeter per second	2,682
miles per minute	feet per second	88
miles per minute	kilometer per minute	1.609
miles per minute	knots per minute	0.8684
miles per minute	miles per hour	60
mil-feet	cubic inches	9.425×10^{-6}
milliers	kilograms	1,000
millimicrons	meters	1×10^{-9}
milligrams	grains	0.01543236
milligrams	grams	0.001
milligrams per liter	parts per million	1.0
millihenries	henries	0.001
milliliters	liters	0.001
millimeters	centimeters	0.1
millimeters	feet	3.281×10^{-3}
millimeters	inches	0.03937
millimeters	kilometers	10^{-6}
millimeters	meters	0.001
millimeters	miles	6.214×10^{-7}
millimeters	mils	39.37
millimeters	yards	1.094×10^{-3}
million gallons per day	cubic foot per second	1.54723
mils	centimeters	2.540×10^{-3}
mils	feet	8.333×10^{-5}
mils	inches	0.001
mils	kilometers	2.540×10^{-8}
mils	yards	2.778×10^{-5}
miner's inches	cubic foot per minute	1.5
minims (British)	cubic centimeter	0.059192
minims (U.S., fluid)	cubic centimeter	0.061612
minutes (angles)	degrees	0.01667
minutes (angles)	quadrants	1.852×10^{-4}
minutes (angles)	radians	2.909×10^{-4}
minutes (angles)	seconds	60.0
myriagrams	kilograms	10.0
myriameters	kilometers	10.0
myriawatts	kilowatts	10.0

N

To Convert	Into	Multiply by
nepers	decibels	8.686
newton	dynes	1×10^5

O

To Convert	Into	Multiply by
ohm (International)	ohm (absolute)	1.0005
ohms	megohms	10^{-6}
ohms	microhms	10^6
ounces	drams	16.0
ounces	grains	437.5
ounces	grams	28.349527
ounces	pounds	0.0625
ounces	ounces (troy)	0.9115
ounces	tons (long)	2.790×10^{-5}
ounces	tons (metric)	2.835×10^{-5}
ounces (fluid)	cu. inches	1.805
ounces (fluid)	liters	0.02957
ounces (troy)	grains	480.0
ounces (troy)	grams	31.103481

To Convert	Into	Multiply by
ounces (troy)	ounces (avoirdupois)	1.09714
ounces (troy)	pennyweights (troy)	20.0
ounces (troy)	pounds (troy)	0.08333
Ounce per square inch	dynes per square centimeter	4,309
ounces per square inch	pounds per square inch	0.0625

P

To Convert	Into	Multiply by
parsec	miles	19×10^{12}
parsec	kilometers	3.084×10^{13}
parts per million	grains per U.S. gallon	0.0584
parts per million	grains per Imperial gallon	0.07016
parts per million	pounds per million gallon	8.345
pecks (British)	cubic inches	554.6
pecks (British)	liters	9.091901
pecks (U.S.)	bushels	0.25
pecks (U.S.)	cubic inches	537.605
pecks (U.S.)	liters	8.809582
pecks (U.S.)	quarts (dry)	8
pennyweights (troy)	grains	24.0
pennyweights (troy)	ounces (troy)	0.05
pennyweights (troy)	grams	1.55517
pennyweights (troy)	pounds (troy)	4.1667×10^{-3}
pints (dry)	cubic inches	33.60
pints (liquid)	cubic centimeter	473.2
pints (liquid)	cubic feet	0.01671
pints (liquid)	cubic inches	28.87
pints (liquid)	cubic meters	4.732×10^{-4}
pints (liquid)	cubic yards	6.189×10^{-4}
pints (liquid)	gallons	0.125
pints (liquid)	liters	0.4732
pints (liquid)	quarts (liquid)	0.5
Planck's quantum	erg-second	6.624×10^{-27}
Poise	Gram per centimeter second	1.00
Pounds (avoirdupois)	ounces (troy)	14.5833
poundals	dynes	13,826
poundals	grams	14.10
poundals	joules per centimeter	1.383×10^{-3}
poundals	joules per meter (newtons)	0.1383
poundals	kilograms	0.01410
poundals	pounds	0.03108
pounds	drams	256
pounds	dynes	44.4823×10^{4}
pounds	grains	7,000
pounds	grams	453.5924
pounds	joules per centimeter	0.04448
pounds	joules per meter (newtons)	4.448
pounds	kilograms	0.4536
pounds	ounces	16.0
pounds	ounces (troy)	14.5833
pounds	poundals	32.17
pounds	pounds (troy)	1.21528
pounds	tons (short)	0.0005
pounds (troy)	grains	5,760
pounds (troy)	grams	373.24177
pounds (troy)	ounces (avoirdupois)	13.1657
pounds (troy)	ounces (troy)	12.0
pounds (troy)	pennyweights (troy)	240.0
pounds (troy)	pounds (avoirdupois)	0.822857
pounds (troy)	tons (long)	3.6735×10^{-4}
pounds (troy)	tons (metric)	3.7324×10^{-4}
pounds (troy)	tons (short)	4.1143×10^{-4}

Conversion Factors

To Convert	Into	Multiply by
pounds of water	cubic foot	0.01602
pounds of water	cubic inches	27.68
pounds of water	gallons	0.1198
pounds of water per minute	cubic foot per second	2.670×10^{-4}
pound-feet	centimeter-dynes	1.356×10^{7}
pound-feet	centimeter-grams	13,825
pound-feet	meter-kilogram	0.1383
pounds per cubic foot	grams per cubic centimeter	0.01602
pounds per cubic foot	kilogram per cubic meter	16.02
pounds per cubic foot	pounds per cubic inch	5.787×10^{-4}
pounds per cubic foot	pounds per mil-foot	5.456×10^{-9}
pounds per cubic inch	gram per cubic centimeter	27.68
pounds per cubic inch	kilogram per cubic meter	2.768×10^{4}
pounds per cubic inch	pounds per cubic foot	1,728
pounds per cubic inch	pounds per mil-foot	9.425×10^{-6}
pounds per foot	kilogram per meter	1.488
pounds per inch	gram per centimeter	178.6
pounds per mil-foot	gram per cubic centimeter	2.306×10^{6}
pounds per square foot	atmospheres	4.725×10^{-4}
pounds per square foot	feet of water	0.01602
pounds per square foot	inches of mercury	0.01414
pounds per square foot	kilogram per square meter	4.882
pounds per square foot	pounds per square inch	6.944×10^{-3}
pounds per square inch	atmospheres	0.06804
pounds per square inch	feet of water	2.307
pounds per square inch	inches of mercury	2.036
pounds per square inch	kilogram per square meter	703.1
pounds per square inch	pounds per square foot	144.0
	Q	
quadrants (angle)	degrees	90.0
quadrants (angle)	minutes	5,400.0
quadrants (angle)	radians	1.571
quadrants (angle)	seconds	3.24×10^{5}
quarts (dry)	cubic inches	67.20
quarts (liquids)	cubic centimeter	946.4
quarts (liquids)	cubic feet	0.03342
quarts (liquids)	cubic inches	57.75
quarts (liquids)	cubic meters	9.464×10^{-4}
quarts (liquids)	cubic yards	1.238×10^{-3}
quarts (liquids)	gallons	0.25
quarts (liquids)	liters	0.9463
	R	
radians	degrees	57.30
radians	minutes	3,438
radians	quadrants	0.6366
radians	seconds	2.063×10^{5}
radians per second	degrees per second	57.30
radians per second	revolutions per minute	9.549
radians per second	revolutions per second	0.1592
radians per second per second	revolution per minute per minute	573.0
radians per second per second	revolution per minute per second	9.549
radians per second per second	revolution per second per second	0.1592
revolutions	degrees	360.0
revolutions	quadrants	4.0
revolutions	radians	6.283
revolutions per minute	degrees per second	6.0
revolutions per minute	radians per second	0.1047
revolutions per minute	revolution per second	0.01667
revolutions per minute per minute	radians per second per second	1.745×10^{-3}
revolutions per minute per minute	revolution per minute per second	0.01667

To Convert	Into	Multiply by
revolutions per minute per minute	revolution per second per second	2.778×10^{-4}
revolutions per second	degrees per second	360.0
revolutions per second	radians per second	6.283
revolutions per second	revolution per minute	60.0
revolutions per second per second	radians per second per second	6.283
revolutions per second per second	revolution per minute per minute	3,600.0
revolutions per second per second	revolution per minute per second	60.0
rod	chain (Gunters)	.25
rod	meters	5.029
rods (surveyors' measure)	yards	5.5
rods	feet	16.5

S

To Convert	Into	Multiply by
scruples	grains	20
seconds (angle)	degrees	2.778×10^{-4}
seconds (angle)	minutes	0.01667
seconds (angle)	quadrants	3.087×10^{-6}
seconds (angle)	radians	4.848×10^{-6}
slug	kilogram	14.59
slug	pounds	32.17
sphere	steradians	12.57
square centimeters	circular mils	1.973×10^{5}
square centimeters	square feet	1.076×10^{-3}
square centimeters	square inches	0.1550
square centimeters	square meters	0.0001
square centimeters	square miles	3.861×10^{-11}
square centimeters	square millimeters	100.0
square centimeters	square yards	1.196×10^{-4}
square feet	acres	2.296×10^{-5}
square feet	circular mils	1.833×10^{8}
square feet	square centimeter	929.0
square feet	square inches	144.0
square feet	square meters	0.09290
square feet	square miles	3.587×10^{-8}
square feet	square millimeters	9.290×10^{4}
square feet	square yards	0.1111
square inches	circular mils	1.273×10^{6}
square inches	square centimeter	6.452
square inches	square feet	6.944×10^{-3}
square inches	square millimeters	645.2
square inches	square mils	10^{6}
square inches	square yards	7.716×10^{-4}
square kilometers	acres	247.1
square kilometers	square centimeter	10^{10}
square kilometers	square foot	10.76×10^{6}
square kilometers	square inches	1.550×10^{9}
square kilometers	square meters	10^{6}
square kilometers	square miles	0.3861
square kilometers	square yards	1.196×10^{6}
square meters	acres	2.471×10^{-4}
square meters	square centimeter	10^{4}
square meters	square feet	10.76
square meters	square inches	1,550
square meters	square miles	3.861×10^{-7}
square meters	square millimeters	10^{6}
square meters	square yards	1.196
square miles	acres	640.0
square miles	square feet	27.88×10^{6}
square miles	square kilometer	2.590
square miles	square meters	2.590×10^{6}
square miles	square yards	3.098×10^{6}

Conversion Factors

To Convert	Into	Multiply by
square millimeters	circular mils	1,973
square millimeters	square centimeter	0.01
square millimeters	square feet	1.076×10^{-5}
square millimeters	square inches	1.550×10^{-3}
square mils	circular mils	1.273
square mils	square centimeter	6.452×10^{-6}
square mils	square inches	10^{-6}
square yards	acres	2.066×10^{-4}
square yards	square centimeter	8,361
square yards	square feet	9.0
square yards	square inches	1,296
square yards	square meters	0.8361
square yards	square miles	3.228×10^{-7}
square yards	square millimeters	8.361×10^5

T

To Convert	Into	Multiply by
temperature (°C) + 273	absolute temperature (°C)	1.0
temperature (°C) + 17.78	temperature (°F)	1.8
temperature(°F) + 460	absolute temperature (°F)	1.0
temperature (°F) − 32	temperature (°C)	5/9
tons (long)	kilograms	1,016
tons (long)	pounds	2,240
tons (long)	tons (short)	1.120
tons (metric)	kilograms	1,000
tons (metric)	pounds	2,205
tons (short)	kilograms	907.1848
tons (short)	ounces	32,000
tons (short)	ounces (troy)	29,166.66
tons (short)	pounds	2,000
tons (short)	pounds (troy)	2,430.56
tons (short)	tons (long)	0.89287
tons (short)	tons (metric)	0.9078
tons (short) per square foot	kilogram per square meter	9,765
tons (short) per square foot	pounds per square inch	2,000
tons of water per 24 hour	pounds of water per hour	83.333
tons of water per 24 hour	gallons per minute	0.16643
tons of water per 24 hour	cubic foot per hour	1.3349

V

To Convert	Into	Multiply by
volt/inch	volt per centimeter	.39370
volt (absolute)	statvolts	.003336

W

To Convert	Into	Multiply by
watts	British thermal unit per hour	3.4129
watts	British thermal unit per minute	0.05688
watts	ergs per second	107
watts	foot-pound per minute	44.27
watts	foot-pound per second	0.7378
watts	horsepower	1.341×10^{-3}
watts	horsepower (metric)	1.360×10^{-3}
watts	kilogram-calories per minute	0.01433
watts	kilowatts	0.001
watts (absolute)	British thermal unit (mean) per minute	0.056884
watts (absolute)	joules per second	1
watt-hours	British thermal unit	3.413
watt-hours	ergs	3.60×10^{10}
watt-hours	foot-pounds	2,656
watt-hours	gram-calories	859.85
watt-hours	horsepower-hour	1.341×10^{-3}
watt-hours	kilogram-calories	0.8605
watt-hours	kilogram-meters	367.2
watt-hours	kilowatt-hour	0.001

To Convert	Into	Multiply by
watt (International)	watt (absolute)	1.0002
webers	maxwells	10^8
webers	kilolines	10^5
webers per square inch	gausses	1.550×10^7
webers per square inch	lines per square inch	10^8
webers per square inch	webers per square centimeter	0.1550
webers per square inch	webers per square meter	1,550
webers per square meter	gausses	10^4
webers per square meter	lines per square inch	6.452×10^4
webers per square meter	webers per square centimeter	10^{-4}
webers per square meter	webers per square inch	6.452×10^{-4}
	Y	
yards	centimeters	91.44
yards	kilometers	9.144×10^{-4}
yards	meters	0.9144
yards	miles (nautical)	4.934×10^{-4}
yards	miles (statute)	5.682×10^{-4}
yards	millimeters	914.4

11.2 Conversion Constants and Multipliers [2]

Recommended Decimal Multiples and Submultiples

Multiples and Submultiples	Prefixes	Symbols	Multiples and Submultiples	Prefixes	Symbols
10^{18}	exa	E	10^{-1}	deci	d
10^{15}	peta	P	10^{-2}	centi	c
10^{12}	tera	T	10^{-3}	milli	m
10^{9}	giga	G	10^{-6}	micro	μ (Greek mu)
10^{6}	mega	M	10^{-9}	nano	n
10^{3}	kilo	k	10^{-12}	pico	p
10^{2}	hecto	h	10^{-15}	femto	f
10	deca	da	10^{-18}	atto	a

Conversion Factors—Metric to English

To Convert	Into	Multiply By
Inches	Centimeters	0.393700787
Feet	Meters	3.280839895
Yards	Meters	1.093613298
Miles	Kilometers	0.6213711922
Ounces	Grams	$3.527396195 \times 10^{-2}$
Pounds	Kilograms	2.204622622
Gallons (U.S. Liquid)	Liters	0.2641720524
Fluid ounces	Milliliters (cc)	$3.381402270 \times 10^{-2}$
Square inches	Square centimeters	0.1550003100
Square feet	Square meters	10.76391042
Square yards	Square meters	1.195990046
Cubic inches	Milliliters (cc)	$6.102374409 \times 10^{-2}$
Cubic feet	Cubic meters	35.31466672
Cubic yards	Cubic meters	1.307950619

[2] Zwillinger, D., ed. 1996. *CRC Standard Mathematical Tables and Formulae*, 30th ed. CRC Press, Boca Raton, FL: Greek alphabet, conversion constants and multipliers (recommended decimal multiples and submultiples, metric to English, English to metric, general, temperature factors).

Conversion Factors—English to Metric [3]

To Convert	Into	Multiply By
Microns	Mils	**25.4**
Centimeters	Inches	**2.54**
Meters	Feet	**0.3048**
Meters	Yards	**0.9144**
Kilometers	Miles	**1.609344**
Grams	Ounces	28.34952313
Kilograms	Pounds	**0.45359237**
Liters	Gallons (U.S. Liquid)	**3.785411784**
Millimeters (cc)	Fluid ounces	29.57352956
Square centimeters	Square inches	**6.4516**
Square meters	Square feet	**0.09290304**
Square meters	Square yards	**0.83612736**
Milliliters (cc)	Cubic inches	**16.387064**
Cubic meters	Cubic feet	$2.831684659 \times 10^{-2}$
Cubic meters	Cubic yards	0.764554858

Conversion Factors—General [3]

To Convert	Into	Multiply By
Atmospheres	Feet of water @ 4°C	2.950×10^{-2}
Atmospheres	Inches of mercury @ 0°C	3.342×10^{-2}
Atmospheres	Pounds per square inch	6.804×10^{-2}
BTU	Foot-pounds	1.285×10^{-3}
BTU	Joules	9.480×10^{-4}
Cubic feet	Cords	**128**
Degree (angle)	Radians	57.2958
Ergs	Foot-pounds	1.356×10^{7}
Feet	Miles	**5280**
Feet of water @ 4°C	Atmospheres	33.90
Foot-pounds	Horsepower-hours	1.98×10^{6}
Foot-pounds	Kilowatt-hours	2.655×10^{6}
Foot-pounds per min	Horsepower	3.3×10^{4}
Horsepower	Foot-pounds per sec	1.818×10^{-3}
Inches of mercury @ 0°C	Pounds per square inch	2.036
Joules	BTU	1054.8
Joules	Foot-pounds	1.35582
Kilowatts	BTU per min	1.758×10^{-2}
Kilowatts	Foot-pounds per min	2.26×10^{-5}
Kilowatts	Horsepower	0.745712
Knots	Miles per hour	0.86897624
Miles	Feet	1.894×10^{-4}
Nautical miles	miles	0.86897624
Radians	Degrees	1.745×10^{-2}
Squares feet	Acres	**43560**
Watts	BTU per min	17.5796

Temperature Factors

°F = 9/5(°C) + 32
Fahrenheit temperature = 1.8(temperature in kelvins) − 459.67
°C = 5/9[(°F) − 32]
Celsius temperature = temperature in kelvin − 273.15
Fahrenheit temperature = 1.8 (Celsius temperature) + 32

---

[3] Boldface numbers are exact; others are given to ten significant figures where so indicated by the multiplier factor.

Conversion of Temperatures

From	To	
°Celsius	°Fahrenheit	$t_F = (t_C \times 1.8) + 32$
	Kelvin	$T_K = t_C + 273.15$
	°Rankine	$T_R = (t_C + 273.15) \times 1.8$
°Fahrenheit	°Celsius	$t_C = \frac{t_F - 32}{1.8}$
	Kelvin	$T_K = \frac{t_F - 32}{1.8} + 273.15$
	°Rankine	$T_R = t_F + 459.67$
Kelvin	°Celsius	$t_C = T_K - 273.15$
	°Rankine	$T_R = T_K \times 1.8$
°Rankine	°Fahrenheit	$t_F = T_R - 459.67$
	Kelvin	$T_K = \frac{T_R}{1.8}$

12
General Mathematical Tables[1]

12.1	Introduction to Mathematics Chapter	242
12.2	Elementary Algebra and Geometry	242
	Fundamental Properties (Real Numbers) • Exponents • Fractional Exponents • Irrational Exponents • Operations with Zero • Logarithms • Factorials • Binomial Theorem • Factors and Expansion • Progression • Complex Numbers • Permutations • Combinations • Algebraic Equations • Geometry	
12.3	Trigonometry	247
	Triangles • Trigonometric Functions of an Angle • Trigonometric Identities • Inverse Trigonometric Functions	
12.4	Series	251
	Bernoulli and Euler Numbers • Series of Functions • Error Function • Series Expansion	
12.5	Differential Calculus	256
	Notation • Slope of a Curve • Angle of Intersection of Two Curves • Radius of Curvature • Relative Maxima and Minima • Points of Inflection of a Curve • Taylor's Formula • Indeterminant Forms • Numerical Methods • Partial Derivatives • Additional Relations with Derivatives	
12.6	Integral Calculus	262
	Indefinite Integral • Definite Integral • Properties • Common Applications of the Definite Integral • Cylindrical and Spherical Coordinates • Double Integration • Surface Area and Volume by Double Integration • Centroid	
12.7	Special Functions	266
	Hyperbolic Functions • Bessel Functions • Bessel Functions of the Second Kind, $Y_n(x)$ (Also Called Neumann Functions or Weber Functions) (Fig. 12.19) • Legendre Polynomials • Laguerre Polynomials • Hermite Polynomials • Orthogonality • Functions with $x^2/a^2 \pm y^2/b^2$ • Functions with $(x^2/a^2 + y^2/b^2 \pm c^2)^{1/2}$	
12.8	Basic Definitions: Linear Algebra Matrices	273
	Algebra of Matrices • Systems of Equations • Vector Spaces • Rank and Nullity • Orthogonality and Length • Determinants • Eigenvalues and Eigenvectors	
12.9	Basic Definitions: Vector Algebra and Calculus	278
	Coordinate Systems • Vector Functions • Gradient, Curl, and Divergence • Integration • Integral Theorems	
12.10	The Fourier Transforms: Overview	282
	Fourier Transforms	

W.F. Ames
Georgia Institute of Technology

George Cain
Georgia Institute of Technology

[1] The material in this chapter was previously published by CRC Press in *The Engineering Handbook*, pp. 2037–2079, 2187–2192, and 2196–2202, 1996.

12.1 Introduction to Mathematics Chapter

Mathematics has been defined as "the logic of drawing unambiguous conclusions from arbitrary assumptions." The assumptions do not come from the mathematics but arise from the engineering or scientific discipline. Mathematical models of real world phenomena have been remarkably useful. The simplest of these express various physical *laws* in mathematical form. For example, Ohms law ($V = IR$), Newton's second law ($F = ma$) and the ideal gas law ($pv = nRT$) provide starting points for many theories. More complicated models employ difference, differential, or integral equations. Whatever the model, the conclusions drawn from it are unambiguous!

In this mathematics chapter we provide for an extensive review of algebra and geometry in the first section. One feature is the collection of common geometric figures together with their areas, volumes, and other data.

The second section reviews the fundamentals of trigonometry and presents the myriad of identities, which are very useful but very easy to forget. The third section presents tables of various series for a number of useful functions, followed by the theory of how these series are calculated.

The next section gives the elements of the differential calculus, discusses the calculation and theory of maxima and minima, and contains a table of derivatives. The fifth section provides a summary of the integral calculus including formulas for calculating such fundamental physical ideas as arclength, area, volume, work, centroids, etc.

Special functions constitute Sec. 12.7. Here are the Bessel functions and various basic polynomials. A special feature found here is the table of three-dimensional quadratic figures.

Because linear algebra and vector calculus are so basic Secs. 12. 8 and 9 are devoted to their summaries. The last section presents the various Fourier transforms and their properties in table form.

Throughout the chapter various references are provided.

12.2 Elementary Algebra and Geometry

Fundamental Properties (Real Numbers)

$a + b = b + a$	Commutative law for addition
$(a + b) + c = a + (b + c)$	Associative law for addition
$a + 0 = 0 + a$	Identity law for addition
$a + (-a) = (-a) + a = 0$	Inverse law for addition
$a(bc) = (ab)c$	Associative law for multiplication
$a(\frac{1}{a}) = (\frac{1}{a})a = 1, \ a \neq 0$	Inverse law for multiplication
$(a)(1) = (1)(a) = a$	Identity law for multiplication
$ab = ba$	Commutative law for multiplication
$a(b + c) = ab + ac$	Distributive law

Division by zero is not defined.

Exponents

For integers m and n,

$$a^n a^m = a^{n+m}$$

$$a^n / a^m = a^{n-m}$$

$$(a^n)^m = a^{nm}$$

$$(ab)^m = a^m b^m$$

$$(a/b)^m = a^m / b^m$$

Fractional Exponents

$$a^{p/q} = (a^{1/q})^p$$

General Mathematical Tables

where $a^{1/q}$ is the positive qth root of a if $a > 0$ and the negative qth root of a if a is negative and q is odd. Accordingly, the five rules of exponents just given (for integers) are also valid if m and n are fractions, provided a and b are positive.

Irrational Exponents

If an exponent is irrational (e.g., $\sqrt{2}$), the quantity, such as $a^{\sqrt{2}}$, is the limit of the sequence $a^{1.4}, a^{1.41}, a^{1.414}, \ldots$.

Operations with Zero

$$0^m = 0 \quad a^0 = 1$$

Logarithms

If x, y, and b are positive and $b \neq 1$,

$$\log_b(xy) = \log_b x + \log_b y$$
$$\log_b(x/y) = \log_b x - \log_b y$$
$$\log_b x^p = p \log_b x$$
$$\log_b(1/x) = -\log_b x$$
$$\log_b b = 1$$
$$\log_b 1 = 0 \quad \text{Note: } b^{\log_b x} = x$$

Change of Base ($a \neq 1$)

$$\log_b x = \log_a x \log_b a$$

Factorials

The factorial of a positive integer n is the product of all of the positive integers less than or equal to the integer n and is denoted $n!$. Thus,

$$n! = 1 \cdot 2 \cdot 3 \cdot \cdots \cdot n$$

Factorial 0 is defined: $0! = 1$.

Stirling's Approximation

$$\lim_{n \to \infty} (n/e)^n \sqrt{2\pi n} = n!$$

Binomial Theorem

For positive integer n

$$(x + y)^n = x^n + nx^{n-1}y + \frac{n(n-1)}{2!}x^{n-2}y^2$$
$$+ \frac{n(n-1)(n-2)}{3!}x^{n-3}y^3 + \cdots + nxy^{n-1} + y^n$$

Factors and Expansion

$$(a + b)^2 = a^2 + 2ab + b^2$$
$$(a - b)^2 = a^2 - 2ab + b^2$$
$$(a + b)^3 = a^3 + 3a^2b + 3ab^2 + b^3$$
$$(a - b)^3 = a^3 - 3a^2b + 3ab^2 - b^3$$
$$(a^2 - b^2) = (a - b)(a + b)$$
$$(a^3 - b^3) = (a - b)(a^2 + ab + b^2)$$
$$(a^3 + b^3) = (a + b)(a^2 - ab + b^2)$$

Progression

An *arithmetic progression* is a sequence in which the difference between any term and the preceding term is a constant (d):

$$a, a + d, a + 2d, \ldots, a + (n - 1)d$$

If the last term is denoted $l\,[= a + (n - 1)d]$, then the sum is

$$s = \frac{n}{2}(a + l)$$

A *geometric progression* is a sequence in which the ratio of any term is a constant r. Thus, for n terms,

$$a, ar, ar^2, \ldots, ar^{n-1}$$

The sum is

$$S = \frac{a - ar^n}{1 - r}$$

Complex Numbers

A complex number is an ordered pair of real numbers (a, b).
    Equality:

$$(a, b) = (c, d) \quad \text{if and only if } a = c \text{ and } b = d$$

    Addition:

$$(a, b) + (c, d) = (a + c, b + d)$$

    Multiplication:

$$(a, b)(c, d) = (ac - bd, ad + bc)$$

The first element (a, b) is called the *real* part, the second the *imaginary* part. An alternative notation for (a, b) is $a + bi$, where $i^2 = (-1, 0)$, and $i = (0, 1)$ or $0 + 1i$ is written for this complex number as a convenience. With this understanding, i behaves as a number, that is, $(2 - 3i)(4 + i) = 8 - 12i + 2i - 3i^2 = 11 - 10i$. The conjugate of $a + bi$ is $a - bi$, and the product of a complex number and its conjugate is $a^2 + b^2$. Thus, *quotients* are computed by multiplying numerator and denominator by the conjugate of the denominator, illustrated as follows:

$$\frac{2 + 3i}{4 + 2i} = \frac{(4 - 2i)(2 + 3i)}{(4 - 2i)(4 + 2i)} = \frac{14 + 8i}{20} = \frac{7 + 4i}{10}$$

General Mathematical Tables

Polar Form

The complex number $x + iy$ may be represented by a plane vector with components x and y:

$$x + iy = r(\cos\theta + i\sin\theta)$$

(See Fig. 12.1.) Then, given two complex numbers $z_1 = r_1(\cos\theta_1 + i\sin\theta_1)$, and $z_2 = r_2(\cos\theta_2 + i\sin\theta_2)$, the product and quotient are as follows.

Product:

$$z_1 z_2 = r_1 r_2 [\cos(\theta_1 + \theta_2) + i\sin(\theta_1 + \theta_2)]$$

Quotient:

$$z_1/z_2 = (r_1/r_2)[\cos(\theta_1 - \theta_2) + i\sin(\theta_1 - \theta_2)]$$

Powers:

$$z^n = [r\cos\theta + i\sin\theta]^n = r^n[\cos n\theta + i\sin n\theta]$$

Roots:

$$z^{1/n} = [r\cos\theta + i\sin\theta]^{1/n}$$
$$= r^{1/n}\left[\cos\frac{\theta + k\cdot 360}{n} + i\sin\frac{\theta + k\cdot 360}{n}\right]$$
$$k = 0, 1, 2, \ldots, n-1$$

Permutations

A permutation is an ordered arrangement (sequence) of all or part of a set of objects. The number of permutations of n objects taken r at a time is

$$p(n, r) = n(n-1)(n-2)\cdots(n-r+1)$$
$$= \frac{n!}{(n-r)!}$$

A permutation of positive integers is *even* or *odd* if the total number of inversions is an even integer or an odd integer, respectively. Inversions are counted relative to each integer j in the permutation by counting the number of integers that follow j and are less than j. These are summed to give the total number of inversions. For example, the permutation 4132 has four inversions: three relative to 4 and one relative to 3. This permutation is therefore even.

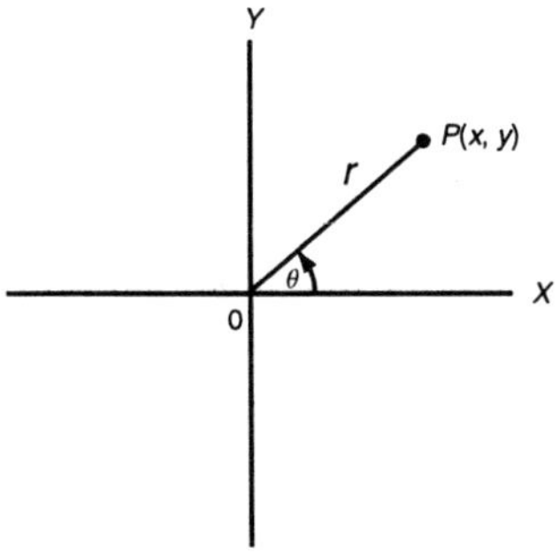

Combinations

A combination is a selection of one or more objects from among a set of objects regardless of order. The number of combinations of n different objects taken r at a time is

$$C(n, r) = \frac{P(n, r)}{r!} = \frac{n!}{r!(n-r)!}$$

FIGURE 12.1 Polar form of complex number.

Algebraic Equations

Quadratic

If $ax^2 + bx + c = 0$, and $a \neq 0$, then roots are

$$x = \frac{-b \pm \sqrt{b^2 - 4ac}}{2a}$$

Cubic

To solve $x^3 + bx^2 + cx + d = 0$, let $x = y - b/3$. Then the *reduced cubic* is obtained,

$$y^3 + py + q = 0$$

where $p = c - (1/3)b^2$ and $q = d - (1/3)bc + (2/27)b^3$. Solutions of the original cubic are then in terms of the reduced cubic roots y_1, y_2, y_3,

$$x_1 = y_1 - (1/3)b \qquad x_2 = y_2 - (1/3)b \qquad x_3 = y_3 - (1/3)b$$

The three roots of the reduced cubic are

$$y_1 = (A)^{1/3} + (B)^{1/3}$$
$$y_2 = W(A)^{1/3} + W^2(B)^{1/3}$$
$$y_3 = W^2(A)^{1/3} + W(B)^{1/3}$$

where

$$A = -\frac{1}{2}q + \sqrt{(1/27)p^3 + \frac{1}{4}q^2}$$

$$B = -\frac{1}{2}q - \sqrt{(1/27)p^3 + \frac{1}{4}q^2}$$

$$W = \frac{-1 + i\sqrt{3}}{2}, \qquad W^2 = \frac{-1 - i\sqrt{3}}{2}$$

When $(1/27)p^3 + (1/4)q^2$ is negative, A is complex; in this case A should be expressed in trigonometric form: $A = r(\cos\theta + i\sin\theta)$ where θ is a first or second quadrant angle, as q is negative or positive. The three roots of the reduced cubic are

$$y_1 = 2(r)^{1/3} \cos(\theta/3)$$
$$y_2 = 2(r)^{1/3} \cos\left(\frac{\theta}{3} + 120°\right)$$
$$y_3 = 2(r)^{1/3} \cos\left(\frac{\theta}{3} + 240°\right)$$

Geometry

Figures 12.2–12.12 are a collection of common geometric figures. Area (A), volume (V), and other measurable features are indicated.

General Mathematical Tables

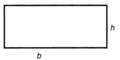

FIGURE 12.2 Rectangle: $A = bh$.

FIGURE 12.3 Parallelogram: $A = bh$.

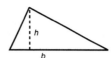

FIGURE 12.4 Triangle: $A = \frac{1}{2}bh$.

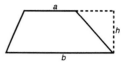

FIGURE 12.5 Trapezoid: $A = \frac{1}{2}(a+b)h$

FIGURE 12.6 Circle: $A = \pi R^2$; circumference $= 2\pi R$; arc length $S = R\theta$ (θ in radians).

FIGURE 12.7 Sector of a circle: $A_{sector} = \frac{1}{2}R^2\theta$; $A_{segment} = \frac{1}{2}R^2(\theta - \sin\theta)$.

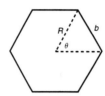

FIGURE 12.8 Regular polygon of n sides: $A = (n/4)b^2 \operatorname{ctn}(\pi/n)$; $R = (b/2)\csc(\pi/n)$.

FIGURE 12.9 Right circular cylinder: $V = \pi R^2 h$; lateral surface area $= 2\pi R h$.

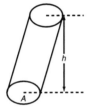

FIGURE 12.10 Cylinder (or prism) with parallel bases: $V = Ah$.

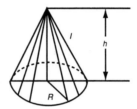

FIGURE 12.11 Right circular cone: $V = \frac{1}{3}\pi R^2 h$; lateral surface area $= \pi R l = \pi R \sqrt{R^2 + h^2}$.

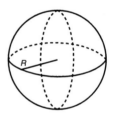

FIGURE 12.12 Sphere: $V = \frac{4}{3}\pi R^3$; surface area $= 4\pi R^2$.

12.3 Trigonometry

Triangles

In any triangle (in a plane) with sides a, b, and c and corresponding opposite angles A, B, and C,

$$\frac{a}{\sin A} = \frac{b}{\sin B} = \frac{c}{\sin C} \quad \text{(Law of sines)}$$

$$a^2 = b^2 + c^2 - 2cb\cos A \quad \text{(Law of cosines)}$$

$$\frac{a+b}{a-b} = \frac{\tan\frac{1}{2}(A+B)}{\tan\frac{1}{2}(A-B)} \quad \text{(Law of Tangents)}$$

$$\sin\frac{1}{2}A = \sqrt{\frac{(s-b)(s-c)}{bc}} \quad \text{where } s = \frac{1}{2}(a+b+c)$$

$$\cos\frac{1}{2}A = \sqrt{\frac{s(s-a)}{bc}}$$

$$\tan \frac{1}{2}A = \sqrt{\frac{(s-b)(s-c)}{s(s-a)}}$$

$$\text{area} = \frac{1}{2}bc \sin A$$

$$= \sqrt{s(s-a)(s-b)(s-c)}$$

If the vertices have coordinates (x_1, y_1), (x_2, y_2), (x_3, y_3), the area is the *absolute value* of the expression

$$\frac{1}{2}\begin{vmatrix} x_1 & y_1 & 1 \\ x_2 & y_2 & 1 \\ x_3 & y_3 & 1 \end{vmatrix}$$

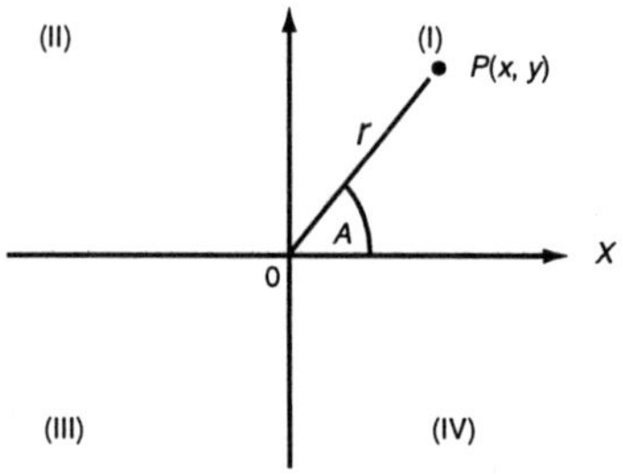

Trigonometric Functions of an Angle

With reference to Fig. 12.13, $P(x, y)$ is a point in any one of the four quadrants and A is an angle whose initial side is coincident with the positive x axis and whose terminal side contains the point $P(x, y)$. The distance from the origin $P(x, y)$ is denoted by r and is positive. The trigonometric functions of the angle A are defined as:

$$\begin{aligned} \sin A &= \text{sine } A &&= y/r \\ \cos A &= \text{cosine } A &&= x/r \\ \tan A &= \text{tangent } A &&= y/x \\ \text{ctn } A &= \text{cotangent } A &&= x/y \\ \sec A &= \text{secant } A &&= r/x \\ \csc A &= \text{cosecant } A &&= r/y \end{aligned}$$

FIGURE 12.13 The trigonometric point: angle A is taken to be positive when the rotation is counterclockwise and negative when the rotation is clockwise. The plane is divided into quadrants as shown.

Angles are measured in degrees or radians; $180° = \pi$ radians; 1 radian $= 180/\pi°$.

The trigonometric functions of 0, 30, and 45°, and integer multiples of these are directly computed.

	0°	30°	45°	60°	90°	120°	135°	150°	180°
sin	0	$\frac{1}{2}$	$\frac{\sqrt{2}}{2}$	$\frac{\sqrt{3}}{2}$	1	$\frac{\sqrt{3}}{2}$	$\frac{\sqrt{2}}{2}$	$\frac{1}{2}$	0
cos	1	$\frac{\sqrt{3}}{2}$	$\frac{\sqrt{2}}{2}$	$\frac{1}{2}$	0	$-\frac{1}{2}$	$-\frac{\sqrt{2}}{2}$	$-\frac{\sqrt{3}}{2}$	-1
tan	0	$\frac{\sqrt{3}}{3}$	1	$\sqrt{3}$	∞	$-\sqrt{3}$	-1	$\frac{\sqrt{3}}{3}$	0
ctn	∞	$\sqrt{3}$	1	$\frac{\sqrt{3}}{3}$	0	$-\frac{\sqrt{3}}{3}$	-1	$-\sqrt{3}$	∞
sec	1	$\frac{2\sqrt{3}}{3}$	$\sqrt{2}$	2	∞	-2	$-\sqrt{2}$	$-\frac{2\sqrt{3}}{3}$	-1
csc	∞	2	$\sqrt{2}$	$\frac{2\sqrt{3}}{3}$	1	$\frac{2\sqrt{3}}{3}$	$\sqrt{2}$	2	∞

Trigonometric Identities

$$\sin A = \frac{1}{\csc A}$$

$$\cos A = \frac{1}{\sec A}$$

General Mathematical Tables

$$\tan A = \frac{1}{\text{ctn } A} = \frac{\sin A}{\cos A}$$

$$\csc A = \frac{1}{\sin A}$$

$$\sec A = \frac{1}{\cos A}$$

$$\text{ctn } A = \frac{1}{\tan A} = \frac{\cos A}{\sin A}$$

$$\sin^2 A + \cos^2 A = 1$$

$$1 + \tan^2 A = \sec^2 A$$

$$1 + \text{ctn}^2 A = \csc^2 A$$

$$\sin(A \pm B) = \sin A \cos B \pm \cos A \sin B$$

$$\cos(A \pm B) = \cos A \cos B \mp \sin A \sin B$$

$$\tan(A \pm B) = \frac{\tan A \pm \tan B}{1 \mp \tan A \tan B}$$

$$\sin 2A = 2 \sin A \cos A$$

$$\sin 3A = 3 \sin A - 4 \sin^3 A$$

$$\sin nA = 2 \sin(n-1)A \cos A - \sin(n-2)A$$

$$\cos 2A = 2 \cos^2 A - 1 = 1 - 2 \sin^2 A$$

$$\cos 3A = 4 \cos^3 A - 3 \cos A$$

$$\cos nA = 2 \cos(n-1)A \cos A - \cos(n-2)A$$

$$\sin A + \sin B = 2 \sin \tfrac{1}{2}(A+B) \cos \tfrac{1}{2}(A-B)$$

$$\sin A - \sin B = 2 \cos \tfrac{1}{2}(A+B) \sin \tfrac{1}{2}(A-B)$$

$$\cos A + \cos B = 2 \cos \tfrac{1}{2}(A+B) \cos \tfrac{1}{2}(A-B)$$

$$\cos A - \cos B = -2 \sin \tfrac{1}{2}(A+B) \sin \tfrac{1}{2}(A-B)$$

$$\tan A \pm \tan B = \frac{\sin(A \pm B)}{\cos A \cos B}$$

$$\text{ctn } A \pm \text{ctn } B = \pm \frac{\sin(A \pm B)}{\sin A \sin B}$$

$$\sin A \sin B = \tfrac{1}{2} \cos(A-B) - \tfrac{1}{2} \cos(A+B)$$

$$\cos A \cos B = \tfrac{1}{2} \cos(A-B) + \tfrac{1}{2} \cos(A+B)$$

$$\sin A \cos B = \tfrac{1}{2} \sin(A+B) + \tfrac{1}{2} \sin(A-B)$$

$$\sin \frac{A}{2} = \pm \sqrt{\frac{1 - \cos A}{2}}$$

$$\cos \frac{A}{2} = \pm \sqrt{\frac{1 + \cos A}{2}}$$

$$\tan\frac{A}{2} = \frac{1-\cos A}{\sin A} = \frac{\sin A}{1+\cos A} = \pm\sqrt{\frac{1-\cos A}{1+\cos A}}$$

$$\sin^2 A = \frac{1}{2}(1-\cos 2A)$$

$$\cos^2 A = \frac{1}{2}(1+\cos 2A)$$

$$\sin^3 A = \frac{1}{4}(3\sin A - \sin 3A)$$

$$\cos^3 A = \frac{1}{4}(\cos 3A + 3\cos A)$$

$$\sin ix = \frac{1}{2}i(e^x - e^{-x}) = i\sinh x$$

$$\cos ix = \frac{1}{2}(e^x + e^{-x}) = \cosh x$$

$$\tan ix = i\frac{(e^x - e^{-x})}{e^x + e^{-x}} = i\tanh x$$

$$e^{x+iy} = e^x(\cos y + i\sin y)$$

$$(\cos x \pm i\sin x)^n = \cos nx \pm i\sin nx$$

Inverse Trigonometric Functions

The inverse trigonometric functions are multiple valued, and this should be taken into account in the use of the following formulas:

$$\sin^{-1} x = \cos^{-1}\sqrt{1-x^2}$$
$$= \tan^{-1}\frac{x}{\sqrt{1-x^2}} = \mathrm{ctn}^{-1}\frac{\sqrt{1-x^2}}{x}$$
$$= \sec^{-1}\frac{1}{\sqrt{1-x^2}} = \csc^{-1}\frac{1}{x}$$
$$= -\sin^{-1}(-x)$$

$$\cos^{-1} x = \sin^{-1}\sqrt{1-x^2}$$
$$= \tan^{-1}\frac{\sqrt{1-x^2}}{x} = \mathrm{ctn}^{-1}\frac{x}{\sqrt{1-x^2}}$$
$$= \sec^{-1}\frac{1}{x} = \csc^{-1}\frac{1}{\sqrt{1-x^2}}$$
$$= \pi - \cos^{-1}(-x)$$

$$\tan^{-1} x = \mathrm{ctn}^{-1}\frac{1}{x}$$
$$= \sin^{-1}\frac{x}{\sqrt{1+x^2}} = \cos^{-1}\frac{1}{\sqrt{1+x^2}}$$
$$= \sec^{-1}\sqrt{1+x^2} = \csc^{-1}\frac{\sqrt{1+x^2}}{x}$$
$$= -\tan^{-1}(-x)$$

12.4 Series

Bernoulli and Euler Numbers

A set of numbers, $B_1, B_3, \ldots, B_{2n-1}$ (Bernoulli numbers) and $B_2, B_4, \ldots, B_{2n}$ (Euler numbers) appear in the series expansions of many functions. A partial listing follows; these are computed from the following equations:

$$B_{2n} - \frac{2n(2n-1)}{2!}B_{2n-2} + \frac{2n(2n-1)(2n-2)(2n-3)}{4!}B_{2n-4} - \cdots + (-1)^n = 0$$

and

$$\frac{2^{2n}(2^{2n}-1)}{2n}B_{2n-1} = (2n-1)B_{2n-2}$$
$$- \frac{(2n-1)(2n-2)(2n-3)}{3!}B_{2n-4} + \cdots + (-1)^{n-1}$$

$B_1 = 1/6$ $B_2 = 1$
$B_3 = 1/30$ $B_4 = 5$
$B_5 = 1/42$ $B_6 = 61$
$B_7 = 1/30$ $B_8 = 1385$
$B_9 = 5/66$ $B_{10} = 50521$
$B_{11} = 691/2730$ $B_{12} = 2\,702\,765$
$B_{13} = 7/6$ $B_{14} = 199\,360\,981$
$\vdots$ $\vdots$

Series of Functions

In the following, the interval of convergence is indicated; otherwise it is all x. Logarithms are to the base e. Bernoulli and Euler numbers (B_{2n-1} and B_{2n}) appear in certain expressions.

$$(a+x)^n = a^n + na^{n-1}x + \frac{n(n-1)}{2!}a^{n-2}x^2 + \frac{n(n-1)(n-2)}{3!}a^{n-3}x^3 + \cdots$$
$$+ \frac{n!}{(n-j)!j!}a^{n-j}x^j + \cdots \qquad [x^2 < a^2]$$

$$(a-bx)^{-1} = \frac{1}{a}\left[1 + \frac{bx}{a} + \frac{b^2x^2}{a^2} + \frac{b^3x^3}{a^3} + \cdots\right] \qquad [b^2x^2 < a^2]$$

$$(1 \pm x)^n = 1 \pm nx + \frac{n(n-1)}{2!}x^2 \pm \frac{n(n-1)(n-2)x^3}{3!} + \cdots \qquad [x^2 < 1]$$

$$(1 \pm x)^{-n} = 1 \mp nx + \frac{n(n+1)}{2!}x^2 \mp \frac{n(n+1)(n+2)}{3!}x^3 + \cdots \qquad [x^2 < 1]$$

$$(1 \pm x)^{\frac{1}{2}} = 1 \pm \frac{1}{2}x - \frac{1}{2\cdot 4}x^2 \pm \frac{1\cdot 3}{2\cdot 4\cdot 6}x^3 - \frac{1\cdot 3\cdot 5}{2\cdot 4\cdot 6\cdot 8}x^4 \pm \cdots \qquad [x^2 < 1]$$

$$(1 \pm x)^{-\frac{1}{2}} = 1 \mp \frac{1}{2}x + \frac{1\cdot 3}{2\cdot 4}x^2 \mp \frac{1\cdot 3\cdot 5}{2\cdot 4\cdot 6}x^3 + \frac{1\cdot 3\cdot 5\cdot 7}{2\cdot 4\cdot 6\cdot 8}x^4 \pm \cdots \qquad [x^2 < 1]$$

$$(1 \pm x^2)^{\frac{1}{2}} = 1 \pm \frac{1}{2}x^2 - \frac{x^4}{2\cdot 4} \pm \frac{1\cdot 3}{2\cdot 4\cdot 6}x^6 - \frac{1\cdot 3\cdot 5}{2\cdot 4\cdot 6\cdot 8}x^8 \pm \cdots \qquad [x^2 < 1]$$

$$(1 \pm x)^{-1} = 1 \mp x + x^2 \mp x^3 + x^4 \mp x^5 + \cdots \qquad [x^2 < 1]$$

$$(1 \pm x)^{-2} = 1 \mp 2x + 3x^2 \mp 4x^3 + 5x^4 \mp \cdots \qquad [x^2 < 1]$$

$$e^x = 1 + x + \frac{x^2}{2!} + \frac{x^3}{3!} + \frac{x^4}{4!} + \cdots$$

$$e^{-x^2} = 1 - x^2 + \frac{x^4}{2!} - \frac{x^6}{3!} + \frac{x^8}{4!} - \cdots$$

$$a^x = 1 + x \log a + \frac{(x \log a)^2}{2!} + \frac{(x \log a)^3}{3!} + \cdots$$

$$\log x = (x-1) - \frac{1}{2}(x-1)^2 + \frac{1}{3}(x-1)^3 - \cdots \qquad [0 < x < 2]$$

$$\log x = \frac{x-1}{x} + \frac{1}{2}\left(\frac{x-1}{x}\right)^2 + \frac{1}{3}\left(\frac{x-1}{x}\right)^3 + \cdots \qquad \left[x > \frac{1}{2}\right]$$

$$\log x = 2\left[\left(\frac{x-1}{x+1}\right) + \frac{1}{3}\left(\frac{x-1}{x+1}\right)^3 + \frac{1}{5}\left(\frac{x-1}{x+1}\right)^5 + \cdots\right] \qquad [x > 0]$$

$$\log(1+x) = x - \frac{1}{2}x^2 + \frac{1}{3}x^3 - \frac{1}{4}x^4 + \cdots \qquad [x^2 < 1]$$

$$\log\left(\frac{1+x}{1-x}\right) = 2\left[x + \frac{1}{3}x^3 + \frac{1}{5}x^5 + \frac{1}{7}x^7 + \cdots\right] \qquad [x^2 < 1]$$

$$\log\left(\frac{x+1}{x-1}\right) = 2\left[\frac{1}{x} + \frac{1}{3}\left(\frac{1}{x}\right)^3 + \frac{1}{5}\left(\frac{1}{x}\right)^5 + \cdots\right] \qquad [x^2 < 1]$$

$$\sin x = x - \frac{x^3}{3!} + \frac{x^5}{5!} - \frac{x^7}{7!} + \cdots$$

$$\cos x = 1 - \frac{x^2}{2!} + \frac{x^4}{4!} - \frac{x^6}{6!} + \cdots$$

$$\tan x = x + \frac{x^3}{3} + \frac{2x^5}{15} + \frac{17x^7}{315}$$
$$+ \cdots + \frac{2^{2n}(2^{2n}-1)B_{2n-1}x^{2n-1}}{2n!} \qquad \left[x^2 < \frac{\pi^2}{4}\right]$$

$$\operatorname{ctn} x = \frac{1}{x} - \frac{x}{3} - \frac{x^3}{45} - \frac{2x^5}{945} - \cdots - \frac{B_{2n-1}(2x)^{2n}}{(2n)!x} - \cdots \qquad [x^2 < \pi^2]$$

$$\sec x = 1 + \frac{x^2}{2!} + \frac{5x^4}{4!} + \frac{61x^6}{6!} + \cdots + \frac{B_{2n}x^{2n}}{(2n)!} + \cdots \qquad \left[x^2 < \frac{\pi^2}{4}\right]$$

$$\csc x = \frac{1}{x} + \frac{x}{3!} + \frac{7x^3}{3 \cdot 5!} + \frac{31x^5}{3 \cdot 7!} + \cdots$$
$$+ \frac{2(2^{2n+1}-1)}{(2n+2)!}B_{2n+1}x^{2n+1} + \cdots \qquad [x^2 < \pi^2]$$

$$\sin^{-1} x = x + \frac{x^3}{6} + \frac{(1 \cdot 3)x^5}{(2 \cdot 4)5} + \frac{(1 \cdot 3 \cdot 5)x^7}{(2 \cdot 4 \cdot 6)7} + \cdots \qquad [x^2 < 1]$$

$$\tan^{-1} x = x - \frac{1}{3}x^3 + \frac{1}{5}x^5 - \frac{1}{7}x^7 + \cdots \qquad [x^2 < 1]$$

$$\sec^{-1} x = \frac{\pi}{2} - \frac{1}{x} - \frac{1}{6x^3} - \frac{1 \cdot 3}{(2 \cdot 4)5x^5} - \frac{1 \cdot 3 \cdot 5}{(2 \cdot 4 \cdot 6)7x^7} - \cdots \qquad [x^2 < 1]$$

$$\sinh x = x + \frac{x^3}{3!} + \frac{x^5}{5!} + \frac{x^7}{7!} + \cdots$$

$$\cosh x = 1 + \frac{x^2}{2!} + \frac{x^4}{4!} + \frac{x^6}{6!} + \frac{x^8}{8!} + \cdots$$

General Mathematical Tables

$$\tanh x = (2^2 - 1)2^2 B_1 \frac{x}{2!} - (2^4 - 1)2^4 B_3 \frac{x^3}{4!} + (2^6 - 1)2^6 B_5 \frac{x^5}{6!} - \cdots \qquad \left[x^2 < \frac{\pi^2}{4}\right]$$

$$\operatorname{ctnh} x = \frac{1}{x}\left(1 + \frac{2^2 B_1 x^2}{2!} - \frac{2^4 B_3 x^4}{4!} + \frac{2^6 B_5 x^6}{6!} - \cdots\right) \qquad [x^2 < \pi^2]$$

$$\operatorname{sech} x = 1 - \frac{B_2 x^2}{2!} + \frac{B_4 x^4}{4!} - \frac{B_6 x^6}{6!} + \cdots \qquad \left[x^2 < \frac{\pi^2}{4}\right]$$

$$\operatorname{csch} x = \frac{1}{x} - (2-1)2B_1 \frac{x}{2!} + (2^3 - 1)2B_3 \frac{x^3}{4!} - \cdots \qquad [x^2 < \pi^2]$$

$$\sinh^{-1} x = x - \frac{1}{2}\frac{x^3}{3} + \frac{1 \cdot 3}{2 \cdot 4}\frac{x^5}{5} - \frac{1 \cdot 3 \cdot 5}{2 \cdot 4 \cdot 6}\frac{x^7}{7} + \cdots \qquad [x^2 < 1]$$

$$\tanh^{-1} x = x + \frac{x^3}{3} + \frac{x^5}{5} + \frac{x^7}{7} + \cdots \qquad [x^2 < 1]$$

$$\operatorname{ctnh}^{-1} x = \frac{1}{x} + \frac{1}{3x^3} + \frac{1}{5x^5} + \cdots \qquad [x^2 > 1]$$

$$\operatorname{csch}^{-1} x = \frac{1}{x} - \frac{1}{2 \cdot 3x^3} + \frac{1 \cdot 3}{2 \cdot 4 \cdot 5x^5} - \frac{1 \cdot 3 \cdot 5}{2 \cdot 4 \cdot 6 \cdot 7x^7} + \cdots \qquad [x^2 > 1]$$

$$\int_0^x e^{-t^2} dt = x - \frac{1}{3}x^3 + \frac{x^5}{5 \cdot 2!} - \frac{x^7}{7 \cdot 3!} + \cdots$$

Error Function

The following function, known as the error function, erf x, arises frequently in applications:

$$\operatorname{erf} x = \frac{2}{\sqrt{\pi}} \int_0^x e^{-t^2} dt$$

The integral cannot be represented in terms of a finite number of elementary functions; therefore, values of erf x have been compiled in tables. The following is the series for erf x:

$$\operatorname{erf} x = \frac{2}{\sqrt{\pi}}\left[x - \frac{x^3}{3} + \frac{x^5}{5 \cdot 2!} - \frac{x^7}{7 \cdot 3!} + \cdots\right]$$

There is a close relation between this function and the area under the standard normal curve. For evaluation it is convenient to use z instead of x; then erf z may be evaluated from the area $F(z)$ by use of the relation

$$\operatorname{erf} z = 2F(\sqrt{2}z)$$

Example

$$\operatorname{erf}(0.5) = 2F[(1.414)(0.5)] = 2F(0.707)$$

By interpolation, $F(0.707) = 0.260$; thus, erf $(0.5) = 0.520$.

Series Expansion

The expression in parentheses following certain series indicates the region of convergence. If not otherwise indicated, it is understood that the series converges for all finite values of x.

Binomial

$$(x+y)^n = x^n + nx^{n-1}y + \frac{n(n-1)}{2!}x^{n-2}y^2 + \frac{n(n-1)(n-2)}{3!}x^{n-3}y^3 + \cdots \qquad (y^2 < x^2)$$

$$(1 \pm x)^n = 1 \pm nx + \frac{n(n-1)x^2}{2!} \pm \frac{n(n-1)(n-2)x^3}{3!} + \cdots \qquad (x^2 < 1)$$

$$(1 \pm x)^{-n} = 1 \mp nx + \frac{n(n+1)x^2}{2!} \mp \frac{n(n+1)(n+2)x^3}{3!} + \cdots \qquad (x^2 < 1)$$

$$(1 \pm x)^{-1} = 1 \mp x + x^2 \mp x^3 + x^4 \mp x^5 + \cdots \qquad (x^2 < 1)$$

$$(1 \pm x)^{-2} = 1 \mp 2x + 3x^2 \mp 4x^3 + 5x^4 \mp 6x^5 + \cdots \qquad (x^2 < 1)$$

Reversion of Series

Let a series be represented by

$$y = a_1 x + a_2 x^2 + a_3 x^3 + a_4 x^4 + a_5 x^5 + a_6 x^6 + \cdots \qquad (a_1 \neq 0)$$

To find the coefficients of the series

$$x = A_1 y + A_2 y^2 + A_3 y^3 + A_4 y^4 + \cdots$$

$$A_1 = \frac{1}{a_1} \qquad A_2 = -\frac{a_2}{a_1^3} \qquad A_3 = \frac{1}{a_1^5}\left(2a_2^2 - a_1 a_3\right)$$

$$A_4 = \frac{1}{a_1^7}\left(5a_1 a_2 a_3 - a_1^2 a_4 - 5a_2^3\right)$$

$$A_5 = \frac{1}{a_1^9}\left(6a_1^2 a_2 a_4 + 3a_1^2 a_3^2 + 14a_2^4 - a_1^3 a_5 - 21 a_1 a_2^2 a_3\right)$$

$$A_6 = \frac{1}{a_1^{11}}\left(7a_1^3 a_2 a_5 + 7a_1^3 a_3 a_4 + 84 a_1 a_2^3 a_3 - a_1^4 a_6 - 28 a_1^2 a_2^2 a_4 - 28 a_1^2 a_2 a_3^2 - 42 a_2^5\right)$$

$$A_7 = \frac{1}{a_1^{13}}\Big(8a_1^4 a_2 a_6 + 8a_1^4 a_3 a_5 + 4a_1^4 a_4^2 + 120 a_1^2 a_2^3 a_4 + 180 a_1^2 a_2^2 a_3^2 + 132 a_2^6 - a_1^5 a_7$$

$$- 36 a_1^3 a_2^2 a_5 - 72 a_1^3 a_2 a_3 a_4 - 12 a_1^3 a_3^3 - 330 a_1 a_2^4 a_3\Big)$$

Taylor

1.

$$f(x) = f(a) + (x-a)f'(a) + \frac{(x-a)^2}{2!} f''(a) + \frac{(x-a)^3}{3!} f'''(a)$$
$$+ \cdots + \frac{(x-a)^n}{n!} f^{(n)}(a) + \cdots$$

2. (Increment form)

$$f(x+h) = f(x) + h f'(x) + \frac{h^2}{2!} f''(x) + \frac{h^3}{3!} f'''(x) + \cdots$$
$$= f(h) + x f'(h) + \frac{x^2}{2!} f''(h) + \frac{x^3}{3!} f'''(h) + \cdots$$

3. If $f(x)$ is a function possessing derivatives of all orders throughout the interval $a \leq x \leq b$, then there is a value X, with $a < X < b$, such that

$$f(b) = f(a) + (b-a)f'(a) + \frac{(b-a)^2}{2!} f''(a) + \cdots$$
$$+ \frac{(b-a)^{n-1}}{(n-1)!} f^{(n-1)}(a) + \frac{(b-a)^n}{n!} f^{(n)}(X)$$

$$f(a+h) = f(a) + hf'(a) + \frac{h^2}{2!}f''(a) + \cdots + \frac{h^{n-1}}{(n-1)!}f^{(n-1)}(a)$$
$$+ \frac{h^n}{n!}f^{(n)}(a+\theta h), \quad b = a+h, \quad 0 < \theta < 1$$

or

$$f(x) = f(a) + (x-a)f'(a) + \frac{(x-a)^2}{2!}f''(a) + \cdots + (x-a)^{n-1}\frac{f^{(n-1)}(a)}{(n-1)!} + R_n$$

where

$$R_n = \frac{f^{(n)}[a + \theta \cdot (x-a)]}{n!}(x-a)^n, \quad 0 < \theta < 1$$

The preceding forms are known as Taylor's series with the remainder term.

4. Taylor's series for a function of two variables: If

$$\left(h\frac{\partial}{\partial x} + k\frac{\partial}{\partial y}\right)f(x,y) = h\frac{\partial f(x,y)}{\partial x} + k\frac{\partial f(x,y)}{\partial y};$$

$$\left(h\frac{\partial}{\partial x} + k\frac{\partial}{\partial y}\right)^2 f(x,y) = h^2\frac{\partial^2 f(x,y)}{\partial x^2} + 2hk\frac{\partial^2 f(x,y)}{\partial x\,\partial y} + k^2\frac{\partial^2 f(x,y)}{\partial y^2}$$

and so forth, and if

$$\left(h\frac{\partial}{\partial x} + k\frac{\partial}{\partial y}\right)^n f(x,y)\bigg|_{\substack{x=a\\y=b}}$$

where the bar and subscripts mean that after differentiation we are to replace x by a and y by b,

$$f(a+h, b+k) = f(a,b) + \left(h\frac{\partial}{\partial x} + k\frac{\partial}{\partial y}\right)f(x,y)\bigg|_{\substack{x=a\\y=b}} + \cdots$$
$$+ \frac{1}{n!}\left(h\frac{\partial}{\partial x} + k\frac{\partial}{\partial y}\right)^n f(x,y)\bigg|_{\substack{x=a\\y=b}} + \cdots$$

MacLaurin

$$f(x) = f(0) + xf'(0) + \frac{x^2}{2!}f''(0) + \frac{x^3}{3!}f'''(0) + \cdots + x^{n-1}\frac{f^{n-1}(0)}{(n-1)!} + R_n$$

where

$$R_n = \frac{x^n f^{(n)}(\theta x)}{n!}, \quad 0 < \theta < 1$$

Exponential

$$e = 1 + \frac{1}{1!} + \frac{1}{2!} + \frac{1}{3!} + \frac{1}{4!} + \cdots$$

$$e^x = 1 + x + \frac{x^2}{2!} + \frac{x^3}{3!} + \frac{x^4}{4!} + \cdots \qquad \text{(all real values of } x\text{)}$$

$$a^x = 1 + x\log_e a + \frac{(x\log_e a)^2}{2!} + \frac{(x\log_e a)^3}{3!} + \cdots$$

$$e^x = e^a\left[1 + (x-a) + \frac{(x-a)^2}{2!} + \frac{(x-a)^3}{3!} + \cdots\right]$$

Logarithmic

$$\log_e x = \frac{x-1}{x} + \frac{1}{2}\left(\frac{x-1}{x}\right)^2 + \frac{1}{3}\left(\frac{x-1}{x}\right)^3 + \cdots \qquad \left(x > \frac{1}{2}\right)$$

$$\log_e x = (x-1) - \frac{1}{2}(x-1)^2 + \frac{1}{3}(x-1)^3 - \cdots \qquad (2 \geq x > 0)$$

$$\log_e x = 2\left[\frac{x-1}{x+1} + \frac{1}{3}\left(\frac{x-1}{x+1}\right)^3 \frac{1}{5}\left(\frac{x-1}{x+1}\right)^5 + \cdots\right] \qquad (x > 0)$$

$$\log_e(1+x) = x - \frac{1}{2}x^2 + \frac{1}{3}x^3 - \frac{1}{4}x^4 + \cdots \qquad (-1 < x \leq 1)$$

$$\log_e(n+1) - \log_e(n-1) = 2\left[\frac{1}{n} + \frac{1}{3n^3} + \frac{1}{5n^5} + \cdots\right]$$

$$\log_e(a+x) = \log_e a + 2\left[\frac{x}{2a+x} + \frac{1}{3}\left(\frac{x}{2a+x}\right)^3 + \frac{1}{5}\left(\frac{x}{2a+x}\right)^5 + \cdots\right]$$
$$(a > 0, -a < x < +\infty)$$

$$\log_e \frac{1+x}{1-x} = 2\left[x + \frac{x^3}{3} + \frac{x^5}{5} + \cdots + \frac{x^{2n-1}}{2n-1} + \cdots\right] \qquad (-1 < x < 1)$$

$$\log_e x = \log_e a + \frac{(x-a)}{a} - \frac{(x-a)^2}{2a^2} + \frac{(x-a)^3}{3a^3} - \cdots \qquad (0 < x \leq 2a)$$

Trigonometric

$$\sin x = x - \frac{x^3}{3!} + \frac{x^5}{5!} - \frac{x^7}{7!} + \cdots \qquad \text{(all real values of } x\text{)}$$

$$\cos x = 1 - \frac{x^2}{2!} + \frac{x^4}{4!} - \frac{x^6}{6!} + \cdots \qquad \text{(all real values of } x\text{)}$$

$$\tan x = x + \frac{x^3}{3} + \frac{2x^5}{15} + \frac{17x^7}{315} + \frac{62x^9}{2835} + \cdots$$
$$+ \frac{(-1)^{n-1}2^{2n}(2^{2n}-1)B_{2n}}{(2n)!}x^{2n-1} + \cdots$$

$(x^2 < \pi^2/4$, and B_n represents the nth Bernoulli number$)$

$$\cot x = \frac{1}{x} - \frac{x}{3} - \frac{x^2}{45} - \frac{2x^5}{945} - \frac{x^7}{4725} - \cdots$$
$$- \frac{(-1)^{n+1}2^{2n}}{(2n)!}B_{2n}x^{2n-1} + \cdots$$

$(x^2 < \pi^2$, and B_n represents the nth Bernoulli number$)$

12.5 Differential Calculus

Notation

For the following equations, the symbols $f(x)$, $g(x)$, and so forth represent functions of x. The value of a function $f(x)$ at $x = a$ is denoted $f(a)$. For the function $y = f(x)$ the derivative of y with respect to x is denoted by one of the following:

$$\frac{dy}{dx}, \quad f'(x), \quad D_x y, \quad y'$$

Higher derivatives are as follows:

$$\frac{d^2y}{dx^2} = \frac{d}{dx}\left(\frac{dy}{dx}\right) = \frac{d}{dx}f'(x) = f''(x)$$

$$\frac{d^3y}{dx^3} = \frac{d}{dx}\left(\frac{d^2y}{dx^2}\right) = \frac{d}{dx}f''(x) = f'''(x)$$

$$\vdots$$

and values of these at $x = a$ are denoted $f''(a)$, $f'''(a)$, and so on (see Table 12.1, Table of Derivatives).

Slope of a Curve

The tangent line at point $P(x, y)$ of the curve $y = f(x)$ has a slope $f'(x)$ provided that $f'(x)$ exists at P. The slope at P is defined to be that of the tangent line at P. The tangent line at $P(x_1, y_1)$ is given by

$$y - y_1 = f'(x_1)(x - x_1)$$

TABLE 12.1 Table of Derivatives*

1. $\dfrac{d}{dx}(a) = 0$

2. $\dfrac{d}{dx}(x) = 1$

3. $\dfrac{d}{dx}(au) = a\dfrac{du}{dx}$

4. $\dfrac{d}{dx}(u + v) = \dfrac{du}{dx} + \dfrac{dv}{dx}$

5. $\dfrac{d}{dx}(uv) = u\dfrac{dv}{dx} + v\dfrac{du}{dx}$

6. $\dfrac{d}{dx}\dfrac{u}{v} = \dfrac{v\dfrac{du}{dx} - u\dfrac{dv}{dx}}{v^2}$

7. $\dfrac{d}{dx}(u^n) = nu^{n-1}\dfrac{du}{dx}$

8. $\dfrac{d}{dx}e^u = e^u\dfrac{du}{dx}$

9. $\dfrac{d}{dx}a^u = (\log_e a)a^u\dfrac{du}{dx}$

10. $\dfrac{d}{dx}\log_e u = \dfrac{1}{u}\dfrac{du}{dx}$

11. $\dfrac{d}{dx}\log_a u = (\log_a e)\dfrac{1}{u}\dfrac{du}{dx}$

12. $\dfrac{d}{dx}u^v = vu^{v-1}\dfrac{du}{dx} + u^v(\log_e u)\dfrac{dv}{dx}$

13. $\dfrac{d}{dx}\sin u = \cos u\dfrac{du}{dx}$

14. $\dfrac{d}{dx}\cos u = -\sin u\dfrac{du}{dx}$

15. $\dfrac{d}{dx}\tan u = \sec^2 u\dfrac{du}{dx}$

16. $\dfrac{d}{dx}\operatorname{ctn} u = -\csc^2 u\dfrac{du}{dx}$

17. $\dfrac{d}{dx}\sec u = \sec u \tan u\dfrac{du}{dx}$

18. $\dfrac{d}{dx}\csc u = -\csc u \operatorname{ctn} u\dfrac{du}{dx}$

19. $\dfrac{d}{dx}\sin^{-1} u = \dfrac{1}{\sqrt{1 - u^2}}\dfrac{du}{dx}$, $\left(-\dfrac{1}{2}\pi \leq \sin^{-1} u \leq \dfrac{1}{2}\pi\right)$

20. $\dfrac{d}{dx}\cos^{-1} u = \dfrac{-1}{\sqrt{1 - u^2}}\dfrac{du}{dx}$, $(0 \leq \cos^{-1} u \leq \pi)$

21. $\dfrac{d}{dx}\tan^{-1} u = \dfrac{1}{1 + u^2}\dfrac{du}{dx}$

22. $\dfrac{d}{dx}\operatorname{ctn}^{-1} u = \dfrac{-1}{1 + u^2}\dfrac{du}{dx}$

23. $\dfrac{d}{dx}\sec^{-1} u = \dfrac{1}{u\sqrt{u^2 - 1}}\dfrac{du}{dx}$,

$\left(-\pi \leq \sec^{-1} u < -\dfrac{1}{2}\pi;\ 0 \leq \sec^{-1} u < \dfrac{1}{2}\pi\right)$

24. $\dfrac{d}{dx}\csc^{-1} u = \dfrac{-1}{u\sqrt{u^2 - 1}}\dfrac{du}{dx}$,

$\left(-\pi \leq \csc^{-1} u \leq -\dfrac{1}{2}\pi;\ 0 < \csc^{-1} u \leq \dfrac{1}{2}\pi\right)$

25. $\dfrac{d}{dx}\sinh u = \cosh u\dfrac{du}{dx}$

26. $\dfrac{d}{dx}\cosh u = \sinh u\dfrac{du}{dx}$

27. $\dfrac{d}{dx}\tanh u = \operatorname{sech}^2 u\dfrac{du}{dx}$

28. $\dfrac{d}{dx}\operatorname{ctnh} u = -\operatorname{csch}^2 u\dfrac{du}{dx}$

29. $\dfrac{d}{dx}\operatorname{sech} u = -\operatorname{sech} u \tanh u\dfrac{du}{dx}$

30. $\dfrac{d}{dx}\operatorname{csch} u = -\operatorname{csch} u \operatorname{ctnh} u\dfrac{du}{dx}$

31. $\dfrac{d}{dx}\sinh^{-1} u = \dfrac{1}{\sqrt{u^2 + 1}}\dfrac{du}{dx}$

32. $\dfrac{d}{dx}\cosh^{-1} u = \dfrac{1}{\sqrt{u^2 - 1}}\dfrac{du}{dx}$

33. $\dfrac{d}{dx}\tanh^{-1} u = \dfrac{1}{1 - u^2}\dfrac{du}{dx}$

34. $\dfrac{d}{dx}\operatorname{ctnh}^{-1} u = \dfrac{-1}{u^2 - 1}\dfrac{du}{dx}$

35. $\dfrac{d}{dx}\operatorname{sech}^{-1} u = \dfrac{-1}{u\sqrt{1 - u^2}}\dfrac{du}{dx}$

36. $\dfrac{d}{dx}\operatorname{csch}^{-1} u = \dfrac{1}{u\sqrt{u^2 + 1}}\dfrac{du}{dx}$

*In this table, a and n are constants, e is the base of the natural logarithms, and u and v denote functions of x.

The *normal line* to the curve at $P(x_1, y_1)$ has slope $-1/f'(x_1)$ and thus obeys the equation

$$y - y_1 = \left[-1/f'(x_1)\right](x - x_1)$$

(The slope of a vertical line is not defined.)

Angle of Intersection of Two Curves

Two curves, $y = f_1(x)$ and $y = f_2(x)$, that intersect at a point $P(X, Y)$ where derivatives $f_1'(X)$, $f_2'(X)$ exist have an angle (α) of intersection given by

$$\tan \alpha = \frac{f_2'(X) - f_1'(X)}{1 + f_2'(X) \cdot f_1'(X)}$$

If $\tan \alpha > 0$, then α is the acute angle; if $\tan \alpha < 0$, then α is the obtuse angle.

Radius of Curvature

The radius of curvature R of the curve $y = f(x)$ at the point $P(x, y)$ is

$$R = \frac{\{1 + [f'(x)]^2\}^{3/2}}{f''(x)}$$

In polar coordinates (θ, r) the corresponding formula is

$$R = \frac{\left[r^2 + \left(\dfrac{dr}{d\theta}\right)^2\right]^{3/2}}{r^2 + 2\left(\dfrac{dr}{d\theta}\right)^2 - r\dfrac{d^2r}{d\theta^2}}$$

The *curvature K* is $1/R$.

Relative Maxima and Minima

The function f has a relative maximum at $x = a$ if $f(a) \geq f(a + c)$ for all values of c (positive or negative) that are sufficiently near zero. The function f has a relative minimum at $x = b$ if $f(b) \leq f(b + c)$ for all values of c that are sufficiently close to zero. If the function f is defined on the closed interval $x_1 \leq x \leq x_2$ and has a relative maximum or minimum at $x = a$, where $x_1 < a < x_2$, and if the derivative $f'(x)$ exists at $x = a$, then $f'(a) = 0$. It is noteworthy that a relative maximum or minimum may occur at a point where the derivative does not exist. Further, the derivative may vanish at a point that is neither a maximum nor a minimum for the function. Values of x for which $f'(x) = 0$ are called *critical values*. To determine whether a critical value of x, say x_c, is a relative maximum or minimum for the function at x_c, one may use the second derivative test:

1. If $f''(x_c)$ is positive, $f(x_c)$ is a minimum.
2. If $f''(x_c)$ is negative, $f(x_c)$ is a maximum.
3. If $f''(x_c)$ is zero, no conclusion may be made.

The sign of the derivative as x advances through x_c may also be used as a test. If $f'(x)$ changes from positive to zero to negative, then a maximum occurs at x_c, whereas a change in $f'(x)$ from negative to zero to positive indicates a minimum. If $f'(x)$ does not change sign as x advances through x_c, then the point is neither a maximum nor a minimum.

Points of Inflection of a Curve

The sign of the second derivative of f indicates whether the graph of $y = f(x)$ is concave upward or concave downward:

$$f''(x) > 0: \text{concave upward}$$
$$f''(x) < 0: \text{concave downward}$$

A point of the curve at which the direction of concavity changes is called a point of inflection (Fig. 12.14). Such a point may occur where $f''(x) = 0$ or where $f''(x)$ becomes infinite. More precisely, if the function $y = f(x)$ and its first derivative $y' = f'(x)$ are continuous in the interval $a \le x \le b$, and if $y'' = f''(x)$ exists in $a < x < b$, then the graph of $y = f(x)$ for $a < x < b$ is concave upward if $f''(x)$ is positive and concave downward if $f''(x)$ is negative.

FIGURE 12.14 Point of inflection.

Taylor's Formula

If f is function that is continuous on an interval that contains a and x, and if its first $(n + 1)$ derivatives are continuous on this interval, then

$$f(x) = f(a) + f'(a)(x - a) + \frac{f''(a)}{2!}(x - a)^2$$
$$+ \frac{f'''(a)}{3!}(x - a)^3 + \cdots + \frac{f^{(n)}(a)}{n!}(x - a)^n + R$$

where R is called the *remainder*. There are various common forms of the remainder.

Lagrange's form:

$$R = f^{(n+1)}(\beta) \cdot \frac{(x - a)^{n+1}}{(n + 1)!}, \quad \beta \text{ between } a \text{ and } x$$

Cauchy's form:

$$R = f^{(n+1)}(\beta) \cdot \frac{(x - B)^n (x - a)}{n!}, \quad \beta \text{ between } a \text{ and } x$$

Integral form:

$$R = \int_a^x \frac{(x - t)^n}{n!} f^{(n+1)}(t) dt$$

Indeterminant Forms

If $f(x)$ and $g(x)$ are continuous in an interval that includes $x = a$, and if $f(a) = 0$ and $g(a) = 0$, the limit $\lim_{x \to a}[f(x)/g(x)]$ takes the form $0/0$, called an *indeterminant form*. L'Hôpital's rule is

$$\lim_{x \to a} \frac{f(x)}{g(x)} = \lim_{x \to a} \frac{f'(x)}{g'(x)}$$

Similarly, it may be shown that if $f(x) \to \infty$ and $g(x) \to \infty$ as $x \to a$, then

$$\lim_{x \to a} \frac{f(x)}{g(x)} = \lim_{x \to a} \frac{f'(x)}{g'(x)}$$

(This holds for $x \to \infty$.)

Examples

$$\lim_{x \to 0} \frac{\sin x}{x} = \lim_{x \to 0} \frac{\cos x}{1} = 1$$

$$\lim_{x \to \infty} \frac{x^2}{e^x} = \lim_{x \to \infty} \frac{2x}{e^x} = \lim_{x \to \infty} \frac{2}{e^x} = 0$$

Numerical Methods

1. *Newton's method* for approximating roots of the equation $f(x) = 0$: A first estimate x_1 of the root is made; then, provided that $f'(x_1) \neq 0$, a better approximation is x_2,

$$x_2 = x_1 - \frac{f(x_1)}{f'(x_1)}$$

The process may be repeated to yield a third approximation, x_3, to the root

$$x_3 = x_2 - \frac{f(x_2)}{f'(x_2)}$$

provided $f'(x_2)$ exists. The process may be repeated. (In certain rare cases the process will not converge.)

2. *Trapezoidal rule for areas* (Fig. 12.15): For the function $y = f(x)$ defined on the interval (a, b) and positive there, take n equal subintervals of width $\Delta x = (b - a)/n$. The area bounded by the curve between $x = a$ and $x = b$ [or definite integral of $f(x)$] is approximately the sum of trapezoidal areas, or

$$A \sim \left(\frac{1}{2} y_0 + y_1 + y_2 + \cdots + y_{n-1} + \frac{1}{2} y_n \right) (\Delta x)^2$$

Estimation of the error (E) is possible if the second derivative can be obtained:

$$E = \frac{b-a}{12} f''(c)(\Delta x)^2$$

where c is some number between a and b.

Functions of Two Variables

For the function of two variables, denoted $z = f(x, y)$, if y is held constant, say at $y = y_1$, then the resulting function is a function of x only. Similarly, x may be held constant at x_1, to give the resulting function of y.

The Gas Laws

A familiar example is afforded by the ideal gas law relating the pressure p, the volume V, and the absolute temperature T of an ideal gas:

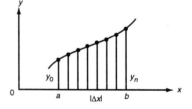

FIGURE 12.15   Trapezoidal rule for area.

$$pV = nRT$$

where n is the number of moles and R is the gas constant per mole, 8.31 ($J \cdot K^{-1} \cdot mole^{-1}$). By rearrangement, any one of the three variables may be expressed as a function of the other two. Further, either one of these two may be held constant. If T is held constant, then we get the form known as Boyle's law:

$$p = kV^{-1} \qquad \text{(Boyle's law)}$$

General Mathematical Tables   261

where we have denoted nRT by the constant k and, of course, $V > 0$. If the pressure remains constant, we have Charles' law:

$$V = bT \qquad \text{(Charles' law)}$$

where the constant b denotes nR/p. Similarly, volume may be kept constant:

$$p = aT$$

where now the constant, denoted a, is nR/V.

Partial Derivatives

The physical example afforded by the ideal gas law permits clear interpretations of processes in which one of the variables is held constant. More generally, we may consider a function $z = f(x, y)$ defined over some region of the xy plane in which we hold one of the two coordinates, say y, constant. If the resulting function of x is differentiable at a point (x, y), we denote this derivative by one of the notations.

$$f_x, \quad \frac{\partial f}{\partial x}$$

called the *partial derivative with respect to x*. Similarly, if x is held constant and the resulting function of y is differentiable, we get the *partial derivative with respect to y*, denoted by one of the following:

$$f_y, \quad \frac{\partial f}{\partial y}$$

Example

Given $z = x^4 y^3 - y \sin x + 4y$, then

$$\frac{\partial z}{\partial x} = 4(xy)^3 - y \cos x$$

$$\frac{\partial z}{\partial y} = 3x^4 y^2 - \sin x + 4$$

In Table 12.1, a and n are constants, e is the base of the natural logarithms, and u and v denote functions of x.

Additional Relations with Derivatives

$$\frac{d}{dt} \int_a^t f(x)\,dx = f(t) \qquad \frac{d}{dt} \int_t^a f(x)\,dx = -f(t)$$

If $x = f(y)$, then

$$\frac{dy}{dx} = \frac{1}{\frac{dx}{dy}}$$

If $y = f(u)$ and $u = g(x)$, then

$$\frac{dy}{dx} = \frac{dy}{du} \cdot \frac{du}{dx} \qquad \text{(chain rule)}$$

If $x = f(t)$ and $y = g(t)$, then

$$\frac{dy}{dx} = \frac{g'(t)}{f'(t)}, \qquad \text{and} \qquad \frac{d^2}{dx^2} = \frac{f'(t)g''(t) - g'(t)f''(t)}{[f'(t)]^3}$$

(*Note*: Exponent in denominator is 3.)

12.6 Integral Calculus

Indefinite Integral

If $F(x)$ is differentiable for all values of x in the interval (a, b) and satisfies the equation $dy/dx = f(x)$, then $F(x)$ is an integral of $f(x)$ with respect to x. The notation is $F(x) = \int f(x)\,dx$ or, in differential form, $dF(x) = f(x)\,dx$.

For any function $F(x)$ that is an integral of $f(x)$, it follows that $F(x) + C$ is also an integral. We thus write

$$\int f(x)\,dx = F(x) + C$$

Definite Integral

Let $f(x)$ be defined on the interval $[a, b]$ which is partitioned by points $x_1, x_2, \ldots, x_j, \ldots, x_{n-1}$ between $a = x_0$ and $b = x_n$. The jth interval has length $\Delta x_j = x_j - x_{j-1}$, which may vary with j. The sum $\sum_{j=1}^{n} f(v_j)\Delta x_j$, where v_j is arbitrarily chosen in the jth subinterval, depends on the numbers $x_0, \ldots, x_n$ and the choice of the v as well as f; but if such sums approach a common value as all Δx approach zero, then this value is the definite integral of f over the interval (a, b) and is denoted $\int_a^b f(x)\,dx$. The *fundamental theorem of integral calculus* states that

$$\int_a^b f(x)\,dx = F(b) - F(a),$$

where F is any continuous indefinite integral of f in the interval (a, b).

Properties

$$\int_a^b [f_1(x) + f_2(x) + \cdots + f_j(x)]\,dx = \int_a^b f_1(x)\,dx + \int_a^b f_2(x)\,dx + \cdots + \int_a^b f_j(x)\,dx$$

$$\int_a^b cf(x)\,dx = c\int_a^b f(x)\,dx, \quad \text{if } c \text{ is a constant}$$

$$\int_a^b f(x)\,dx = -\int_b^a f(x)\,dx$$

$$\int_a^b f(x)\,dx = \int_a^c f(x)\,dx + \int_c^b f(x)\,dx$$

Common Applications of the Definite Integral

Area (Rectangular Coordinates)

Given the function $y = f(x)$ such that $y > 0$ for all x between a and b, the area bounded by the curve $y = f(x)$, the x axis, and the vertical lines $x = a$ and $x = b$ is

$$A = \int_a^b f(x)\,dx$$

Length of Arc (Rectangular Coordinates)

Given the smooth curve $f(x, y) = 0$ from point (x_1, y_1) to point (x_2, y_2), the length between these points is

$$L = \int_{x_1}^{x_2} \sqrt{1 + \left(\frac{dy}{dx}\right)^2}\,dx$$

$$L = \int_{y_1}^{y_2} \sqrt{1 + \left(\frac{dx}{dy}\right)^2} \, dy$$

Mean Value of a Function

The mean value of a function $f(x)$ continuous on $[a, b]$ is

$$\frac{1}{(b-a)} \int_a^b f(x) \, dx$$

Area (Polar Coordinates)

Given the curve $r = f(\theta)$, continuous and nonnegative for $\theta_1 \leq \theta \leq \theta_2$, the area enclosed by this curve and the radial lines $\theta = \theta_1$ and $\theta = \theta_2$ is given by

$$A = \int_{\theta_1}^{\theta_2} \frac{1}{2}[f(\theta)]^2 \, d\theta$$

Length of Arc (Polar Coordinates)

Given the curve $r = f(\theta)$ with continuous derivative $f'(\theta)$ on $\theta_1 \leq \theta \leq \theta_2$, the length of arc from $\theta = \theta_1$ to $\theta = \theta_2$ is

$$L = \int_{\theta_1}^{\theta_2} \sqrt{[f(\theta)]^2 + [f'(\theta)]^2} \, d\theta$$

Volume of Revolution

Given a function $y = f(x)$ continuous and nonnegative on the interval (a, b), when the region bounded by $f(x)$ between a and b is revolved about the x axis, the volume of revolution is

$$V = \pi \int_a^b [f(x)]^2 \, dx$$

Surface Area of Revolution (Revolution about the x axis, between a and b)

If the portion of the curve $y = f(x)$ between $x = a$ and $x = b$ is revolved about the x axis, the area A of the surface generated is given by the following:

$$A = \int_a^b 2\pi f(x) \{1 + [f'(x)]^2\}^{\frac{1}{2}} \, dx$$

Work

If a variable force $f(x)$ is applied to an object in the direction of motion along the x axis between $x = a$ and $x = b$, the work done is

$$W = \int_a^b f(x) \, dx$$

Cylindrical and Spherical Coordinates

1. Cylindrical coordinates (Fig. 12.16):

$$x = r \cos \theta$$
$$y = r \sin \theta$$

Element of volume $dV = r \, dr \, d\theta \, dz$.

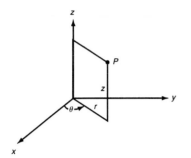

FIGURE 12.16 Cylindrical coordinates.

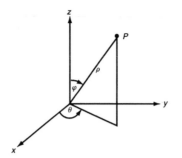

FIGURE 12.17 Spherical coordinates.

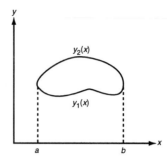

FIGURE 12.18 Region R bounded by $y_2(x)$ and $y_1(x)$.

2. Spherical coordinates (Fig. 12.17):

$$x = \rho \sin \phi \cos \theta$$
$$y = \rho \sin \phi \sin \theta$$
$$z = \rho \cos \phi$$

Element of volume $dV = \rho^2 \sin \phi \, d\rho \, d\phi \, d\theta$.

Double Integration

The evaluation of a double integral of $f(x, y)$ over a plane region R,

$$\iint_R f(x, y) \, dA$$

is practically accomplished by iterated (repeated) integration. For example, suppose that a vertical straight line meets the boundary of R in at most two points so that there is an upper boundary, $y = y_2(x)$, and a lower boundary, $y = y_1(x)$. Also, it is assumed that these functions are continuous from a to b (see Fig. 12.18). Then

$$\iint_R f(x, y) \, dA = \int_a^b \left(\int_{y_1(x)}^{y_2(y)} f(x, y) \, dy \right) dx$$

If R has left-hand boundary $x = x_1(y)$ and a right-hand boundary $x = x_2(y)$, which are continuous from c to d (the extreme values of y in R), then

$$\iint_R f(x, y) \, dA = \int_c^d \left(\int_{x_1(y)}^{x_2(y)} f(x, y) \, dx \right) dy$$

Such integrations are sometimes more convenient in polar coordinates, $x = r \cos \theta$, $y = r \sin \theta$, $dA = r \, dr \, d\theta$.

Surface Area and Volume by Double Integration

For the surface given by $z = f(x, y)$, which projects onto the closed region R of the xy plane, one may calculate the volume V bounded above by the surface and below by R, and the surface area S by the following:

$$V = \iint_R z \, dA = \iint_R f(x, y) \, dx \, dy$$
$$S = \iint_R [1 + (\partial z/\partial x)^2 + (\partial z/\partial y)^2]^{\frac{1}{2}} \, dx \, dy$$

[In polar coordinates, (r, θ), we replace dA by $r \, dr \, d\theta$.]

Centroid

The centroid of a region R of the xy plane is a point (x', y') where

$$x' = \frac{1}{A} \iint_R x \, dA, \quad y' = \frac{1}{A} \iint_R y \, dA$$

and A is the area of the region.

Example

For the circular sector of angle 2α and radius R, the area A is αR^2; the integral needed for x', expressed in polar coordinates, is

$$\iint x \, dA = \int_{-\alpha}^{\alpha} \int_0^R (r\cos\theta) r \, dr \, d\theta = \left[\frac{R^3}{3}\sin\theta\right]_{-\alpha}^{+\alpha} = \frac{2}{3}R^3 \sin\alpha$$

and thus,

$$x' = \frac{\frac{2}{3}R^3 \sin\alpha}{\alpha R^2} = \frac{2}{3}R\frac{\sin\alpha}{\alpha}$$

Centroids of some common regions are shown in Table 12.2

TABLE 12.2 Centroids

	Area	x'	y'
Rectangle	bh	$b/2$	$h/2$
Isosceles triangle	$bh/2$	$b/2$	$h/3$
($y' = h/3$ for any triangle of altitude h.)			
Semicircle	$\pi R^2/2$	R	$4R/3\pi$
Quarter circle	$\pi R^2/4$	$4R/3\pi$	$4R/3\pi$
Circular sector	$R^2 A$	$2R\sin A/3A$	0

12.7 Special Functions

Hyperbolic Functions

$$\sinh x = \frac{e^x - e^{-x}}{2}$$

$$\cosh x = \frac{e^x + e^{-x}}{2}$$

$$\tanh x = \frac{e^x - e^{-x}}{e^x + e^{-x}}$$

$$\sinh(-x) = -\sinh x$$

$$\cosh(-x) = \cosh x$$

$$\tanh(-x) = -\tanh x$$

$$\tanh x = \frac{\sinh x}{\cosh x}$$

$$\cosh^2 x - \sinh^2 x = 1$$

$$\sinh^2 x = \frac{1}{2}(\cosh 2x - 1)$$

$$\operatorname{csch}^2 x - \operatorname{sech}^2 x = \operatorname{csch}^2 x \operatorname{sech}^2 x$$

$$\operatorname{csch} x = \frac{1}{\sinh x}$$

$$\operatorname{sech} x = \frac{1}{\cosh x}$$

$$\operatorname{ctnh} x = \frac{1}{\tanh x}$$

$$\operatorname{ctnh}(-x) = -\operatorname{ctnh} x$$

$$\operatorname{sech}(-x) = \operatorname{sech} x$$

$$\operatorname{csch}(-x) = -\operatorname{csch} x$$

$$\operatorname{ctnh} x = \frac{\cosh x}{\sinh x}$$

$$\cosh^2 x = \frac{1}{2}(\cosh 2x + 1)$$

$$\operatorname{ctnh}^2 x - \operatorname{csch}^2 x = 1$$

$$\tanh^2 x + \operatorname{sech}^2 x = 1$$

$$\sinh(x + y) = \sinh x \cosh y + \cosh x \sinh y$$

$$\cosh(x + y) = \cosh x \cosh y + \sinh x \sinh y$$

$$\sinh(x - y) = \sinh x \cosh y - \cosh x \sinh y$$

$$\cosh(x - y) = \cosh x \cosh y - \sinh x \sinh y$$

$$\tanh(x + y) = \frac{\tanh x + \tanh y}{1 + \tanh x \tanh y}$$

$$\tanh(x - y) = \frac{\tanh x - \tanh y}{1 - \tanh x \tanh y}$$

Bessel Functions

Bessel functions, also called cylindrical functions, arise in many physical problems as solutions of the differential equation

$$x^2 y'' + x y' + (x^2 - n^2) y = 0$$

which is known as Bessel's equation. Certain solutions, known as *Bessel functions of the first kind of order n*, are given by

$$J_n(x) = \sum_{k=0}^{\infty} \frac{(-1)^k}{k!\,\Gamma(n + k + 1)} \left(\frac{x}{2}\right)^{n+2k}$$

$$J_{-n}(x) = \sum_{k=0}^{\infty} \frac{(-1)^k}{k!\,\Gamma(-n + k + 1)} \left(\frac{x}{2}\right)^{-n+2k}$$

see Fig. 12.19a.

In the preceding it is noteworthy that the gamma function must be defined for the negative argument q : $\Gamma(q) = \Gamma(q+1)/q$, provided that q is not a negative integer. When q is a negative integer, $1/\Gamma(q)$ is defined to be zero. The functions $J_{-n}(x)$ and $J_n(x)$ are solutions of Bessel's equation for all real n. It is seen, for $n = 1, 2, 3, \ldots$, that

$$J_{-n}(x) = (-1)^n J_n(x)$$

General Mathematical Tables

and, therefore, these are not independent; hence, a linear combination of these is not a general solution. When, however, n is not a positive integer, a negative integer, or zero, the linear combination with arbitrary constants c_1 and c_2,

$$y = c_1 J_n(x) + c_2 J_{-n}(x)$$

is the general solution of the Bessel differential equation.

The zero-order function is especially important as it arises in the solution of the heat equation (for a long cylinder):

$$J_0(x) = 1 - \frac{x^2}{2^2} + \frac{x^4}{2^2 4^2} - \frac{x^6}{2^2 4^2 6^2} + \cdots$$

whereas the following relations show a connection to the trigonometric functions:

$$J_{1/2}(x) = \left[\frac{2}{\pi x}\right]^{1/2} \sin x$$

$$J_{-1/2}(x) = \left[\frac{2}{\pi x}\right]^{1/2} \cos x$$

The following recursion formula gives $J_{n+1}(x)$ for any order in terms of lower order functions:

$$\frac{2n}{x} J_n(x) = J_{n-1}(x) + J_{n+1}(x)$$

Bessel Functions of the Second Kind, $Y_n(x)$
(Also Called *Neumann Functions* or *Weber Functions*) (Fig. 12.19b)

Domain: $[x > 0]$

Recurrence relation:

$$Y_{n+1}(x) = \frac{2n}{x} Y_n(x) - Y_{n-1}(x), \quad n = 0, 1, 2, \ldots$$

Symmetry: $Y_{-n}(x) = (-1)^n Y_n(x)$

Legendre Polynomials

If Laplace's equation, $\nabla^2 V = 0$, is expressed in spherical coordinates, it is

$$r^2 \sin\theta \frac{\delta^2 V}{\delta r^2} + 2r \sin\theta \frac{\delta V}{\delta r} + \sin\theta \frac{\delta^2 V}{\delta \theta^2} + \cos\theta \frac{\delta V}{\delta \theta} + \frac{1}{\sin\theta} \frac{\delta^2 V}{\delta \phi^2} = 0$$

and any of its solutions, $V(r, \theta, \phi)$, are known as *spherical harmonics*. The solution as a product

$$V(r, \theta, \phi) = R(r)\Theta(\theta)$$

which is independent of ϕ, leads to

$$\sin^2\theta \Theta'' + \sin\theta \cos\theta \Theta' + [n(n+1) \sin^2\theta]\Theta = 0$$

Rearrangement and substitution of $x = \cos\theta$ leads to

$$(1 - x^2)\frac{d^2\Theta}{dx^2} - 2x\frac{d\Theta}{dx} + n(n+1)\Theta = 0$$

known as *Legendre's equation*. Important special cases are those in which n is zero or a positive integer, and, for

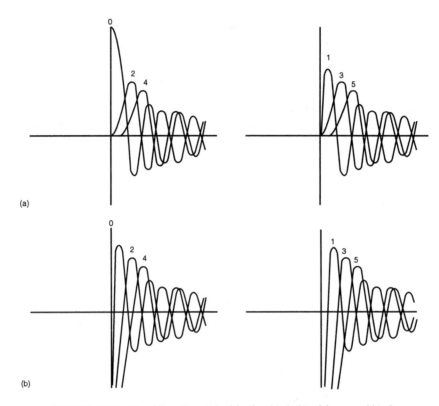

FIGURE 12.19 Bessel Functions: (a) of the first kind, (b) of the second kind.

such cases, Legendre's equation is satisfied by polynomials called Legendre polynomials, $P_n(x)$. A short list of Legendre polynomials, expressed in terms of x and $\cos\theta$, is given next. These are given by the following general formula:

$$P_n(x) = \sum_{j=0}^{L} \frac{(-1)^j (2n-2j)!}{2^n j!(n-j)!(n-2j)!} x^{n-2j}$$

where $L = n/2$ if n is even and $L = (n-1)/2$ if n is odd,

$$P_0(x) = 1$$
$$P_1(x) = x$$
$$P_2(x) = \frac{1}{2}(3x^2 - 1)$$
$$P_3(x) = \frac{1}{2}(5x^3 - 3x)$$
$$P_4(x) = \frac{1}{8}(35x^4 - 30x^2 + 3)$$
$$P_5(x) = \frac{1}{8}(63x^5 - 70x^3 + 15x)$$
$$P_0(\cos\theta) = 1$$
$$P_1(\cos\theta) = \cos\theta$$
$$P_2(\cos\theta) = \frac{1}{4}(3\cos 2\theta + 1)$$

General Mathematical Tables

$$P_3(\cos\theta) = \frac{1}{8}(5\cos 3\theta + 3\cos\theta)$$

$$P_4(\cos\theta) = \frac{1}{64}(35\cos 4\theta + 20\cos 2\theta + 9)$$

Additional Legendre polynomials may be determined from the *recursion formula*

$$(n+1)P_{n+1}(x) - (2n+1)xP_n(x) + nP_{n-1}(x) = 0 \quad (n = 1, 2, \ldots)$$

or the *Rodrigues formula*

$$P_n(x) = \frac{1}{2^n n!}\frac{d^n}{dx^n}(x^2 - 1)^n$$

Laguerre Polynomials

Laguerre polynomials, denoted $L_n(x)$, are solutions of the differential equation

$$xy'' + (1-x)y' + ny = 0$$

and are given by

$$L_n(x) = \sum_{j=0}^{n} \frac{(-1)^j}{j!} C_{(n,j)} x^j \quad (n = 0, 1, 2, \ldots)$$

Thus,

$$L_0(x) = 1$$

$$L_1(x) = 1 - x$$

$$L_2(x) = 1 - 2x + \frac{1}{2}x^2$$

$$L_3(x) = 1 - 3x + \frac{3}{2}x^2 - \frac{1}{6}x^3$$

Additional Laguerre polynomials may be obtained from the recursion formula

$$(n+1)L_{n+1}(x) - (2n+1-x)L_n(x) + nL_{n-1}(x) = 0$$

Hermite Polynomials

The Hermite polynomials, denoted $H_n(x)$, are given by

$$H_0 = 1, \quad H_n(x) = (-1)^n e^{x^2}\frac{d^n e^{-x^2}}{dx^n}, \quad (n = 1, 2, \ldots)$$

and are solutions of the differential equation

$$y'' - 2xy' + 2ny = 0 \quad (n = 0, 1, 2, \ldots)$$

The first few Hermite polynomials are

$$H_0 = 1 \qquad H_1(x) = 2x$$
$$H_2(x) = 4x^2 - 2 \qquad H_3(x) = 8x^3 - 12x$$
$$H_4(x) = 16x^4 - 48x^2 + 12$$

Additional Hermite polynomials may be obtained from the relation

$$H_{n+1}(x) = 2xH_n(x) - H_n'(x)$$

where prime denotes differentiation with respect to x.

Orthogonality

A set of functions $\{f_n(x)\}(n = 1, 2, \ldots)$ is orthogonal in an interval (a, b) with respect to a given weight function $w(x)$ if

$$\int_a^b w(x) f_m(x) f_n(x) dx = 0 \qquad \text{when } m \neq n$$

The following polynomials are orthogonal on the given interval for the given $w(x)$.

Legendre polynomials:

$$P_n(x) \; w(x) = 1$$
$$a = -1, \quad b = 1$$

Leguerre polynomials:

$$L_n(x) \; w(x) = \exp(-x)$$
$$a = 0, \quad b = \infty$$

Hermite polynomials:

$$H_n(x) w(x) = \exp(-x^2)$$
$$a = -\infty, \quad b = \infty$$

The Bessel functions of *order n*, $J_n(\lambda_1 x)$, $J_n(\lambda_2 x)$, ..., are orthogonal with respect to $w(x) = x$ over the interval $(0, c)$ provided that the λ_i are the positive roots of $J_n(\lambda c) = 0$,

$$\int_0^c x J_n(\lambda_j x) J_n(\lambda_k x) \, dx = 0 \quad (j \neq k)$$

where n is fixed and $n \geq 0$.

Functions with $x^2/a^2 \pm y^2/b^2$

Elliptic Paraboloid (Fig. 12.20)

$$z = c(x^2/a^2 + y^2/b^2)$$
$$x^2/a^2 + y^2/b^2 - z/c = 0$$

Hyperbolic Paraboloid (Commonly Called *Saddle*) (Fig. 12.21):

$$z = c(x^2/a^2 - y^2/b^2)$$
$$x^2/a^2 - y^2/b^2 - z/c = 0$$

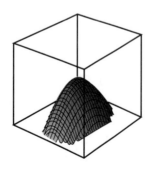
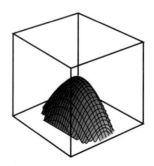

FIGURE 12.20 Elliptic paraboloid: (a) $a = 0.5, b = 1.0, c = -1.0$, viewpoint $= (5, -6, 4)$; (b) $a = 1.0, b = 1.0, c = -2.0$, viewpoint $= (5, -6, 4)$.

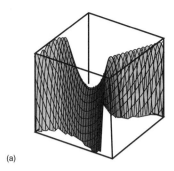 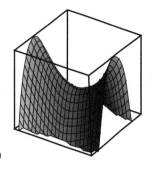

FIGURE 12.21 Hyperbolic paraboloid: (a) $a = 0.50, b = 0.5, c = 1.0$, viewpoint $= (4, -6, 4)$; (b) $a = 1.00, b = 0.5, c = 1.0$, viewpoint $= (4, -6.4)$.

Elliptic Cylinder (Fig. 12.22):

$$1 = x^2/a^2 + y^2/b^2$$
$$x^2/a^2 + y^2/b^2 - 1 = 0$$

Hyperbolic Cylinder (Fig. 12.23)

$$1 = x^2/a^2 - y^2/b^2$$
$$x^2/a^2 - y^2/b^2 - 1 = 0$$

Functions with $(x^2/a^2 + y^2/b^2 \pm c^2)^{1/2}$

Sphere (Fig. 12.24)

$$z = (1 - x^2 - y^2)^{1/2}$$
$$x^2 + y^2 + z^2 - 1 = 0$$

Ellipsoid (Fig. 12.25)

$$z = c(1 - x^2/a^2 - y^2/b^2)^{1/2}$$
$$x^2/a^2 + y^2/b^2 + z^2/c^2 - 1 = 0$$

Special cases:

$$a = b > c \text{ gives oblate spheroid}$$
$$a = b < c \text{ gives prolate spheroid}$$

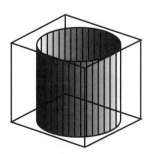 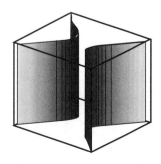 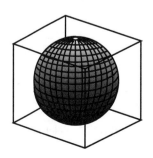

FIGURE 12.22 Elliptic cylinder. $a = 1.0, b = 1.0$, viewpoint $= (4, -5, 2)$.

FIGURE 12.23 Hyperbolic cylinder. $a = 1.0, b = 1.0$, viewpoint $= (4, -6, 3)$.

FIGURE 12.24 Sphere: viewpoint $= (4, -5, 2)$.

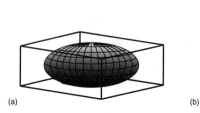

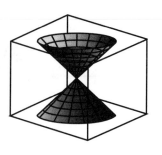

FIGURE 12.25 Ellipsoid: (a) $a = 1.00, b = 1.00, c = 0.5$, viewpoint = $(4, -5, 2)$; (b) $a = 0.50, b = 0.50, c = 1.0$, viewpoint = $(4, -5, 2)$.

FIGURE 12.26 Cone: viewpoint = $(4, -5, 2)$.

Cone (Fig. 12.26)

$$z = (x^2 + y^2)^{1/2}$$
$$x^2 + y^2 - z^2 = 0$$

Elliptic Cone (Circular Cone if $a = b$) (Fig. 12.27)

$$z = c(x^2/a^2 + y^2/b^2)^{1/2}$$
$$x^2/a^2 + y^2/b^2 - z^2/c^2 = 0$$

Hyperboloid of One Sheet (Fig. 12.28)

$$z = c(x^2/a^2 + y^2/b^2 - 1)^{1/2}$$
$$x^2/a^2 + y^2/b^2 - z^2/c^2 - 1 = 0$$

Hyperboloid of Two Sheets (Fig. 12.29)

$$z = c(x^2/a^2 + y^2/b^2 + 1)^{1/2}$$
$$x^2/a^2 + y^2/b^2 - z^2/c^2 + 1 = 0$$

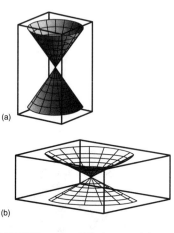
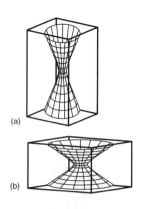
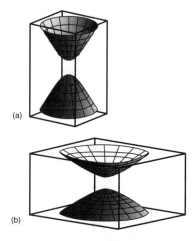

FIGURE 12.27 Elliptic cone: (a) $a = 0.5, b = 0.5, c = 1.00$, viewpoint = $(4, -5, 2)$; (b) $a = 1.0, b = 1.0, c = 0.50$, viewpoint = $(4, -5, 2)$.

FIGURE 12.28 Hyperboloid of one sheet: (a) $a = 0.1, b = 0.1, c = 0.2, \pm z = c\sqrt{15}$, viewpoint = $(4, -5, 2)$; (b) $a = 0.2, b = 0.2, c = 0.2, \pm z = c\sqrt{15}$, viewpoint = $(4, -5, 2)$.

FIGURE 12.29 Hyperboloid of two sheets: (a) $a = 0.125, b = 0.125, c = 0.2, \pm z = c\sqrt{17}$, viewpoint = $(4, -5, 2)$; (b) $a = 0.25, b = 0.25, c = 0.2, \pm z = c\sqrt{17}$, viewpoint = $(4, -5, 2)$.

12.8 Basic Definitions: Linear Algebra Matrices

A *matrix* A is a rectangular array of numbers (real or complex):

$$A = \begin{bmatrix} a_{11} & a_{12} & \cdots & a_{1m} \\ a_{21} & a_{22} & \cdots & a_{2m} \\ \vdots & & & \\ a_{n1} & a_{n2} & \cdots & a_{nm} \end{bmatrix}$$

The *size* of the matrix is said to be $n \times m$. The $1 \times m$ matrices $[a_{i1}\ a_{i2}\ \cdots\ a_{im}]$ are called *rows* of A, and the $n \times 1$ matrices

$$\begin{bmatrix} a_{1j} \\ a_{2j} \\ \vdots \\ a_{nj} \end{bmatrix}$$

are called *columns* of A. An $n \times m$ matrix thus consists of n rows and m columns; a_{ij} denotes the *element*, or *entry*, of A in the ith row and jth column. A matrix consisting of just one row is called a *row vector*, whereas a matrix of just one column is called a *column vector*. The elements of a vector are frequently called *components* of the vector. When the size of the matrix is clear from the context, we sometimes write $A = (a_{ij})$.

A matrix with the same number of rows as columns is a *square* matrix, and the number of rows and columns is the *order* of the matrix. The diagonal of an $n \times n$ square matrix A from a_{11} to a_{nn} is called the *main*, or *principal*, *diagonal*. The word *diagonal* with no modifier usually means the main diagonal. The *transpose* of a matrix A is the matrix that results from interchanging the rows and columns of A. It is usually denoted by A^T. A matrix A such that $A = A^T$ is said to be *symmetric*. The *conjugate transpose* of A is the matrix that results from replacing each element of A^T by its complex conjugate, and is usually denoted by A^H. A matrix such that $A = A^H$ is said to be *Hermitian*.

A square matrix $A = (a_{ij})$ is *lower triangular* if $a_{ij} = 0$ for $j > i$ and is *upper triangular* if $a_{ij} = 0$ for $j < i$. A matrix that is both upper and lower triangular is a *diagonal* matrix. The $n \times n$ *identity matrix* is the $n \times n$ diagonal matrix in which each element of the main diagonal is 1. It is traditionally denoted I_n, or simply I when the order is clear from the context.

Algebra of Matrices

The sum and difference of two matrices A and B are defined whenever A and B have the same size. In that case $C = A \pm B$ is defined by $C = (c_{ij}) = (a_{ij} \pm b_{ij})$. The product tA of a scalar t (real or complex number) and a matrix A is defined by $tA = (ta_{ij})$. If A is an $n \times m$ matrix and B is an $m \times p$ matrix, the product $C = AB$ is defined to be the $n \times p$ matrix $C = (c_{ij})$ given by $c_{ij} = \sum_{k=1}^{m} a_{ik} b_{kj}$. Note that the product of an $n \times m$ matrix and an $m \times p$ matrix is an $n \times p$ matrix, and the product is defined only when the number of column of the first factor is the same as the number of rows of the second factor. Matrix multiplication is, in general, associative: $A(BC) = (AB)C$. It also distributes over addition (and subtraction):

$$A(B + C) = AB + AC \quad \text{and} \quad (A + B)C = AC + BC$$

It is, however, not in general true that $AB = BA$, even in case both products are defined. It is clear that $(A + B)^T = A^T + B^T$ and $(A + B)^H = A^H + B^H$. It is also true, but not so obvious perhaps, that $(AB)^T = B^T A^T$ and $(AB)^H = B^H A^H$.

The $n \times n$ identity matrix I has the property that $IA = AI = A$ for every $n \times n$ matrix A. If A is square, and if there is a matrix B such that $AB = BA = I$, then B is called the *inverse* of A and is denoted A^{-1}. This terminology and notation are justified by the fact that a matrix can have at most one inverse. A matrix having an inverse is said to be *invertible*, or *nonsingular*, whereas a matrix not having an inverse is said to be *noninvertible*, or *singular*. The product of two invertible matrices is invertible and, in fact, $(AB)^{-1} = B^{-1}A^{-1}$. The sum of two invertible matrices is, obviously, not necessarily invertible.

Systems of Equations

The system of n linear equations in m unknowns

$$a_{11}x_1 + a_{12}x_2 + a_{13}x_3 + \cdots + a_{1m}x_m = b_1$$
$$a_{21}x_1 + a_{22}x_2 + a_{23}x_3 + \cdots + a_{2m}x_m = b_2$$
$$\vdots$$
$$a_{n1}x_1 + a_{n2}x_2 + a_{n3}x_3 + \cdots + a_{nm}x_m = b_n$$

may be written $Ax = b$, where $A = (a_{ij})$, $x = [x_1 \quad x_2 \quad \cdots \quad x_m]^T$ and $b = [b_1 \quad b_2 \quad \cdots \quad b_n]^T$. Thus, A is an $n \times m$ matrix, and x and b are column vectors of the appropriate sizes.

The matrix A is called the *coefficient matrix* of the system. Let us first suppose the coefficient matrix is square; that is, there are an equal number of equations and unknowns. If A is upper triangular, it is quite easy to find all solutions of the system. The ith equation will contain only the unknowns $x_i, x_{i+1}, \ldots, x_n$, and one simply solves the equations in reverse order: the last equation is solved for x_n; the result is substituted into the $(n-1)$st equation, which is then solved for x_{n-1}; these values of x_n and x_{n-1} are substituted in the $(n-2)$th equation, which is solved for x_{n-2}, and so on. This procedure is known as *back substitution*.

The strategy for solving an arbitrary system is to find an upper-triangular system equivalent with it and solve this upper-triangular system using back substitution. First, suppose the element $a_{11} \neq 0$. We may rearrange the equations to ensure this, unless, of course the first column of A is all 0s. In this case proceed to the next step, to be described later. For each $i \geq 2$ let $m_{i1} = a_{i1}/a_{11}$. Now replace the ith equation by the result of multiplying the first equation by m_{i1} and subtracting the new equation from the ith equation. Thus,

$$a_{i1}x_1 + a_{i2}x_2 + a_{i3}x_3 + \cdots + a_{im}x_m = b_i$$

is replaced by

$$0 \cdot x_1 + (a_{i2} + m_{i1}a_{12})x_2 + (a_{i3} + m_{i1}a_{13})x_3 + \cdots + (a_{im} + m_{i1}a_{1m})x_m = b_i + m_{i1}b_1$$

After this is done for all $i = 2, 3, \ldots, n$, there results the equivalent system

$$a_{11}x_1 + a_{12}x_2 + a_{13}x_3 + \cdots + a_{1n}x_n = b_1$$
$$0 \cdot x_1 + a'_{22}x_2 + a'_{23}x_3 + \cdots + a'_{2n}x_n = b'_2$$
$$0 \cdot x_1 + a'_{32}x_2 + a'_{33}x_3 + \cdots + a'_{3n}x_n = b'_3$$
$$\vdots$$
$$0 \cdot x_1 + a'_{n2}x_2 + a'_{n3}x_3 + \cdots + a'_{nn}x_n = b'_n$$

in which all entries in the first column below a_{11} are 0. (Note that if all entries in the first column were 0 to begin with, then $a_{11} = 0$ also.) This procedure is now repeated for the $(n-1) \times (n-1)$ system

$$a'_{22}x_2 + a'_{23}x_3 + \cdots + a'_{2n}x_n = b'_2$$
$$a'_{32}x_2 + a'_{33}x_3 + \cdots + a'_{3n}x_n = b'_3$$
$$\vdots$$
$$a'_{n2}x_2 + a'_{n3}x_3 + \cdots + a'_{nn}x_n = b'_n$$

to obtain an equivalent system in which all entries of the coefficient matrix below a'_{22} are 0. Continuing, we obtain an upper-triangular system $Ux = c$ equivalent with the original system. This procedure is known as *Gaussian elimination*. The numbers m_{ij} are known as the *multipliers*.

Essentially the same procedure may be used in case the coefficient matrix is not square. If the coefficient matrix is not square, we may make it square by appending either rows or columns of 0s as needed. Appending rows of 0s and appending 0s to make b have the appropriate size is equivalent to appending equations $0 = 0$ to the system. Clearly the new system has precisely the same solutions as the original system. Appending columns of 0s and adjusting the size of x appropriately yields a new system with additional unknowns, each appearing only with coefficient 0, thus not affecting the solutions of the original system. In either case we may assume the coefficient matrix is square, and apply the Gauss elimination procedure.

Suppose the matrix A is invertible. Then if there were no row interchanges in carrying out the above Gauss elimination procedure, we have the *LU factorization* of the matrix A:

$$A = LU$$

where U is the upper-triangular matrix produced by elimination and L is the lower-triangular matrix given by

$$L = \begin{bmatrix} 1 & 0 & \cdots & \cdots & 0 \\ m_{21} & 1 & 0 & \cdots & 0 \\ \vdots & & \ddots & & \\ m_{n1} & m_{n2} & \cdots & & 1 \end{bmatrix}$$

A *permutation* P_{ij} matrix is an $n \times n$ matrix such that $P_{ij}A$ is the matrix that results from exchanging row i and j of the matrix A. The matrix P_{ij} is the matrix that results from exchanging row i and j of the identity matrix. A product P of such matrices P_{ij} is called a *permutation* matrix. If row interchanges are required in the Gauss elimination procedure, then we have the factorization

$$PA = LU$$

where P is the permutation matrix giving the required row exchanges.

Vector Spaces

The collection of all column vectors with n real components is *Euclidean n-space*, and is denoted R^n. The collection of column vectors with n complex components is denoted C^n. We shall use *vector space* to mean either R^n or C^n. In discussing the space R^n, the word *scalar* will mean a real number, and in discussing the space C^n, it will mean a complex number. A subset S of a vector space is a *subspace* such that if u and v are vectors in S, and if c is any scalar, then $u + v$ and cu are in S. We shall sometimes use the word *space* to mean a subspace. If $B = \{v_1, v_2, \ldots, v_k\}$ is a collection of vectors in a vector space, then the set S consisting of all vectors $c_1v_1 + c_2v_2 + \cdots + c_mv_m$ for all scalars $c_1, c_2, \ldots, c_m$ is a subspace, called the *span* of B. A collection $\{v_1, v_2, \ldots, v_m\}$ of vectors $c_1v_1 + c_2v_2 + \cdots + c_mv_m\}$ is a *linear combination* of B. If S is a subspace and $B = \{v_1, v_2, \ldots, v_m\}$ is a subset of S such that S is the space of B, then B is said to *span S*.

A collection $\{v_1, v_2, \ldots, v_m\}$ of n vectors is *linearly dependent* if there exist scalars $c_1, c_2, \ldots, c_m$, not all zero, such that $c_1v_1 + c_2v_2 + \cdots + c_mv_m = 0$. A collection of vectors that is not linearly dependent is said to be *linearly independent*. The modifier *linearly* is frequently omitted, and we speak simply of dependent and independent collections. A linearly independent collection of vectors in a space S that spans S is a *basis* of S. Every basis of a space S contains the same number of vectors; this number is the *dimension* of S. The dimension of the space consisting of only the zero vector is 0. The collection $B = \{e_1, e_2, \ldots, e_n\}$, where $e_1 = [1, 0, 0, \ldots, 0]^T$, $e_2 = [0, 1, 0, \ldots, 0]^T$, and so forth ($e_i$ has 1 as its ith component and zero for all other components) is a basis for the spaces R^n and C^n. This is the *standard basis* for these spaces. The dimension of these spaces is thus n. In a space S of dimension n, no collection of fewer than n vectors can span S, and no collection of more than n vectors in S can be independent.

Rank and Nullity

The *column space* of an $n \times m$ matrix A is the subspace of R^n or C^n spanned by the columns of A. The *row space* is the subspace of R^m or C^m spanned by the rows of A. Note that for any vector $x = [x_1 x_2 \cdots x_m]^T$,

$$Ax = x_1 \begin{bmatrix} a_{11} \\ a_{21} \\ \vdots \\ a_{n1} \end{bmatrix} + x_2 \begin{bmatrix} a_{12} \\ a_{22} \\ \vdots \\ a_{n2} \end{bmatrix} + \cdots + x_m \begin{bmatrix} a_{1m} \\ a_{2m} \\ \vdots \\ a_{nm} \end{bmatrix}$$

so that the column space is the collection of all vectors Ax, and thus the system $Ax = b$ has a solution if and only if b is a member of the column space of A.

The dimension of the column space is the *rank* of A. The row space has the same dimension as the column space. The set of all solutions of the system $Ax = 0$ is a subspace called the *null space* of A, and the dimension of this null space is the *nullity* of A. A fundamental result in matrix theory is the fact that, for an $n \times m$ matrix A,

$$\text{rank } A + \text{nullity } A = m$$

The difference of any two solutions of the linear system $Ax = b$ is a member of the null space of A. Thus this system has at most one solution if and only if the nullity of A is zero. If the system is square (that is, if A is $n \times n$), then there will be a solution for every right-hand side b if and only if the collection of columns of A is linearly independent, which is the same as saying the rank of A is n. In this case the nullity must be zero. Thus, for any b, the square system $Ax = b$ has exactly one solution if and only if rank $A = n$. In other words the $n \times n$ matrix A is invertible if and only if rank $A = n$.

Orthogonality and Length

The *inner product* of two vectors x and y is the scalar $x^H y$. The *length*, or norm, $\|x\|$, of the vector x is given by $\|x\| = \sqrt{x^H x}$. A *unit vector* is a vector of norm 1. Two vectors x and y are *orthogonal* if $x^H y = 0$. A collection of vectors $\{v_1, v_2, \ldots, v_m\}$ in a space S is said to be an *orthonormal* collection if $v_i^H v_j = 0$ for $i \neq j$ and $v_i^H v_i = 1$. An orthonormal collection is necessarily linearly independent. If S is a subspace (or R^n or C^n) spanned by the orthonormal collection $\{v_1, v_2, \ldots, v_m\}$, then the *projection* of a vector x onto S is the vector

$$\text{proj}(x; S) = (x^H v_1)v_1 + (x^H v_2)v_2 + \cdots + (x^H v_m)v_m$$

The projection of x onto S minimizes the function $f(y) = \|x - y\|^2$ for $y \in S$. In other words the projection of x onto S is the vector in S that is closest to x.

If b is a vector and A is an $n \times m$ matrix, then a vector x minimizes $\|b - Ax\|^2$ if and only if it is a solution of $A^H Ax = A^H b$. This system of equations is called the *system of normal equations* for the least-squares problem of minimizing $\|b - Ax\|^2$.

If A is an $n \times m$ matrix and rank $A = k$, then there is an $n \times k$ matrix Q whose columns form an orthonormal basis for the column space of A and a $k \times m$ upper-triangular matrix R of rank k such that

$$A = QR$$

This is called the *QR factorization* of A. It now follows that x minimizes $\|b - Ax\|^2$ if and only if it is a solution of the upper-triangular system $Rx = Q^H b$.

If $\{w_1, w_2, \ldots, w_m\}$ is a basis for a space S, the following procedure produces an orthonormal basis $\{v_1, v_2, \ldots, v_m\}$ for S:

- Set $v_1 = w_1/\|w_1\|$
- Let $\tilde{v}_2 = w_2 - \text{proj}(w_2; S_1)$, where S_1 is the span of $\{v_1\}$; set $v_2 = \tilde{v}_2/\|\tilde{v}_2\|$
- Next, let $\tilde{v}_3 = w_3 - \text{proj}(w_3; s_2)$, where S_2 is the span of $\{v_1, v_2\}$; set $v_3 = \tilde{v}_3/\|\tilde{v}_3\|$.

And so on: $\tilde{v}_i = w_i - \text{proj}(w_i; S_{i-1})$, where S_{i-1} is the span of $\{v_1, v_2, \ldots, v_{i-1}\}$ set $v_i = \tilde{v}_i / \|\tilde{v}_i\|$ This is the *Gram–Schmidt procedure*.

If the collection of columns of a square matrix is an orthonormal collection, the matrix is called a *unitary matrix*. In case the matrix is a real matrix, it is usually called an *orthogonal matrix*. A unitary matrix U is invertible, and $U^{-1} = U^H$. (In the real case an orthogonal matrix Q is invertible, and $Q^{-1} = Q^T$.)

Determinants

The *determinant* of a square matrix is defined inductively. First, suppose the determinant det A has been defined for all square matrices of order $< n$. Then

$$\det A = a_{11}C_{11} + a_{12}C_{12} + \cdots + a_{1n}C_{1n}$$

where the numbers C_{ij} are *cofactors* of the matrix A,

$$C_{ij} = (-1)^{i+j} \det M_{ij}$$

where M_{ij} is the $(n-1) \times (n-1)$ matrix obtained by deleting the ith row and jth column of A. Now det A is defined to be the only entry of a matrix of order 1. Thus, for a matrix of order 2, we have

$$\det \begin{bmatrix} a & b \\ c & d \end{bmatrix} = ad - bc$$

There are many interesting but not obvious properties of determinants. It is true that

$$\det A = a_{i1}C_{i1} + a_{i2}C_{i2} + \cdots + a_{in}C_{in}$$

for any $1 \leq i \leq n$. It is also true that det A = det A^T, so that we have

$$\det A = a_{1j}C_{1j} + a_{2j}C_{2j} + \cdots + a_{nj}C_{nj}$$

for any $1 \leq j \leq n$.

If A and B are matrices of the same order, then det AB = (det A) (det B), and the determinant of any identity matrix is 1. Perhaps the most important property of the determinant is the fact that a matrix is invertible if and only if its determinant is not zero.

Eigenvalues and Eigenvectors

If A is a square matrix, and $Av = \lambda v$ for a scalar λ and a nonzero v, then λ is an *eigenvalue* of A and v is an *eigenvector* of A that *corresponds* to λ. Any nonzero linear combination of eigenvectors corresponding to the same eigenvalue λ is also an eigenvector corresponding to λ. The collection of all eigenvectors corresponding to a given eigenvalue λ is thus a subspace, called an *eigenspace* of A. A collection of eigenvectors corresponding to different eigenvalues is necessarily linear independent. It follows that a matrix of order n can have at most n distinct eigenvectors. In fact, the eigenvalues of A are the roots of the nth degree polynomial equation

$$\det(A - \lambda I) = 0$$

called the *characteristic equation* of A. (Eigenvalues and eigenvectors are frequently called *characteristic values* and *characteristic vectors*.)

If the nth order matrix A has an independent collection of n eigenvectors, then A is said to have a *full set* of eigenvectors. In this case there is a set of eigenvectors of A that is a basis for R^n or, in the complex case, C^n. In case there are n distinct eigenvalues of A, then, of course, A has a full set of eigenvectors. If there are fewer than

n distinct eigenvalues, then A may or may not have a full set of eigenvectors. If there is a full set of eigenvectors, then

$$D = S^{-1} A S \quad \text{or} \quad A = S D S^{-1}$$

where D is a diagonal matrix the eigenvalues of A on the diagonal, and S is a matrix whose columns are the full set of eigenvectors. If A is symmetric, there are n real distinct eigenvalues of A and the corresponding eigenvectors are orthogonal. There is thus an orthonormal collection of eigenvectors that span R^n, and we have

$$A = Q D Q^T \quad \text{and} \quad D = Q^T A Q$$

where Q is a real orthogonal matrix and D is diagonal. For the complex case, if A is Hermitian, we have

$$A = U D U^H \quad \text{and} \quad D = U^H A U$$

where U is a unitary matrix and D is a *real* diagonal matrix. (A Hermitian matrix also has n distinct real eigenvalues.)

12.9 Basic Definitions: Vector Algebra and Calculus

A vector is a directed line segment, with two vectors being equal if they have the same length and the same direction. More precisely, a *vector* is an equivalence class of directed line segments, where two directed segments are equivalent if they have the same length and the same direction. The *length* of a vector is the common length of its directed segments, and the *angle between* vectors is the angle between any of their segments. The length of a vector u is denoted $|u|$. There is defined a distinguished vector having zero length, which is usually denoted $\mathbf{0}$. It is frequently useful to visualize a directed segment as an arrow; we then speak of the nose and the tail of the segment. The *sum* $u + v$ of two vectors u and v is defined by taking directed segments from u and v and placing the tail of the segment representing v at the nose of the segment representing u and defining $u + v$ to be the vector determined by the segment from the tail of the u representative to the nose of the v representative. It is easy to see that $u + v$ is well defined and that $u + v = v + u$. Subtraction is the inverse operation of addition. Thus the *difference* $u - v$ of two vectors is defined to be the vector that when added to v gives u. In other words, if we take a segment from u and a segment from v and place their tails together, the difference is the segment from the nose of v to the nose of u. The zero vector behaves as one might expect: $u + 0 = u$, and $u - u = 0$. Addition is associative: $u + (v + w) = (u + v) + w$.

To distinguish them from vectors, the real numbers are called *scalars*. The product tu of a scalar t and a vector u is defined to be the vector having length $|t|\,|u|$ and direction the same as u if $t > 0$, the opposite direction if $t < 0$. If $t = 0$, then tu is defined to be the zero vector. Note that $t(\mathbf{u} + \mathbf{v}) = t\mathbf{u} + t\mathbf{v}$, and $(t+s)\mathbf{u} = t\mathbf{u} + s\mathbf{u}$. From this it follows that $u - v = u + (-1)v$.

The *scalar product* $u \cdot v$ of two vectors is $|u|\,|v|\cos\theta$, where θ is the angle between u and v. The scalar product is frequently called the *dot product*. The scalar product distributes over addition,

$$u \cdot (v + w) = u \cdot v + u \cdot w$$

and it is clear that $(t\mathbf{u}) \cdot \mathbf{v} = t(\mathbf{u} \cdot \mathbf{v})$. The *vector product* $u \times v$ of two vectors is defined to be the vector perpendicular to both u and v and having length $|u|\,|v|\sin\theta$, where θ is the angle between u and v. The direction of $u \times v$ is the direction a right-hand threaded bolt advances if the vector u is rotated to v. The vector product is frequently called the *cross product*. The vector product is both associative and distributive, but not commutative: $u \times v = -v \times u$.

General Mathematical Tables

Coordinate Systems

Suppose we have a right-handed Cartesian coordinate system is space. For each vector u, we associate a point in space by placing the tail of a representative of u at the origin and associating with u the point at the nose of the segment. Conversely, associated with each point in space is the vector determined by the directed segment from the origin to that point. There is thus a one-to-one correspondence between the points in space and all vectors. The origin corresponds to the zero vector. The coordinates of the point associated with a vector u are called *coordinates* of u. One frequently refers to the vector u and writes $u = (x, y, z)$, which is, strictly speaking, incorrect, because the left side of this equation is a vector and the right side gives the coordinates of a point in space. What is meant is that (x, y, z) are the coordinates of the point associated with u under the correspondence described. In terms of coordinates, for $u = (u_1, u_2, u_3)$ and $v = (v_1, v_2, v_3)$, we have

$$u + v = (u_1 + v_1, u_2 + v_2, u_3 + v_3)$$

$$tu = (tu_1, tu_2, tu_3)$$

$$u \cdot v = u_1 v_1 + u_2 v_2 + u_3 v_3$$

$$u \times v = (u_2 v_3 - v_2 u_3, u_3 v_1 - v_3 u_1, u_1 v_2 - v_1 u_2)$$

The *coordinate vectors* i, j, and k are the unit vectors $i = (1, 0, 0)$, $j = (0, 1, 0)$, and $k = (0, 0, 1)$. Any vector $u = (u_1, u_2, u_3)$ is thus a linear combination of these coordinate vectors: $u = u_1 i + u_2 j + u_3 k$. A convenient form for the vector product is the formal determinant

$$u \times v = \det \begin{bmatrix} i & j & k \\ u_1 & u_2 & u_3 \\ v_1 & v_2 & v_2 \end{bmatrix}$$

Vector Functions

A *vector function F of one variable* is a rule that associates a vector $F(t)$ with each real number t is some set, called the *domain* of F. The expression $\lim_{t \to t_0} F(t) = a$ means that for any $\varepsilon > 0$, there is a $\delta > 0$ such that $|F(t) - a| < \varepsilon$ whenever $0 < |t - t_0| < \delta$. If $F(t) = [x(t), y(t), z(t)]$ and $a = (a_1, a_2, a_3)$, then $\lim_{t \to t_0} F(t) = a$ if and only if

$$\lim_{t \to t_0} x(t) = a_1$$

$$\lim_{t \to t_0} y(t) = a_2$$

$$\lim_{t \to t_0} z(t) = a_3$$

A vector function F is *continuous* at t_0 if $\lim_{t \to t_0} F(t) = F(t_0)$. The vector function F is continuous at t_0 if and only if each of the coordinates $x(t)$, $y(t)$, and $z(t)$ is continuous at t_0.

The function F is *differentiable* at t_0 if the limit

$$\lim_{h \to 0} \frac{1}{h}[F(t + h) - F(t)]$$

exists. This limit is called the *derivative* of F at t_0 and is usually written $F'(t_0)$, or $(dF/dt)(t_0)$. The vector function F is differentiable at t_0 if and only if each of its coordinate functions is differentiable at t_0. Moreover, $(dF/dt)(t_0) = [(dx/dt)(t_0), (dy/dt)(t_0), (dz/dt)(t_0)]$. The usual rules for derivatives of real valued functions all hold for vector functions. Thus, if F and G are vector functions and s is a scalar function, then

$$\frac{d}{dt}(\mathbf{F} + \mathbf{G}) = \frac{d\mathbf{F}}{dt} + \frac{d\mathbf{G}}{dt}$$

$$\frac{d}{dt}(s\mathbf{F}) = s\frac{d\mathbf{F}}{dt} + \frac{ds}{dt}\mathbf{F}$$

$$\frac{d}{dt}(\mathbf{F} \cdot \mathbf{G}) = \mathbf{F} \cdot \frac{d\mathbf{G}}{dt} + \frac{d\mathbf{F}}{dt} \cdot \mathbf{G}$$

$$\frac{d}{dt}(\mathbf{F} \times \mathbf{G}) = \mathbf{F} \times \frac{d\mathbf{G}}{dt} + \frac{d\mathbf{F}}{dt} \times \mathbf{G}$$

If $\mathbf{R}$ is a vector function defined for t is some interval, then, as t varies, with the tail of $\mathbf{R}$ at the origin, the nose traces out some object C in space. For nice functions $\mathbf{R}$, the object C is a *curve*. If $\mathbf{R}(t) = [x(t), y(t), z(t)]$, then the equations

$$x = x(t)$$
$$y = y(t)$$
$$z = z(t)$$

are called *parametric equations* of C. At points where $\mathbf{R}$ is differentiable, the derivative $d\mathbf{R}/dt$ is a vector *tangent* to the curve. The unit vector $\mathbf{T} = (d\mathbf{R}/dt)/|d\mathbf{R}/dt|$ is called the *unit tangent vector*. If $\mathbf{R}$ is differentiable and if the length of the arc of curve described by $\mathbf{R}$ between $\mathbf{R}(a)$ and $\mathbf{R}(t)$ is given by $s(t)$, then

$$\frac{ds}{dt} = \left|\frac{d\mathbf{R}}{dt}\right|$$

Thus the length L of the arc from $\mathbf{R}(t_0)$ to $\mathbf{R}(t_1)$ is

$$L = \int_{t_0}^{t_1} \frac{ds}{dt} dt = \int_{t_0}^{t_1} \left|\frac{d\mathbf{R}}{dt}\right| dt$$

The vector $d\mathbf{T}/ds = (d\mathbf{T}/dt)/(ds/dt)$ is perpendicular to the unit tangent $\mathbf{T}$, and the number $\kappa = |d\mathbf{T}/ds|$ is the *curvature* of C. The unit vector $\mathbf{N} = (1/\kappa)(d\mathbf{T}/ds)$ is the *principal normal*. The vector $\mathbf{B} = \mathbf{T} \times \mathbf{N}$ is the *binormal*, and $d\mathbf{B}/ds = -\tau \mathbf{N}$. The number τ is the *torsion*. Note that C is a plane curve if and only if τ is zero for all t.

A *vector function* $\mathbf{F}$ *of two variables* is a rule that assigns a vector $\mathbf{F}(s, t)$ to each point (s, t) is some subset of the plane, called the *domain* of $\mathbf{F}$. If $\mathbf{R}(s, t)$ is defined for all (s, t) in some region D of the plane, then as the point (s, t) varies over D, with its rail at the origin, the nose of $\mathbf{R}(s, t)$ traces out an object in space. For a nice function $\mathbf{R}$, this object is a *surface*, S. The partial derivatives $(\partial \mathbf{R}/\partial s)(s, t)$ and $(\partial \mathbf{R}/\partial t)(s, t)$ are tangent to the surface at $\mathbf{R}(s, t)$ and the vector $(\partial \mathbf{R}/\partial s) \times (\partial \mathbf{R}/\partial t)$ is thus *normal* to the surface. Of course, $(\partial \mathbf{R}/\partial t) \times (\partial \mathbf{R}/\partial s) = -(\partial \mathbf{R}/\partial s) \times (\partial \mathbf{R}/\partial t)$ is also normal to the surface and points in the direction opposite that of $(\partial \mathbf{R}/\partial s) \times (\partial \mathbf{R}/\partial t)$. By electing one of these normals, we are choosing an *orientation* of the surface. A surface can be oriented only if it has two sides, and the process of orientation consists of choosing which side is positive and which is negative.

Gradient, Curl, and Divergence

If $f(x, y, z)$ is a scalar field defined in some region D, the *gradient* of f is the vector function

$$\operatorname{grad} f = \frac{\partial f}{\partial x}\mathbf{i} + \frac{\partial f}{\partial y}\mathbf{j} + \frac{\partial f}{\partial z}\mathbf{k}$$

If $\mathbf{F}(x, y, z) = F_1(x, y, z)\mathbf{i} + F_2(x, y, z)\mathbf{j} + F_3(x, y, z)\mathbf{k}$ is a vector field defined in some region D, then the *divergence* of $\mathbf{F}$ is the scalar function

$$\text{div } \mathbf{F} = \frac{\partial F_1}{\partial x} + \frac{\partial F_2}{\partial y} + \frac{\partial F_3}{\partial z}$$

The curl is the vector function

$$\text{curl } \mathbf{F} = \left(\frac{\partial F_3}{\partial y} - \frac{\partial F_2}{\partial z}\right)\mathbf{i} + \left(\frac{\partial F_1}{\partial z} - \frac{\partial F_3}{\partial x}\right)\mathbf{j} + \left(\frac{\partial F_2}{\partial x} - \frac{\partial F_1}{\partial y}\right)\mathbf{k}$$

In terms of the vector operator *del*, $\nabla = \mathbf{i}(\partial/\partial x) + \mathbf{j}(\partial/\partial y) + \mathbf{k}(\partial/\partial z)$, we can write

$$\text{grad } f = \nabla f$$
$$\text{div } \mathbf{F} = \nabla \cdot \mathbf{F}$$
$$\text{curl } \mathbf{F} = \nabla \times \mathbf{F}$$

The *Laplacian operator* is div (grad) $= \nabla \cdot \nabla = \nabla^2 = (\partial^2/\partial x^2) + (\partial^2/\partial y^2) + (\partial^2/\partial z^2)$.

Integration

Suppose C is a curve from the point (x_0, y_0, z_0) to the point (x_1, y_1, z_1) are is described by the vector function $\mathbf{R}(t)$ for $t_0 \leq t \leq t_1$. If f is a scalar function (sometimes called a *scalar field*) defined on C, then the integral of f over C is

$$\int_C f(x, y, z) \, ds = \int_{t_0}^{t_1} f[\mathbf{R}(t)] \left|\frac{d\mathbf{R}}{dt}\right| dt$$

If $\mathbf{F}$ is a vector function (sometimes called a *vector field*) defined on C, then the integral of $\mathbf{F}$ over C is

$$\int_C \mathbf{F}(x, y, z) \cdot d\mathbf{R} = \int_{t_0}^{t_1} \mathbf{F}[\mathbf{R}(t)] \cdot \frac{d\mathbf{R}}{dt} dt$$

These integrals are called *line integrals*.

In case there is a scalar function f such that $\mathbf{F} = \text{grad } f$, then the line integral

$$\int_C \mathbf{F}(x, y, z) \cdot d\mathbf{R} = f[\mathbf{R}(t_1)] - f[\mathbf{R}(t_0)]$$

The value of the integral thus depends only on the end points of the curve C and not on the curve C itself. The integral is said to be *path independent*. The function f is called a *potential function* for the vector field $\mathbf{F}$, and $\mathbf{F}$ is said to be a *conservative field*. A vector field $\mathbf{F}$ with domain D is conservative if and only if the integral of $\mathbf{F}$ around every closed curve in D is zero. If the domain D is simply connected (that is, every closed curve in D can be continuously deformed in D to a point), then $\mathbf{F}$ is conservative if and only if curl $\mathbf{F} = 0$ in D.

Suppose S is a surface described by $\mathbf{R}(s, t)$ for (s, t) in a region D of the plane. If f is a scalar function defined on D, then the integral of f over S is given by

$$\iint_S f(x, y, z) \, dS = \iint_D f[\mathbf{R}(s, t)] \left|\frac{\partial \mathbf{R}}{\partial s} \times \frac{\partial \mathbf{R}}{\partial t}\right| ds \, dt$$

If $\mathbf{F}$ is a vector function defined on S, and if an orientation for S is chosen, then the integral of $\mathbf{F}$ over S, sometimes called the *flux* of **F** through S, is

$$\iint_S \mathbf{F}(x,y,z) \cdot d\mathbf{S} = \iint_D \mathbf{F}[\mathbf{R}(s,t)] \cdot \left(\frac{\partial \mathbf{R}}{\partial s} \times \frac{\partial \mathbf{R}}{\partial t}\right) ds\, dt$$

Integral Theorems

Suppose **F** is a vector field with a closed domain D bounded by the surface S oriented so that the normal points out from D. Then the *divergence theorem* states that

$$\iiint_D \operatorname{div} \mathbf{F}\, dV = \iint_S \mathbf{F} \cdot d\mathbf{S}$$

If S is an orientable surface bounded by a closed curve C, the orientation of the closed curve C is chosen to be consistent with the orientation of the surface S. Then we have *Stokes's theorem*:

$$\iint_S (\operatorname{curl} \mathbf{F}) \cdot d\mathbf{S} = \oint_C \mathbf{F} \cdot d\mathbf{s}$$

12.10 The Fourier Transforms: Overview

For a piecewise continuous function $F(x)$ over a finite interval $0 \le x \le \pi$, the *finite Fourier cosine transform* of $F(x)$ is

$$f_c(n) = \int_0^\pi f(x) \cos nx\, dx \quad (n = 0, 1, 2, \ldots) \tag{12.1}$$

If x ranges over the interval $0 \le x \le L$, the substitution $x' = \pi x / L$ allows the use of this definition also. The inverse transform is written

$$\bar{F}(x) = \frac{1}{\pi} f_c(0) + \frac{2}{\pi} \sum_{n=1}^\infty f_c(n) \cos nx \quad (0 < x < \pi) \tag{12.2}$$

where $\bar{F}(x) = [F(x+0) + F(x-0)]/2$. We observe that $\bar{F}(x) = F(x)$ at points of continuity. The formula

$$\begin{aligned} f_c^{(2)}(n) &= \int_0^\pi F''(x) \cos nx\, dx \\ &= -n^2 f_c(n) - F'(0) + (-1)^n F'(\pi) \end{aligned} \tag{12.3}$$

makes the finite Fourier cosine transform useful in certain boundary-value problems.

Analogously, the *finite Fourier sine transform* of $F(x)$ is

$$f_s(n) = \int_0^\pi F(x) \sin nx\, dx \quad (n = 1, 2, 3, \ldots) \tag{12.4}$$

and

$$\bar{F}(x) = \frac{2}{\pi} \sum_{n=1}^\infty f_s(n) \sin\, nx \quad (0 < x < \pi) \tag{12.5}$$

Corresponding to Eq. (12.3), we have

$$f_s^{(2)}(n) = \int_0^\pi F''(x) \sin nx\, dx \tag{12.6}$$

General Mathematical Tables

$$= -n^2 f_s(n) - nF(0) - n(-1)^n F(\pi)$$

Fourier Transforms

If $F(x)$ is defined for $x \geq 0$ and is piecewise continuous over any finite interval, and if

$$\int_0^\infty F(x)\,dx$$

is absolutely convergent, then

$$f_c(\alpha) = \sqrt{\frac{2}{\pi}} \int_0^\infty f(x) \cos(\alpha x)\,dx \qquad (12.7)$$

is the *Fourier cosine transform of F(x)*. Furthermore,

$$\bar{F}(x) = \sqrt{\frac{2}{\pi}} \int_0^\infty f_c(\alpha) \cos(\alpha x)\,d\alpha \qquad (12.8)$$

If $\lim_{x \to \infty} d^n F/dx^n = 0$, an important property of the Fourier cosine transform,

$$f_c^{(2r)}(\alpha) = \sqrt{\frac{2}{\pi}} \int_0^\infty \left(\frac{d^{2r} F}{dx^{2r}}\right) \cos(\alpha x)\,dx$$

$$= -\sqrt{\frac{2}{\pi}} \sum_{n=0}^{r-1} (-1)^n a_{2r-2n-1} \alpha^{2n} + (-1)^r \alpha^{2r} f_c(\alpha) \qquad (12.9)$$

where $\lim_{x \to 0} d^r F/dx^r = a_r$, makes it useful in the solution of many problems.

Under the same conditions,

$$f_s(\alpha) = \sqrt{\frac{2}{\pi}} \int_0^\infty F(x) \sin(\alpha x)\,dx \qquad (12.10)$$

defines the *Fourier sine transform of F(x)*, and

$$\bar{F}(x) = \sqrt{\frac{2}{\pi}} \int_0^\infty f_s(\alpha) \sin(\alpha x)\,d\alpha \qquad (12.11)$$

Corresponding to Eq. (12.9) we have

$$f_s^{(2r)}(\alpha) = \sqrt{\frac{2}{\pi}} \int_0^\infty \frac{d^{2r} F}{dx^{2r}} \sin(\alpha x)\,dx$$

$$= -\sqrt{\frac{2}{\pi}} \sum_{n=1}^{r} (-1)^n \alpha^{2n-1} a_{2r-2n} + (-1)^{r-1} \alpha^{2r} f_s(\alpha) \qquad (12.12)$$

Similarly, if $F(x)$ is defined for $-\infty < x < \infty$, and if $\int_{-\infty}^\infty F(x)\,dx$ is absolutely convergent, then

$$f(\alpha) = \frac{1}{\sqrt{2\pi}} \int_{-\infty}^\infty F(x) e^{i\alpha x}\,dx \qquad (12.13)$$

is the *Fourier transform of F(x)*, and

$$\bar{F}(x) = \frac{1}{\sqrt{2\pi}} \int_{-\infty}^\infty f(\alpha) e^{-i\alpha x}\,d\alpha \qquad (12.14)$$

Also, if
$$\lim_{|x|\to\infty} \left|\frac{d^n F}{dx^n}\right| = 0 \quad (n = 1, 2, \ldots, r-1)$$
then
$$f^{(r)}(\alpha) = \frac{1}{\sqrt{2\pi}} \int_{-\infty}^{\infty} F^{(r)}(x) e^{i\alpha x}\, dx = (-i\alpha)^r f(\alpha) \tag{12.15}$$

TABLE 12.3 Finite Sine Transforms

$f_s(n)$	$F(x)$
1. $f_s(n) = \int_0^\pi F(x)\sin nx\, dx$ $(n=1,2,3,\ldots)$	$F(x)$
2. $(-1)^{n+1} f_s(n)$	$F(\pi - x)$
3. $\dfrac{1}{n}$	$\dfrac{\pi - x}{\pi}$
4. $\dfrac{(-1)^{n+1}}{n}$	$\dfrac{x}{\pi}$
5. $\dfrac{1-(-1)^n}{n}$	1
6. $\dfrac{2}{n^2}\sin\dfrac{n\pi}{2}$	$\begin{cases} x & \text{when } 0 < x < \pi/n \\ \pi - x & \text{when } \pi/2 < x < \pi \end{cases}$
7. $\dfrac{(-1)^{n+1}}{n^3}$	$\dfrac{x(\pi^2 - x^2)}{6\pi}$
8. $\dfrac{1-(-1)^n}{n^3}$	$\dfrac{x(\pi - x)}{2}$
9. $\dfrac{\pi^2(-1)^{n-1}}{n} - \dfrac{2[1-(-1)^n]}{n^3}$	x^2
10. $\pi(-1)^n\left(\dfrac{6}{n^3} - \dfrac{\pi^2}{n}\right)$	x^3
11. $\dfrac{n}{n^2 + c^2}[1-(-1)^n e^{c\pi}]$	e^{cx}
12. $\dfrac{n}{n^2 + c^2}$	$\dfrac{\sinh c(\pi - x)}{\sinh c\pi}$
13. $\dfrac{n}{n^2 - k^2}$ $(k \neq 0, 1, 2, \ldots)$	$\dfrac{\sinh k(\pi - x)}{\sin k\pi}$
14. $\begin{cases} \dfrac{\pi}{2} & \text{when } n = m \\ 0 & \text{when } n \neq m \end{cases}$ $(m = 1, 2, \ldots)$	$\sin mx$
15. $\dfrac{n}{n^2 - k^2}[1-(-1)^n \cos k\pi]$ $(k \neq 1, 2, \ldots)$	$\cos kx$
16. $\begin{cases} \dfrac{n}{n^2 - m^2}[1-(-1)^{n+m}] & \text{when } n \neq m = 1, 2, \ldots \\ 0 & \text{when } n = m \end{cases}$	$\cos mx$
17. $\dfrac{n}{(n^2 - k^2)^2}$ $(k \neq 0, 1, 2, \ldots)$	$\dfrac{\pi \sin kx}{2k \sin^2 k\pi} - \dfrac{x \cos k(\pi - x)}{2k \sin k\pi}$
18. $\dfrac{b^n}{n}$ $(|b| \le 1)$	$\dfrac{2}{\pi}\arctan\dfrac{b \sin x}{1 - b \cos x}$
19. $\dfrac{1-(-1)^n}{n}b^n$ $(|b| \le 1)$	$\dfrac{2}{\pi}\arctan\dfrac{2b \sin x}{1 - b^2}$

TABLE 12.4 Finite Cosine Transforms

$f_c(n)$	$F(x)$
1. $f_c(n) = \int_0^\pi F(x) \cos nx \, dx$ $(n = 0, 1, 2, \ldots)$	$F(x)$
2. $(-1)^n f_c(n)$	$F(\pi - x)$
3. 0 when $n = 1, 2, \ldots$; $f_c(0) = \pi$	1
4. $\dfrac{2}{\pi} \sin \dfrac{n\pi}{2}$; $f_c(0) = 0$	$\begin{cases} 1 & \text{when } 0 < x < \pi/2 \\ -1 & \text{when } \pi/2 < x < \pi \end{cases}$
5. $-\dfrac{1 - (-1)^n}{n^2}$; $f_c(0) = \dfrac{\pi^2}{2}$	x
6. $\dfrac{(-1)^n}{n^2}$; $f_c(0) = \dfrac{\pi^2}{6}$	$\dfrac{x^2}{2\pi}$
7. $\dfrac{1}{n^2}$; $f_c(0) = 0$	$\dfrac{(\pi - x)^2}{2\pi} - \dfrac{\pi}{6}$
8. $3\pi^2 \dfrac{(-1)^n}{n^2} - 6\dfrac{1 - (-1)^n}{n^4}$; $f_c(0) = \dfrac{\pi^4}{4}$	x^3
9. $\dfrac{(-1)^n e^c \pi - 1}{n^2 + c^2}$	$\dfrac{1}{c} e^{cx}$
10. $\dfrac{1}{n^2 + c^2}$	$\dfrac{\cosh c(\pi - x)}{c \sinh c\pi}$
11. $\dfrac{k}{n^2 - k^2}[(-1)^n \cos \pi k - 1]$ $(k \neq 0, 1, 2, \ldots)$	$\sin kx$
12. $\dfrac{(-1)^{n+m} - 1}{n^2 - m^2}$	$\dfrac{1}{m} \sin mx$
13. $\dfrac{1}{n^2 - k^2}$ $(k \neq 0, 1, 2, \ldots)$	$-\dfrac{\cos k(\pi - x)}{k \sin k\pi}$
14. 0 when $n = 1, 2, \ldots$;	$\cos mx$

TABLE 12.5 Fourier Sine Transforms

$F(x)$	$f_s(\alpha)$
1. $\begin{cases} 1 & (0 < x < a) \\ 0 & (x > a) \end{cases}$	$\sqrt{\dfrac{2}{\pi}} \left[\dfrac{1 - \cos \alpha}{\alpha} \right]$
2. x^{p-1} $(0 < p < 1)$	$\sqrt{\dfrac{2}{\pi}} \dfrac{\Gamma(p)}{\alpha^p} \sin \dfrac{p\pi}{2}$
3. $\begin{cases} \sin x & (0 < x < a) \\ 0 & (x > a) \end{cases}$	$\dfrac{1}{\sqrt{2\pi}} \left[\dfrac{\sin[a(1 - \alpha)]}{1 - \alpha} - \dfrac{\sin[a(1 + \alpha)]}{1 + \alpha} \right]$
4. e^{-x}	$\sqrt{\dfrac{2}{\pi}} \left[\dfrac{\alpha}{1 + \alpha^2} \right]$
5. $xe^{-x^2/2}$	$\alpha e^{-\alpha^2/2}$
6. $\cos \dfrac{x^2}{2}$	$\sqrt{2} \left[\sin \dfrac{\alpha^2}{2} C\left(\dfrac{\alpha^2}{2}\right) - \cos \dfrac{\alpha^2}{2} S\left(\dfrac{\alpha^2}{2}\right) \right]^a$
7. $\sin \dfrac{x^2}{2}$	$\sqrt{2} \left[\cos \dfrac{\alpha^2}{2} C\left(\dfrac{\alpha^2}{2}\right) + \sin \dfrac{\alpha^2}{2} S\left(\dfrac{\alpha^2}{2}\right) \right]^a$

[a] $C(y)$ and $S(y)$ are the Fresnel integrals

$$C(y) = \dfrac{1}{\sqrt{2\pi}} \int_0^y \dfrac{1}{\sqrt{t}} \cos t \, dt$$

$$S(y) = \dfrac{1}{\sqrt{2\pi}} \int_0^y \dfrac{1}{\sqrt{t}} \sin t \, dt$$

TABLE 12.6 Fourier Cosine Transforms

$F(x)$	$f_c(\alpha)$
1. $\begin{cases} 1 & (0 < x < a) \\ 0 & (x < a) \end{cases}$	$\sqrt{\dfrac{2}{\pi}} \dfrac{\sin a\alpha}{\alpha}$
2. x^{p-1} $(0 < p < 1)$	$\sqrt{\dfrac{2}{\pi}} \dfrac{\Gamma(p)}{\alpha^p} \cos \dfrac{p\pi}{2}$
3. $\begin{cases} \cos x & (0 < x < a) \\ 0 & (x > a) \end{cases}$	$\dfrac{1}{\sqrt{2\pi}} \left[\dfrac{\sin[a(1-\alpha)]}{1-\alpha} + \dfrac{\sin[a(1+\alpha)]}{1+\alpha} \right]$
4. e^{-x}	$\sqrt{\dfrac{2}{\pi}} \left(\dfrac{1}{1+\alpha^2} \right)$
5. $e^{-x^2/2}$	$e^{-\alpha^2/2}$
6. $\cos \dfrac{x^2}{2}$	$\cos \left(\dfrac{\alpha^2}{2} - \dfrac{\pi}{4} \right)$
7. $\sin \dfrac{x^2}{2}$	$\cos \left(\dfrac{\alpha^2}{2} - \dfrac{\pi}{4} \right)$

TABLE 12.7 Fourier Transforms

$F(x)$	$f(\alpha)$
1. $\dfrac{\sin ax}{x}$	$\begin{cases} \sqrt{\dfrac{\pi}{2}} & |\alpha| < a \\ 0 & |\alpha| > a \end{cases}$
2. $\begin{cases} e^{iwx} & (p < x < q) \\ 0 & (x < p, x > q) \end{cases}$	$\dfrac{i}{\sqrt{2\pi}} \dfrac{e^{ip(w+\alpha)} - e^{iq(w+\alpha)}}{(w+\alpha)}$
3. $\begin{cases} e^{-cx+iwx} & (x > 0) \\ 0 & (x < 0) \end{cases}$ $(c > 0)$	$\dfrac{i}{\sqrt{2\pi}(w+\alpha+ic)}$
4. e^{-px^2} $R(p) > 0$	$\dfrac{1}{\sqrt{2p}} e^{-\alpha^2/4p}$
5. $\cos px^2$	$\dfrac{1}{\sqrt{2p}} \cos \left[\dfrac{\alpha^2}{4p} - \dfrac{\pi}{4} \right]$
6. $\sin px^2$	$\dfrac{1}{\sqrt{2p}} \cos \left[\dfrac{\alpha^2}{4p} + \dfrac{\pi}{4} \right]$
7. $|x|^{-p}$ $(0 < p < 1)$	$\sqrt{\dfrac{2}{\pi}} \dfrac{\Gamma(1-p) \sin \dfrac{p\pi}{2}}{|\alpha|^{(1-p)}}$
8. $\dfrac{e^{-a|x|}}{\sqrt{|x|}}$	$\dfrac{\sqrt{\sqrt{(a^2+\alpha^2)}+a}}{\sqrt{a^2+\alpha^2}}$
9. $\dfrac{\cosh ax}{\cosh \pi x}$ $(-\pi < a < \pi)$	$\sqrt{\dfrac{2}{\pi}} \dfrac{\cos \dfrac{a}{2} \cosh \dfrac{\alpha}{2}}{\cosh \alpha + \cos a}$
10. $\dfrac{\sinh ax}{\sinh \pi x}$ $(-\pi < a < \pi)$	$\dfrac{1}{\sqrt{2\pi}} \dfrac{\sin a}{\cosh \alpha + \cos a}$
11. $\begin{cases} \dfrac{1}{\sqrt{a^2-x^2}} & (|x| < a) \\ 0 & (|x| > a) \end{cases}$	$\sqrt{\dfrac{\pi}{2}} J_0(a\alpha)$
12. $\dfrac{\sin[b\sqrt{a^2+x^2}]}{\sqrt{a^2+x^2}}$	$\begin{cases} 0 & (|\alpha| > b) \\ \sqrt{\dfrac{\pi}{2}} J_0(a\sqrt{b^2-\alpha^2}) & (|\alpha| < b) \end{cases}$

General Mathematical Tables

TABLE 12.7 Fourier Transforms (*Continued*)

	$F(x)$	$f(\alpha)$
13.	$\begin{cases} P_n(x) & (|x| < 1) \\ 0 & (|x| > 1) \end{cases}$	$\dfrac{i^n}{\sqrt{\alpha}} J_{n+1/2}(\alpha)$
14.	$\begin{cases} \dfrac{\cos[b\sqrt{a^2 - x^2}]}{\sqrt{a^2 - x^2}} & (|x| < a) \\ 0 & (|x| > a) \end{cases}$	$\sqrt{\dfrac{\pi}{2}} J_0(a\sqrt{a^2 + b^2})$
15.	$\begin{cases} \dfrac{\cosh[b\sqrt{a^2 - x^2}]}{\sqrt{a^2 - x^2}} & (|x| < a) \\ 0 & (|x| > a) \end{cases}$	$\sqrt{\dfrac{\pi}{2}} J_0(a\sqrt{a^2 - b^2})$

TABLE 12.8 Functions Among Transforms Tables Entries

Funtion	Definition	Name
$\text{Ei}(x)$	$\int_{-\infty}^{x} \dfrac{e^v}{v} dv;$ or sometimes defined as $-Ei(-x) = \int_{x}^{\infty} \dfrac{e^{-v}}{v} dv$	Exponential integral function
$\text{Si}(x)$	$\int_0^x \dfrac{\sin v}{v} dv$	Sine integral function
$\text{Ci}(x)$	$\int_\infty^x \dfrac{\cos v}{v} dv;$ or sometimes defined as negative of this integral	Cosine integral function
$\text{erf}(x)$	$\dfrac{2}{\sqrt{\pi}} \int_0^x e^{-v^2} dv$	Error function
$\text{erfc}(x)$	$1 - \text{erf}(x) = \dfrac{2}{\pi} \int_x^\infty e^{-v^2} dv$	Complementary function to error function
$L_n(x)$	$\dfrac{e^x}{n!} \dfrac{d^n}{dx^n}(x^n e^{-x}), \quad n = 0, 1, \ldots$	Laguerre polynomial of degree n

References

Daniel, J.W. and Noble, B. 1988. *Applied Linear Algebra*. Prentice Hall, Englewood Cliffs, NJ.
Davis, H.F. and Snider, A.D. 1991. *Introduction to Vector Analysis,* 6th ed. Wm. C. Brown, Dubuque, IA.
Strang, G. 1993. *Introduction to Linear Algebra*. Wellesley–Cambridge Press, Wellesley, MA.
Wylie, C.R. 1975. *Advanced Engineering Mathematics,* 4th ed. McGraw–Hill, New York.

Further Information

More advanced topics leading into the theory and applications of tensors may be found in J.G. Simmonds, *A Brief on Tensor Analysis* (1982, Springer–Verlag, New York).

13

Communications Terms: Abbreviations[1]

A	288
B	289
C	290
D	291
E	292
F	293
G	294
H	294
I	294
J	295
K	295
L	295
M	296
N	297
O	297
P	298
Q	299
R	299
S	299
T	301
U	301
V	301
W	302
X	302
Y	302
Z	302

A

A	ampere
Å	Angstrom
A-to-D converter	analog to digital converter
AALU	arithmetic and logical unit
AAMPS	advanced mobile phone system
AC	access control
AC	alternating current

[1] Courtesy of Intertec Publishing, Overland Park, KS.

ACC	automatic color correction
ADC	analog to digital converter
ADPCM	adaptive differential pulse-code modulation
ADR	automatic dialog replacement
AES	Audio Engineering Society
AFC	automatic frequency control
AFCEA	Armed Forces Communications and Electronics Association
AFP	AppleTalk filing protocol
AFRTS	Armed Forces Radio and Television Service
AFV	audio-follow-video
AGC	automatic gain control
AI	artificial intelligence
AIN	advanced intelligent network
ALAP	AppleTalk link access protocol
ALGOL	algorithmic language, algorithmic oriented language
AM	amplitude modulation
AMI	alternate mark inversion
ANI	automatic number identification
ANSI	American National Standards Institute
APD	avalanche photodiode
API	application program interface
APL	average picture level
APPC	advanced program-to-program communications
ARIS	access request information system
ARP	address resolution protocol
ASCII	American standard code for information interchange
ASI	adapter support interface
ASIC	application specific integrated circuit
ASK	amplitude-shift keying
ASR	access service request
ATE	automatic test equipment
ATM	asynchronous transfer mode
ATR	audio tape recorder
ATSC	Advanced Television Systems Committee
ATV	advanced television
AUI	attachment unit interface
AVK	audio/video kernel
AVL	automatic vehicle location
AWG	American wire gauge

B

B-link	bridge link
B8ZS	bipolar with eight zeros substitution
BBC	British Broadcasting Corporation
BCC	background color cancellation
BCC	Bellcore Client Company
BCD	binary coded decimal
BCS	background color suppression
BER	bit error rate
BER	bit error ratio
BETRS	Basic Exchange Telecommunications Radio Service

BEXR	Basic Exchange Radio Service
BF	burst flag
BG	burst gate
BIP-N	bit interleaved parity-N
BISDN	broadband integrated services digital network
BITS	building integrated timing supply
BKGD	background
BLKG	blanking
BNC	bayonet Neill–Concelman
BNZS	bipolar with N zeros substitution
BOC	Bell Operating Company
BORSCHT	battery feed, overvoltage protection, ringing, supervision, coding/decoding, hybrid, and testing
BT	British Telecom
BTSC	Broadcast Television Systems Committee
BVB	black-video-black

C

C/N	carrier-to-noise ratio
CABSC	Canadian Advanced Broadcast Systems Committee
CAD/CAM	computer-aided design/computer-aided manufacture
CAMA	centralized automatic message accounting
CARS	community antenna relay service
CATV	community antenna television
CAV	component analog video
CBD	central business district
CBU	Caribbean Broadcasting Union
CC	calling channel
CCC	clear channel capability
CCD	charge coupled device
CCIR	Comité Consultatif International de Radiocommunications (International Radio Consultative Committee)
CCITT	Comité Consultatif International Télégraphique et Téléphonique (Consultative Committee for International Telephone and Telegraph)
CCS	centum call seconds
CCU	camera control unit
CD	compact disc
CD-ROM	compact disc-read only memory
CDI	compact disc-interactive
CDMA	code division multiple access
CEPT	Conference of European Postal and Telecommunications Administrations
CEV	controlled environmental vault
CG	character generator
CGSA	cellular geographic service area
CIBER	cellular intercarrier billing exchange roamer
CIC	circuit identification code
CIE	Commission Internationale de l'Eclairageclear
CIF	common intermediate format
CIMAP/CC	circuit installation and maintenance assistance package/control center
CIMAP/SSC	circuit installation and maintenance assistance package/special service center
CLONES	common language on-line entry system

CMDS	centralized message data system
CMOS	complementary metal oxide semiconductor
CMR	common mode rejection
CMRR	common mode rejection ratio
CMRS	Cellular Mobile Radiotelephone Service
CO	central office
COMSAT	Communications Satellite Corporation
CPE	customer premise equipment
CPU	central processing unit
CRC	cyclical redundancy check
CRCC	cyclic redundancy check code
CRT	cathode ray tube
CS	composite sync
CSMA	carrier-sense multiple access
CSMA/CD	carrier sense multiple access with collision detection
CSU	channel service unit
CTIA	Cellular Telecommunications Industry Association
CUCRIT	capital utilization criteria
CVGB	cable vault ground bar
CWA	Communications Workers of America
CWCG	copper wire counterpoise ground

D

D-to-A converter	digital to analog converter
D-to-D	digital to digital transfer
D/I	drop and insert
DA	distribution amplifier
DAC	digital to analog converter
DARPA	Defense Advanced Research Projects Agency
DAS	data acquisition system
dB	decibel
dBi	decibels relative to an isotropic antenna
dBk	decibels relative to 1 kilowatt
dBm	decibels relative to 1 milliwatt
dBmv	decibels relative to 1 millivolt
dBrn	decibels above reference noise
DBS	direct-broadcast satellite
dBV	decibels relative to 1 volt
dBW	decibels relative to 1 watt
DC	direct current
DCE	data communications equipment
DCPSK	differentially coherent phase-shift keying
DCT	discrete cosine transform
DDD	direct distance dialing
DDS	digital data system
DES	data encryption standard
DID	direct inward dialing
DILEP	Digital Line Engineering Program
DIP	dual in-line package
DIR/ECT	directory project
DLC	data-link control

DMA	direct memory access
dmW	digital milliwatt
DNS	domain name service
DOD	direct outward dialing
DOMSAT	domestic satellite
DOS	disk operating system
DPCM	differential pulse-code modulation
DPSK	differential phase-shift keying
DRAM	dynamic random access memory
DSB	double sideband (AM)
DSE	data switching exchange
DSK	downstream keyer
DSSC	double-sideband suppressed carrier
DSX	digital signal cross connect
DTE	data terminal equipment
DTL	diode-transistor logic
DTMF	dual tone multifrequency
DVE	digital video effects
DVI	digital video interactive
DVTR	digital videotape recorder
DWG	drilled well ground

E

E-link	extension link
e-mail	electronic mail
EADAS	Engineering and Administrative Data Acquisition System
EAROM	electrically alterable read-only memory
EBCDIC	extended binary-coded decimal interchange code
EBU	European Broadcasting Union
EC	European Community
ECC	error correcting code
ECCS	economic CCS
ECL	emitter-coupled logic
ECS	European communication satellite
EDFA	erbium doped fiber amplifier
EDI	electronic data interchange
EDL	edit decision list
EDTV	extended definition television
EEPROM	electrically erasable programmable read-only memory
EFM	eight to fourteen modulation
EFP	electronic field production
efx	effects
EHF	extremely high frequency
EIA	Electronic Industries Association
EIRP	effective isotropic radiated power
EISA	expanded industry standard architecture
ELF	extremely low frequency
EME	economic modular evaluation
EMF	electromotive force
EMI	electromagnetic interference
EMP	electromagnetic pulse

ENG	electronic news gathering
EPLD	erasable programmable logic device
EPROM	erasable programmable read-only memory
EQ	equalization
ERP	effective radiated power
ESA	European Space Agency
ESD	electrostatic discharge
ESDI	enhanced small device interface
ESI	equivalent step index
ESN	electronic service number
ESPRIT	European Strategic Program for Research in Information Technology
ETM	eight to ten modulation
ETSI	European Telecommunications Standards Institute

F

F-link	fully associated link
FACTS	fully automated collect and third-number service
FCC	Federal Communications Commission
FCD	frame continuity date
FD	frequency distance
FDDI	fiber distributed data interface
FDM	frequency division multiplexing
FDMA	frequency division multiple access
FDR	frequency dependent rejection
FDRL	filed data of regions and LECS
FEC	forward error correction
FEO	foreign exchange office
FET	field effect transistor
FFSK	fast frequency shift keying
FID	field identifier
FIFO	first-in-first-out
FIR filter	finite impulse response filter
FIT	failure in time
FIVE	format independent visual exchange
FM	frequency modulation
FNPA	foreign numbering plan area
FO	fiber optics
FOMS/FUSA	frame operations management system/frame user switch access system
FPGA	field programmable gate array
FPLA	field programmable logic array
FPLF	field programmable logic family
FPLS	field programmable logic sequence
fs	femtosecond
FSK	frequency shift keying
FSS	fixed satellite service
FTA	fault tree analysis
FTAM	file transfer, access, and management
FTB	fade-to-black
fx	effects
FX	foreign exchange
FXS	foreign exchange station

G

G/T	gain-over-noise temperature
GaAs	gallium arsenide
GADS	generic advisory diagnostic system
GBR	green, blue, red
GHz	gigahertz
GIGO	garbage-in-garbage-out
GMT	Greenwich mean time
GND	ground
GOS	grade of service
GOSIP	government open systems interconnection profile
GPI	general purpose interface
GSM	Global System for Mobile Communications

H

H	horizontal
HCS fiber	hard clad silica fiber
HDTV	high definition television
HF	high frequency
HLL	high level language
HVAC	heating, ventilation, and air conditioning
Hz	hertz

I

I/O	input/output
IACC	interaural cross correlation
IBA	Independent Broadcasting Authority
IBG	interblock gap
IC	integrated circuit
IC	interexchange carrier
IDTV	improved definition television
IEC	International Electrotechnical Commission
IEE	Institution of Electrical Engineers
IEEE	Institute of Electrical and Electronics Engineers
IFRB	International Frequency Registration Board
IIR	infinite impulse response
ILD	injection laser diode
ILF	infra low frequency
IM	intensity modulation
IMD	intermodulation distortion
IMTS	improved mobile telephone service
IN	intelligent network
INA	information networking architecture
INA	integrated network access
INFORMS	integrated forecasting management system
Inmarsat	International Maritime Satellite Organization
INREFS	integrated reference system
INTELSAT	International Telecommunications Satellite Consortium
INWATS	inward WATS

IOC	integrated optical circuit
IP	Internet protocol
IPC	interprocess communications
IPX	internetwork packet exchange
IPX/SPX	internetwork packet exchange/sequenced packet exchange
IR	infrared
IRQ	interrupt request
ISA	industry standard architecture
ISD	international subscriber dialing
ISDN	Integrated Services Digital Network
ISO	International Standards Organization
ITC	Independent Television Commission
ITDG	initial-time-delay gap
ITS	International Teleproduction Society
ITU	International Telecommunication Union

J

JF	junction frequency
JPEG	Joint Photographic Experts Group

K

K	Kelvin
k/s	kilobits per second
kHz	kilohertz
kV	kilovolt

L

LAN	local area network
LAP	link access protocol
LAPB	link access protocol—balanced
LAPD	link access protocol on D channel
LAT	local area transport
LATA	local access and transport area
LCD	liquid crystal display
LCRIS	loop cable record inventory system
LEC	light energy converter
LEC	local exchange carrier (company)
LED	light emitting diode
LF	low frequency
LFACS	loop facilities assignment and control system
LIFO	last-in-first-out
LLC	logical link control
LMOS	loop maintenance operations system
LNA	launch numerical aperture
LNA	low noise amplifier
LOMS	loop assignment center operations management system
LOS	line of sight
LP	linearly polarized
LPC	linear predictive coding

LPIE	loop plant improvement evaluator
LPTV	low power television
LRC	longitudinal redundancy check
LSB	least significant bit
LSB	lower sideband
LSI	large-scale integration
LSL	link support layer
LTC	longitudinal time code
LU	logical unit
LU 6.2	logical unit 6.2
LUM	luminance

M

M/E	mix/effects
mA	milliamperes
MAC	multiplexed analog components
MACS	major apparatus and cable system
MAP	manufacturing automation protocol
master SPG	master reference synchronizing pulse generator
MATV	master antenna television
MAU	media access unit
MAU	multistation access unit
MAVEN	mapping and access for valid equipment nomenclature
mb/s	megabits per second
MCA	media control architecture
MCA	microchannel architecture
MCI	media control interface
MDS	multipoint distribution system
MESFET	metal semiconductor field effect transistor
MF	medium frequency
MFJ	modification of final judgment
MFSK	multiple frequency shift keying
MHz	megahertz
MIB	management information base
MICR	magnetic ink character recognition
MIDI	musical instrument digital interface
MIPS	millions of instructions per second
MITI	Ministry of International Trade and Industry
MLID	multiple link interface driver
MNOS	metal, nitride, oxide semiconductor
MOPS	millions of operations per second
MOS	metal-oxide semiconductor
MPCD	minimum perceptible color difference
MPEG	Motion Picture Experts Group
MPT	Ministry of Posts and Telecommunications
ms	millisecond
MSA	metropolitan statistical area
MSB	most significant bit
MSU	medium-scale integration
MTBF	mean time between failures

MTSO	mobile telephone switching office
MTTF	mean time to failure
MTTR	mean time to repair
MUF	maximum usable frequency
MUSA	multiple unit steerable antenna
MUX	multiplex
mV	millivolt
MW	medium wave
mW	milliwatt

N

NAM	negative nonadditive mix
NAM	numeric assignment module
NANP	North American Numbering Plan
NARUC	National Association of Regulatory Utility Commissioners
NASC	Number Administration and Service Center
NBP	name binding protocol
NCP	network control point
NCS	National Communications System
NCTA	National Telephone Cooperative Association
NEBS	network equipment building system
NEC	National Electrical Code
NECA	National Exchange Carriers Association
NEP	noise equivalent power
NF	noise factor
NHK	Nippon Hoso Kyokai
NIC	network interface card
NICAM	near instantaneously companded audio multiplex
NIST	National Institute of Standards and Technology
NLM	network loadable module
NOS	network operating system
NPA	number plan area
NPR	noise power ratio
NRZ	nonreturn-to-zero
ns	nanosecond
NSDB	network and services database
NSEP	National Security Emergency Preparedness
NSIS	network server interface specification
NT1	network termination 1
NTC/C	network configuration
NTIA	National Telecommunications and Information Administration
NTL	National Transcommunications Limited
NTSC	National Television System Committee
NTT	Nippon Telegraph and Telephone

O

OCR	optical character recognition
ODI	open data-link interface
OEM	original equipment manufacturer
ONI	operator number identification

OPASTCO	Organization for the Protection and Advancement of Small Telephone Companies
OPS/INE	operations process system/intelligent network elements
OSI	open systems interconnection
OTDR	optical time domain reflectometer
OTF	optimum traffic frequency
OXO	ovenized crystal oscillator

P

p-p	peak-to-peak
PAD	packet assembler/disassembler
PAL	phase alternate each line
PAL	programmable array logic
PAM	pulse amplitude modulation
PAP	printer access protocol
PBX	private branch exchange
PC board	printed circuit board
PCF	physical control fields
PCM	pulse code modulation
PCN	personal communications network
PCS	personal communications services
PCSA	personal computing system architecture
PDM	pulse duration modulation
PDN	public data network
PDP	plasma display panel
PERT	program evaluation and review technique
PFM	pulse frequency modulation
PGM	program
PICS/DCPR	plug-in inventory control system/detailed continuing property record
PIN	personal identification number
PIN	positive-intrinsic-negative
PLD	programmable logic device
PLL	phase locked loop
PLV	production level video
PM	phase modulation
PM	pulse modulation
PMR	public mobile radio
POSIX	portable operating system interface
POTS	plain old telephone service
PPM	pulse position modulation
PPSN	public packet switched network
PPSS	public packet switched service
PRBS	pseudorandom bit stream
programmable GPI	programmable general purpose interface
PROM	programmable read-only memory
PSDS	public switched digital service
PSK	phase-shift keying
PSTN	public switched telephone network
PTM	pulse time modulation
PTT	post, telephone, and telegraph
PTT	push to talk
PUC	public utilities commission

pulse DA	pulse distribution amplifier
pulse delay DA	pulse delay distribution amplifier
PVC	polyvinylchloride

Q

Q	quality factor
QA	quality assurance
QBE	query by example
QC	quality control
QPSK	quadrature phase-shift keying
QUIL	quad-in-line

R

R-Y	red minus luminance
RAA	rural allocation area
RACE	Research in Advanced Communications in Europe
RAM	random access memory
RAR	read after read
RARP	reverse address resolution protocol
RC	resistor–capacitor
RCC	radio common carrier
REA	Rural Electrification Administration
RF	radio frequency
RFI	radio frequency interference
RGB	red, green, blue
RIFF	resource interchange file format
RISC	reduced instruction set computer
RIT	rate of information transfer
RMAS	Remote Memory Administration System
RMS	root means square
ROM	read-only memory
RPC	remote procedure call
RPG	report program generator
RSA	rural service area
RSC	Reed Solomon code
RSL	received signal level
RTC	real time clock
RTL	resistor-transistor logic
RU	rack unit
RZ	return to zero

S

(S+N)/N	signal-plus-noise to noise ratio
S/N	signal-to-noise ratio
SAP	secondary audio program
SAP	service access point
SAT	supervisory audio tone
SAW	surface acoustic wave
SC	subcarrier

SC/H phase	subcarrier to horizontal phase
SCCP	signaling connection control part
SCP	service control point
SCP/800	service control point/800
SCPC	single-channel-per-carrier
SCSI	small computer systems interface
SDLC	synchronous data link control
SECAM	sequential couleur avec memoire
SEF	source explicit forwarding
SGML	Standard Generalized Markup Language
SHF	super high frequency
SID	sudden ionospheric disturbance
SID	system identification
SINAD	signal-to-noise and distortion
SMB	server message block
SMDS	switched multimegabit data service
SMPTE	Society of Motion Picture and Television Engineers
SMRS	Specialized Mobile Radio Service
SMSA	standard metropolitan statistical area
SMT	surface mount technology
SMTP	simple mail transfer protocol
SNA	systems network architecture
SNET	Southern New England Telecommunications Corporation
SNMP	simple network management protocol
SNR	signal-to-noise ratio
SOAC	service order analysis and control
SOH	start-of-heading
SONET	synchronous optical network
SOS	silicon on sapphire
SPC	stored-program control
SPDT	single-pole double-throw
SPG	sync pulse generator
SPL	sound pressure level
SPP	sequenced packet protocol
SPST	single-pole single-throw
SQL	structured query language
SS6	signaling system 6
SS7	signaling system 7
SSB	single sideband
SSBSC	single-sideband suppressed carrier
SSC	special services center
SSP	service switching point
SST	single-sideband transmission
STA	spanning tree algorithm
STL	studio-transmitter link
STP	signaling transfer point
STSL	synchronous transport signal level
STX	start-of-text
SW	short wave
SWR	standing wave ratio

T

TA	terminal adapter
TAP	test access port
TASI	time-assignment speech interpolation
TBC	time base corrector
TCP	transmission control protocol
TCP/IP	transmission control protocol/internet protocol
TCXO	temperature compensated crystal oscillator
TDD	telecommunications device for the deaf
TDM	time division multiplexing
TDMA	time division multiple access
TE	transverse electric
TEM	transverse electromagnetic
THD	total harmonic distortion
THF	tremendously high frequency
TIE	terminal interface equipment
TM	transverse magnetic
TMDA	time-division multiple access
TND	telephone network for the deaf
TNDS/TK	total network data system/trunking
TOP	technical and office protocols
TRF	tuned radio frequency
TTL	transistor-transistor logic
TVI	television interference
TVRO	television receive-only

U

UART	universal asynchronous receiver/transmitter
UHF	ultrahigh frequency
UPS	uninterruptible power supply
USART	universal synchronous/asynchronous receiver/transmitter
USOA	uniform system of accounts
USOAR	uniform system of accounts rewrite
USOC	uniform service order code
USTA	United States Telephone Association
USTSA	United States Telephone Suppliers Association
UTC	coordinated universal time

V

VA	volt-amperes
VCR	video cassette recorder
VCXO	voltage controlled crystal oscillator
VDRV	variable data rate video
VDT	video display terminal
VFD	vacuum fluorescent display
VFO	variable frequency oscillator
VHF	very high frequency
VIA-D	voice interface access-disabled
VIR	vertical interval reference

VITC	vertical interval time code
VITS	vertical interval test signal
VLF	very low frequency
VLSI	very large-scale integration
VOM	volt-ohm-milliammeter
vox	voice-operated relay
VSAT	very small aperture terminal
VSWR	voltage standing wave ratio
VT	virtual tributary
VTR	videotape recorder
VU	volume unit

W

WAN	wide area network
WARC	World Administrative Radio Conference
WATS	wide area telecommunications service
WDM	wavelength division multiplexing
WF monitor	waveform monitor
WORD	work order record and details
WORM	write once read many
WYSIWYG	what you see is what you get

X

XFMR	transformer
XMTR	transmitter
XOR	exclusive OR
XTALK	crosstalk

Y

YAG	yttrium-aluminum garnet
YIG	yttrium-iron garnet

Z

Z	impedance

Index

AC signal, 4
accepted interference, 203
active bits, 16
active satellite, 203
active sensor, 203
addition and cancellation of distortion components, 10
additional filters, 9
additional relations with derivatives, 261
aeronautical
    earth station, 203
    fixed service, 203
    fixed station, 203
    mobile off-route service, 203
    mobile route service, 203
    mobile-satellite off-route service, 203
    mobile-satellite route service, 203
    mobile-satellite service, 203
    mobile service, 203
    radionavigation-satellite service, 203
    radionavigation service, 203
    station, 203
aircraft earth station, 203
aircraft station, 203
algebra, 242
algebra of matrices, 273
algebraic equations, 246
aliasing, 73, 79
allocation of a frequency band, 203
allotment of a radio frequency, 204
alloy, 116
alphabet, greek, 216
alternative trigonometric form, 74
altitude of the apogee or perigee, 204
amateur-satellite service, 204
amateur service, 204
amateur station, 204
amplitude modulation effects of linear distortion, 49
amplitude response, 87
analysis
    audio frequency distortion mechanisms and, 3
    computer-based signal, 82
    digital audio, 18
    FFT, 14
    Fourier waveform, 70, 79
    functions, statistical data, 84, 88
    mathematical preliminaries for Fourier, 71
    multitone, 11
    radio frequency distortion mechanisms and, 38
    signal, 65
    signal generation and, 82
    tools, 18
    using the DFT/FFT in Fourier, 77
    video display distortion mechanisms and, 19
analyzer
    distortion, 8, 9
    logic, 57
    notch filter, 8
    protocol, 62
    signature, 60
    spectrum, 8, 9
    state, 57
    timing, 57
angle of intersection of two curves, 258
aperiodic, 71
aperiodic waveform, 79
application considerations, 34
application of the zone plate signal, 27
applications for computer-controlled testing, 63
applications of the definite integral, 262
approximation, Stirling's, 243
area, 263
area rectangular coordinates, 262
arithmetic progression, 244
assessment of color reproduction, 24
assigned frequency, 204
assigned frequency band, 204
assignment of a radio frequency, 204
atomic structure, 92
audio distortion mechanisms, 18
audio frequency distortion mechanisms and analysis, 3
audio measurements, 3
audio signal, 6
audio system, 7
automated test instruments, 62
autoranging, 65
average-response measurements, 4

backdriving, 62
band properties of semiconductors, 104
bandlimited, 73, 79
bandpass filter, 40
bar edge measurement, 35

base earth station, 204
base station, 204
bed of nails, 62, 68
Bernoulli numbers, 251
Bessel functions, 242, 266
Bessel functions of the second kind, 267
binomial, 253
binomial theorem, 243
bravais lattice, 116
broadcasting-satellite service, 204
broadcasting service, 204
broadcasting station, 204
bus state triggering, 58
bus state triggering mode, 68

capture effect, 55
carrier power of a radio transmitter, 204
cast permanent magnetic alloys, 108
categories of services, 118
cathode ray tube guns, 23
CCITT test, 10
centroid, 265
CGPM, 214
change of base $a = /1$, 243
characteristic frequency, 204
chemical composition, 92
chemical properties of engineering materials, 95
chromatic adaptation and white balance, 24
chromaticity, 23
chrominance, 20
class of emission, 204
classical mechanics, 217
clipping, 15, 17
coast earth station, 204
coast station, 204
code violations, 15
coherent demodulation of AM signal, 45
color differences, 25
color displays, 23
color-matching functions, 23
color reproduction
    assessment, 24
    colorimetric, 24, 25
    corresponding, 24, 36
    equivalent, 24
    exact, 24
    preferred, 24
    spectral, 24, 36
colorimetric color reproduction, 24, 25
command window, 82, 87
common applications of the definite integral, 262
common geometric figures, 246
communications terms abbreviations, 288
community reception, 204
complex exponential form, 74, 76
complex numbers, 244
complex patterns, 30
complex signal, 7
computer-based signal analysis, 82
computer-instrument interface, 63

cone, 272
connectivity libraries, 62
considerations for digital and audio systems, 15
constants, conversion, 238
constants involving e, 217
continuous-time, 85
    aperiodic functions, 75
    fixed linear systems, 87
    periodic functions, 74
    waveform, 70, 79
contrast ratio, 26
contrast ratio divisions, 26
conversion constants and multipliers, 238
conversion factors, 222
    english to metric, 239
    general, 239
    metric to english, 238
conversion of temperatures, 240
coordinate systems, 279
coordinated universal time, 205
coordination area, 205
coordination contour, 205
coordination distance, 205
corresponding color reproduction, 24, 36
Costas receiver, 47
crest factor, 17
crosstalk, 40, 55
CRT, 23
CRT measurement techniques, 32
crystalline, 116
crystalline solids, 92
cubic, 246
curl, 280
cursor measurement, 65
curve fitting, 83
cylindrical and spherical coordinates, 263

data window, 58
DC offset, 16
decibel measurements, 6
decimal multiples and submultiples, 238
deep space, 205
definite integral, 262
definitions of SI base units, 214
demodulation of angle modulated waves, 49
density, 116
determinants, 277
device under test, 11
DFTs, 70
dielectric constants of ceramics, 102
dielectric constants of glasses, 102
dielectric constants of solids, 101
differential calculus, 242, 256
digital and audio systems, 15
digital audio analysis, 18
digital oscilloscope, 63
digital storage oscilloscope features, 65
digital test equipment and measurement systems, 57
direct reception, 204
direct sequence systems, 205

Index

Dirichlet conditions, 71
disassemblers, 59
discrete
    Fourier transforms, 70
    frequency, 34
    frequency method, 35
    time aperiodic functions, 76
    time fixed linear systems, 87
    time linear systems, 85
    time periodic functions, 76
    time waveform, 70, 79
    tone testing, 13
    tones, 11
dispersion, 56
display resolution and pixel format, 26
distortion
    amplitude modulation effects of linear, 49
    analyzer, 8, 9
    audio frequency, 3
    components, 10
    due to time-variant multipath channels, 41
    harmonic, 8
    in angle modulation, phase, 52
    intermodulation, 9
    linear, 39
    linear and nonlinear, 49
    measurements, 3, 15
    measures, total harmonic, 78
    mechanism, interference as a radio frequency, 53
    mechanisms and analysis, 3
    mechanisms and analysis, radio frequency, 38
    mechanisms, nonlinear, 8
    nonlinear, 40
    total harmonic, 3
    types of, 39
    video display, 19
divergence, 280
dot pitch, 26, 36
double integration, 264
double-sideband suppressed-carrier (DSB-SC) demodulation errors, 46
DSO, 65
DSP-based technology, 14
duplex operation, 205
DUT, 11

earth exploration-satellite service, 205
earth station, 205
edge triggering, 58
edge triggering mode, 68
effective radiated power in a given direction, 205
effects of amplitude nonlinearities on angle modulated waves, 51
EHF, 42
eigenvalues, 277
eigenvectors, 277
electrical and optical properties of engineering materials, 95
electrical resistivity of pure metals, 96
electrical resistivity of selected alloys, 99
electrical resistivity of selected materials, 96
electricity and magnetism, 218
electromagnetic radiation, 219
electromotive force series, 95, 116
electronic test patterns, 28
electronics workbench, 87
elementary algebra and geometry, 242
elements of the differential calculus, 242
ellipsoid, 271
elliptic cone, 272
elliptic cylinder, 271
elliptic paraboloid, 270
emergency position-indicating radiobeacon station, 205
emission, 205
emulative tester, 61
energy spectral density, 76
engineering strain, 93, 116
engineering stress, 93, 116
enhanced triggering modes, 67
envelope mode, 64
enveloping, 68
equivalent color reproduction, 24
equivalent isotropically radiated power, 205
equivalent monopole radiated power, 205
equivalent satellite link noise temperature, 205
equivalent-time sampling, 64, 68
error function, 253
Euler numbers, 251
even function, 71, 79
exact color reproduction, 24
example applications of Fourier waveform techniques, 77
experimental station, 205
exponential series, 255
exponents, 242
extra high frequency, 42
extremes, 68

facsimile, 205
factor, Kell, 34
factorials, 243
factors and expansion, 244
factors, conversion, 222
factors, temperature, 239
factors that affect the appearance of the image, 25
fast Fourier transform, 70, 79
FDM, 40
feeder link, 206
FFT, 70, 79
FFT analysis, 14
finite cosine transforms, 285
finite sine transforms, 284
fixed-satellite service, 206
fixed service, 206
fixed station, 206
Fourier
    analysis, 71
    cosine transforms, 286
    methods, 70
    series, 8, 21, 74
    series for continuous-time periodic functions, 74
    series for discrete-time periodic functions, 76

sine transforms, 285
transform for continuous-time aperiodic functions, 75
transform for discrete-time aperiodic functions, 76
transform method, 35
transforms, 34, 242, 286
transforms overview, 282
waveform analysis, 70, 79
waveform techniques, 77
Fourier's law, 95, 116
fractional exponents, 242
frequency
   allocations, U.S. table of, 118
   allotment of a radio, 204
   and time signal-satellite service, standard, 212
   and time signal service, standard, 212
   and time signal station, standard, 212
   assigned, 204
   assignment of a radio, 204
   band, allocation of a, 203
   band, assigned, 204
   bands and assignments, 118
   channel, radio, 204
   characteristic, 204
   discrete, 34, 35
   distortion mechanism, interference as a radio, 53
   distortion mechanisms and analysis, radio, 38
   distortion mechanisms and analysis, audio, 3
   division multiplexing, 40
   energy, radio, 207
   extra high, 42
   fundamental, 71, 79
   harmonic, 80
   hopping systems, 206
   line-harmonic, 22
   maximum camera, 21
   maximum modulating, 21
   maximum video, 21
   minimum video, 20
   Nyquist, 73, 80
   reference, 211
   sample, 78
   shift telegraphy, 206
   sweep signal, 27
   tolerance, 206
   very low, 42
full carrier single-sideband emission, 206
function, 87
   among transforms, 287
   Bessel, 242, 266
   color-matching, 23
   continuous-time aperiodic, 75
   continuous-time periodic, 74
   discrete-time aperiodic, 76
   discrete-time periodic, 76
   error, 253
   even, 71, 79
   Fourier series for continuous-time periodic, 74
   Fourier series for discrete-time periodic, 76
   Fourier transform for continuous-time aperiodic, 75
   Fourier transform for discrete-time aperiodic, 76
   hyperbolic, 266

   inverse trigonometric, 250
   mean value of a, 263
   modulation transfer, 32
   Neumann, 267
   odd, 80
   of two variables, 260
   polyfit, 83
   series of, 251
   sine, 29
   single-valued, 71
   special, 266
   statistical data analysis, 84, 88
   transfer, 85, 88
   trigonometric, 70, 71
   vector, 279
   vertical dimension sine, 29
   Weber, 267
functional, 62
functions among transforms, 287
fundamental frequency, 71, 79
fundamental properties, 242

gain of an antenna, 206
gamma requirements, 25
Gaussian elimination, 275
Gaussian variates, 85, 87
general conference on weights and measures, 214
general mathematical tables, 242
general purpose interface bus, 63
general purpose mobile service, 206
geometric figures, 246
geometric progression, 244
geometry, 242, 246
geostationary satellite, 206
geostationary satellite orbit, 206
geosynchronous satellite, 206
GFI, 61
Gibbs phenomenon, 71, 79
glitch, 58, 68
glitch capture mode, 64
GPIB, 63
gradient, 280
greek alphabet, 216
ground reflected waves, 43
ground waves, 43
group delay, 7, 17
guided-fault isolation, 61

half-power width, 34
half-power width method, 34
half-wave symmetry, 71, 80
harmful interference, 206
harmonic distortion, 8
harmonic frequency, 80
Hermite polynomials, 269
hertzian waves, 211
Hewlett Packard interface bus, 63
high-pass filter, 40

Index

high-permeability magnetic alloys, 107
high silicon transformer steels, 110
highest bar reading, 15
highest true peak, 15
highlight luminance, 34
Hilbert transform, 47
histogram, 87
horizontal resolution, 21
HPIB, 63
hybrid spread spectrum systems, 206
hyperbolic cylinder, 271
hyperbolic functions, 266
hyperbolic paraboloid, 270
hyperboloid of one sheet, 272
hyperboloid of two sheets, 272

ILS, 207
IM, 9
IM measurements, 9
impulse Fourier transform, 34
in-circuit, 62
inclination of an orbit, 207
indefinite integral, 262
indeterminant forms, 259
index of refraction, 96, 116
individual reception, 207
instrument landing system, 207
instrument landing system glide path, 207
instrument landing system localizer, 207
integral calculus, 242, 262
integral theorems, 282
integrated circuits, 60
integration, 281
interference, 207
    accepted, 203
    as a radio frequency distortion mechanism, 53
    harmful, 206
    in amplitude modulation, 53
    in angle modulation, 55
    in DSB-SC AM, 54
    in SSB-SC, 54
    permissible, 203, 209
intermodulation distortion, 9
international standards and constants, 214
international system of units, 214
international telecommunication union, 118
interpolation algorithms, 31
intersatellite service, 207
intersection of two curves, 258
introduction to mathematics, 242
invalid samples, 15
inverse assemblers, 59
inverse trigonometric functions, 250
ionosphere, 41, 56
ionospheric scatter, 207
irrational exponents, 243
ITU, 118

Kell factor, 34
knife edge Fourier transform, 34

Laguerre polynomials, 269
land earth station, 207
land mobile earth station, 207
land mobile-satellite service, 207
land mobile service, 207
land mobile station, 207
land station, 207
large metamerism, 23
leakage, 14
learning mode, 61
least-mean square-error fit, 83
least-square-error, 71
Least-square-error fit, 88
left-hand (or anticlockwise) polarized wave, 207
Legendre polynomials, 267
length of arc, 262
letter designations of microwave bands, 221
level measurements, 4
level triggering, 58
level triggering mode, 68
line A, 207
line B, 207
line C, 208
line D, 208
line-harmonic frequency, 22
line integrals, 281
line-of-sight waves, 43
line width, 32
line width method, 33
linear algebra matrices, 273
linear and nonlinear distortion, 49
linear coefficient of thermal expansion, 116
linear distortion, 39
logarithmic series, 256
logarithms, 243
logic analyzer, 57
lower-sideband, 47
LSB, 47
luminance, 20, 23, 33, 34

M-file, 88
m-file, 82
MacLaurin series, 255
magnetic properties of transformer steels, 110
magnitude, 74
manual probe diagnosis, 60
maritime mobile-satellite service, 208
maritime mobile service, 208
maritime radionavigation-satellite service, 208
maritime radionavigation service, 208
marker beacon, 208
materials, properties of, 92
materials science and engineering, 92
Mathcad, 88
Mathematica, 88

mathematical preliminaries for Fourier analysis, 71
mathematical tables, 242
mathematics, introduction to, 242
MATLAB, 82, 88
maxima and minima, 258
maximum camera frequency, 21
maximum modulating frequency, 21
maximum video frequency, 21
mean power of a radio transmitter, 208
mean value of a function, 263
measurements
    audio, 3
    average-response, 4
    bar edge, 35
    CRT, 32
    cursor, 65
    decibel, 6
    distortion, 3
    IM, 9
    level, 4
    line-width, 33
    noise, 6
    noise and distortion, 15
    objective CRT, 34
    of color displays, 23
    peak-response, 5
    phase, 7
    root-mean-square, 4
    shrinking raster, 33
    signal, 3
    spectroradiometric, 23
    subjective CRT, 32
    systems, digital test equipment, 57
mechanical properties of engineering materials, 93
melting point, 116
meteorological aids service, 208
meteorological-satellite service, 208
method
    CCITT, 10
    discrete frequency, 35
    Fourier transform, 35
    Fourier's, 70
    half-power width, 34
    line width, 33
    Newton's, 260
    numerical, 260
    of measuring distortion, 8
    shrinking raster, 33
    SMPTE, 9
    to measure intermodulation, 9
    TV limiting resolution, 33
microphotometer, 35
microscopic scale structure, 92
microwave bands, 221
minimum video frequency, 20
mobile earth station, 208
mobile-satellite
    off-route (OR) service, aeronautical, 203
    route (R) service, aeronautical, 203
    service, 208
    service, aeronautical, 203
    service, land, 207
    service, maritime, 208
mobile service, 208
mobile station, 208
mode
    bus state triggering, 68
    edge triggering, 68
    enhanced triggering, 67
    envelope, 64
    glitch capture, 64
    learning, 61
    level triggering, 68
    peak accumulation, 64, 68
    triggering, 66
modulation transfer function, 32
modulus of elasticity, 116
motion detection, 31
MTF, 32, 35
multiburst signal, 27
multidimensional sweep, 30
multipliers, 238, 275
multisatellite link, 208
multitone
    analysis advantages, 11
    audio testing, 11
    fundamentals, 15
    original, 11
    signals, 11
    testing, 13, 17
    vs. discrete tones, 12

names and symbols for the SI base units, 215
National Television Systems Committee, 21
necessary bandwidth, 208
Neumann functions, 267
Newton's method, 260
noise, 9
noise measurement, 6, 15
nonlinear distortion, 40
nonlinear distortion mechanisms, 8
notch filter analyzer, 8
NTSC, 21
nullity, 276
number of clips, 15
number of mutes, 15
numerical constants, 217
numerical methods, 260
Nyquist frequency, 73, 80
Nyquist rate, 73, 80
Nyquist sampling theorem, 73

objective CRT measurements, 34
occupied bandwidth, 208
odd function, 80
odd harmonics, 71
Ohm's law, 96, 116
onboard communication station, 208
operational considerations, 14

operations with zero, 243
optical horizon, 43
orbit, 209
original multitone, 11
orthogonality, 270
orthogonality and length, 276
out-of-band emission, 209
overall gamma requirements, 25

parametric equations, 280
parity errors, 15
partial derivatives, 261
passive sensor, 209
peak accumulation mode, 64, 68
peak envelope power of a radio transmitter, 209
peak-equivalent sine, 6, 17
peak luminance, 34
peak-response measurements, 5
perception of color differences, 25
period of a satellite, 209
periodic, 71
periodic waveform, 80
permeability of high purity iron, 110
permissible interference, 203, 209
permitted services, 119
permutation matrix, 275
permutations, 245
personality modules, 61
phase distortion in angle modulation, 52
phase measurement, 7
phase response, 7, 88
phase spectra, 74
photoelectric colorimeter, 23
physical constants, 216
physical factors, 25
physical properties of engineering materials, 93
π constants, 217
pilot carrier, 47, 56
pixel, 36
  definition, 26
  density, 26, 36
  format, 26, 36
  size, 26
points of inflection of a curve, 259
polar coordinates, 263
polar form, 245
polyfit, 83
polyfit function, 83
polynomial fit, 88
polyval, 88
port operations service, 209
port station, 209
power, 209
power spectral density, 75
PPM behavior, 15
PRBS, 60
preferred color reproduction, 24
primary radar, 209
primary services, 119
printed wiring board, 57

producing the zone plate signal, 29
progression, 244
properties, 262
properties of antiferromagnetic compounds, 109
properties of magnetic alloys, 106, 110
properties of materials, 92
properties of semiconductors, 103
property, 116
protection ratio, 209
protocol analyzer, 62
pseudorandom binary sequence, 60
pseudorandom sequence, 209
PSpice, 88
psychophysical factors, 25
public correspondence, 209
pulsed FM systems, 209
PWB, 57

QR factorization, 276
quadratic, 246
quarter-wave symmetry, 80

RACON, 210
radar, 209
radar beacon, 210
radiation, 210
radio, 210
  altimeter, 210
  astronomy, 210
  astronomy service, 210
  astronomy station, 210
  direction-finding, 210
  direction-finding station, 210
  frequency, allotment of a, 204
  frequency, assignment of a, 204
  frequency channel, 204
  frequency distortion mechanisms and analysis, 38
  frequency energy, 207
  horizon, 43, 56
  regulations, 118, 206
  spectrum, 118
  transmitter, carrier power of a, 204
  transmitter, mean power of a, 208
  transmitter, peak envelope power of a, 209
  waves, 211
radiobeacon station, 210
radiocommunication, 210
radiocommunication service, 210
radiodetermination, 210
radiodetermination-satellite service, 210
radiodetermination service, 210
radiodetermination station, 210
radiolocation, 210
radiolocation land station, 210
radiolocation mobile station, 210
radiolocation service, 210
radionavigation, 210
radionavigation land station, 210

radionavigation mobile station, 210
radionavigation-satellite service, 210
radionavigation service, 210
radiosonde, 210
radiotelegram, 210
radiotelemetry, 210
radiotelephone call, 210
radiotelex call, 211
radius of curvature, 258
rank and nullity, 276
real numbers, 242
recommended decimal multiples and submultiples, 238
rectangular coordinates, 262
reduced carrier single-sideband emission, 211
reference frequency, 211
reference memory, 66
reflecting satellite, 211
refractive index of selected polymers, 96
relative maxima and minima, 258
resistance of wires, 113
resistivity, 116
resistivity of selected ceramics, 101
resistivity of semiconducting minerals, 105
resolution, 26
resolution, horizontal, 21
resolution parameter, 26
resolution signals, 34
resolution, spurious, 33
reversion of series, 254
right-hand polarized wave, 211
rms
    RMS, 4
roof/floor, 68
root-mean-square measurements, 4

saddle, 270
safety service, 211
sample frequencies, 78
sample rate, 16
sampling rate, 73, 80
satellite, 211
satellite, active, 203
satellite link, 211
satellite network, 211
satellite system, 211
saturation constants for magnetic substances, 110
scalar, 275
scanning, 21
secondary radar, 211
secondary services, 119
semi-duplex operation, 211
semiconductors, 103
sensor, active, 203
sequence term, 59
series, 251
    Bernoulli, 251
    electromotive force, 95, 116
    Euler numbers, 251
    expansion, 253
    exponential, 255

    Fourier, 8, 21
    logarithmic, 256
    MacLaurin, 255
    of functions, 251
    reversion of, 254
    table of, 242
    Taylor, 254
    trigonometric, 256
services
    aeronautical fixed, 203
    aeronautical mobile, 203
    aeronautical mobile off-route (OR), 203
    aeronautical mobile route (R), 203
    aeronautical mobile-satellite, 203
    aeronautical mobile-satellite off-route (OR), 203
    aeronautical mobile-satellite route (R), 203
    aeronautical radionavigation, 203
    aeronautical radionavigation-satellite, 203
    amateur, 204
    amateur-satellite, 204
    broadcasting, 204
    broadcasting-satellite, 204
    categories of, 118
    earth exploration-satellite, 205
    fixed, 206
    fixed-satellite, 206
    general purpose mobile, 206
    intersatellite, 207
    land mobile, 207
    land mobile-satellite, 207
    maritime mobile, 208
    maritime mobile-satellite, 208
    maritime radionavigation, 208
    maritime radionavigation-satellite, 208
    meteorological aids, 208
    meteorological-satellite, 208
    mobile, 208
    mobile-satellite, 208
    permitted, 119
    port operations, 209
    primary, 119
    radio astronomy, 210
    radiocommunication, 210
    radiodetermination, 210
    radiodetermination-satellite, 210
    radiolocation, 210
    radionavigation, 210
    radionavigation-satellite, 210
    safety, 211
    secondary, 119
    ship movement, 211
    space operation, 211
    space research, 211
    special, 212
    standard frequency and time signal, 212
    standard frequency and time signal-satellite, 212
ship earth station, 211
ship movement service, 211
ship station, 211
Ship's emergency transmitter, 211
shrinking raster, 32

Index

shrinking raster method, 33
SI, 214
SI base units, 214
SI derived units with special names and symbols, 215
signal
    AC, 4
    amplitude, 6
    analysis, 65
    analysis, computer-based, 82
    applications of the zone plate, 27
    audio, 6
    coherent demodulation of AM, 45
    complex, 7
    frequency sweep, 27
    generation, 82
    generation and analysis, 82
    impurity, 8
    level, 9
    measurement, 3
    multiburst, 27
    multitone, 11
    processing, 85
    processing toolbox, 85, 88
    producing the zone plate, 29
    resolution, 34
    satellite service, standard frequency and time, 212
    service, standard frequency and time, 212
    single-sine-wave, 4
    spectra, video, 19
    spectrum, 21, 27
    station, standard frequency and time, 212
    test, 27
    to-noise ratio, 3
    vertical, 30
    voltage, 29
signature analyzer, 60
simple zone plate patterns, 27
simplex operation, 211
sine function, 29
sine waves, 4
single-hop reflection, 45
single-sideband emission, 211
single-sideband suppressed-carrier (SSB-SC) demodulation
    errors, 47
single-valued, 80
sky waves, 44
slope of a curve, 257
SMPTE IM, 9, 17
SNR, 3
software, 63
solid state, 220
solid-state materials, 92
space
    deep, 205
    operation service, 211
    radiocommunication, 211
    research service, 211
    station, 211
    system, 212
    telecommand, 212
    telemetry, 212
    tracking, 212
spacecraft, 211
special functions, 266
special service, 212
spectral color reproduction, 24, 36
spectroradiometric measurement, 23
spectrum analyzer, 8, 9
sphere, 271
spherical coordinates, 263
spline, 83
spread spectrum systems, 212
spurious emission, 212
spurious resolution, 33
SPW, 88
standard frequency and time signal-satellite service, 212
standard frequency and time signal service, 212
standard frequency and time signal station, 212
standard serial interface, 63
state, 59, 68
state analyzer, 57
state qualified triggering, 67
station, 212
    aeronautical, 203
    aeronautical earth, 203
    aeronautical fixed, 203
    aircraft, 203
    aircraft earth, 203
    amateur, 204
    base, 204
    base earth, 204
    broadcasting, 204
    coast, 204
    coast earth, 204
    earth, 205
    emergency position-indicating radiobeacon, 205
    experimental, 205
    fixed, 206
    land, 207
    land earth, 207
    land mobile, 207
    land mobile earth, 207
    mobile, 208
    mobile earth, 208
    onboard communication, 208
    port, 209
    radio astronomy, 210
    radio direction-finding, 210
    radiobeacon, 210
    radiodetermination, 210
    radiolocation land, 210
    radiolocation mobile, 210
    radionavigation land, 210
    radionavigation mobile, 210
    ship, 211
    ship earth, 211
    space, 211
    standard frequency and time signal, 212
    survival craft, 212
    terrestrial, 213
statistical data analysis functions, 84, 88
step response, 88

Stirling's approximation, 243
Stoke's theorem, 282
subjective CRT measurements, 32
suppressed carrier single-sideband emission, 212
surface area and volume by double integration, 264
surface area of revolution, 263
survival craft station, 212
symbolic mathematics, 86
symbols and terminology for physical and chemical quantities, 217
systems
    audio, 7
    digital and audio, 15
    direct sequence, 205
    frequency hopping, 206
    hybrid spread spectrum, 206
    of equations, 274
    pulsed FM, 209
    spread spectrum, 212
    telephone, 7
    time hopping, 213
SystemView, 88

table of derivatives, 242
Taylor series, 254
Taylor's formula, 259
TDM, 40
telecommand, 212
telecommunication, 212
telegram, 212
telegraphy, 212
telemetry, 212
telephone systems, 7
telephony, 213
television, 213
television lines, 26
temperatures, conversion of, 240
temperature factors, 239
tensile strength, 93, 116
terrestrial radiocommunication, 213
terrestrial station, 213
test, CCITT, 10
test signals, 27
testing, multitone audio, 11
THD, 3, 78
theorem, binomial, 243
theorems, integral, 282
theory of maxima and minima, 242
thermal conductivity, 95, 116
thermal expansion of alloy cast irons, 95
time dimension, 31
time-division multiplexing, 40
time hopping systems, 213
timing analyzer, 57
toolboxes, 85, 88
topology database, 61
total harmonic distortion, 3
total harmonic distortion and noise, 9
total harmonic distortion measures, 78
total tristimulus, 23

trace point, 58
trace quality, 65
transfer function, 85, 88
transforms, finite cosine, 285
transforms, finite sine, 284
transponder, 213
trapezoidal rule for areas, 260
triangles, 247
trigger flexibility, 65
triggering mode, 66
trigonometric form, 74
trigonometric function, 70, 71
trigonometric functions of an angle, 248
trigonometric identities, 248
trigonometric series, 256
trigonometry, 242, 247
troposphere, 41, 56
tropospheric scatter, 213
tropospheric waves, 42, 44
TV limiting resolution, 32
TV limiting resolution method, 33
types of distortion, 39

uniform, 85
uniform variates, 88
unitary matrix, 277
units in use together with the SI, 216
unwanted emissions, 213
upper-sideband, 47
U.S. table of frequency allocations, 118
USB, 47
using the DFT/FFT in Fourier analysis, 77
UTC, 205

vector algebra and calculus, 278
vector functions, 279
vector spaces, 275
vertical dimension sine function, 29
vertical signal, 30
vertical testing patterns, 28
very low frequency, 42
video
    display distortion mechanisms and analysis, 19
    frequencies arising from scanning, 21
    frequency, maximum, 21
    frequency, minimum, 20
    signal spectra, 19
VLF, 42
volume of revolution, 263
volume unit, 15
von Kries model, 25
VU, 15

waveform delta, 68
waves
    demodulation of angle modulated, 49
    effects of amplitude nonlinearities on angle modulated, 51

Index

    ground, 43
    ground reflected, 43
    hertzian, 211
    line-of-sight, 43
    radio, 211
    right-hand polarized, 211
    sine, 4
    sky, 44
    tropospheric, 42, 44
Weber functions, 267
weighting filter, 17
white balance, 24
windowing, 80
wireless radio channel, 42
work, 263

Young's modulus, 94, 116

zone plate, 36
zone plate signal, 27